COLUMN GENERATION

GERAD 25th Anniversary Series

- **Essays and Surveys in Global Optimization**
  Charles Audet, Pierre Hansen, and Gilles Savard, editors

- **Graph Theory and Combinatorial Optimization**
  David Avis, Alain Hertz, and Odile Marcotte, editors

- **Numerical Methods in Finance**
  Hatem Ben-Ameur and Michèle Breton, editors

- **Analysis, Control and Optimization of Complex Dynamic Systems**
  El-Kébir Boukas and Roland Malhamé, editors

- **Column Generation**
  Guy Desaulniers, Jacques Desrosiers, and Marius M. Solomon, editors

- **Statistical Modeling and Analysis for Complex Data Problems**
  Pierre Duchesne and Bruno Rémillard, editors

- **Performance Evaluation and Planning Methods for the Next Generation Internet**
  André Girard, Brunilde Sansò, and Félisa Vázquez-Abad, editors

- **Dynamic Games: Theory and Applications**
  Alain Haurie and Georges Zaccour, editors

- **Logistics Systems: Design and Optimization**
  André Langevin and Diane Riopel, editors

- **Energy and Environment**
  Richard Loulou, Jean-Philippe Waaub, and Georges Zaccour, editors

COLUMN GENERATION

Edited by
GUY DESAULNIERS
GERAD and École Polytechnique de Montréal

JACQUES DESROSIERS
GERAD and HEC Montréal

MARIUS M. SOLOMON
GERAD and Northeastern University, Boston

 Springer

Guy Desaulniers
GERAD & École Polytechnique de Montréal
Montréal, Canada

Jacques Desrosiers
GERAD and HEC
Montréal, Canada

Marius M. Solomon
GERAD and Northeastern University
Boston, MA, USA

Library of Congress Cataloging-in-Publication Data

Column generation / edited by Guy Desaulniers, Jacques Desrosiers, Marius M. Solomon.
     p. cm. – (GERAD 25th anniversary series)
    Includes bibliographical references.

     ISBN-13: 978-1-4419-3799-5

     ISBN-13: 978-0-387-25486-9 (e-book)
     1. Production scheduling. I. Desaulniers, Guy. II. Desrosiers, Jacques. III. Solomon, Marius M. IV. Series.

    TS 157.5.C64 2005
    658.5'3—dc22

                     2005043438

Printed in the United States of America.

9 8 7 6 5 4 3 2 1

springeronline.com

Foreword

GERAD celebrates this year its 25th anniversary. The Center was created in 1980 by a small group of professors and researchers of HEC Montréal, McGill University and of the École Polytechnique de Montréal. GERAD's activities achieved sufficient scope to justify its conversion in June 1988 into a Joint Research Centre of HEC Montréal, the École Polytechnique de Montréal and McGill University. In 1996, the Université du Québec à Montréal joined these three institutions. GERAD has fifty members (professors), more than twenty research associates and post doctoral students and more than two hundreds master and Ph.D. students.

GERAD is a multi-university center and a vital forum for the development of operations research. Its mission is defined around the following four complementarily objectives:

- The original and expert contribution to all research fields in GERAD's area of expertise;

- The dissemination of research results in the best scientific outlets as well as in the society in general;

- The training of graduate students and post doctoral researchers;

- The contribution to the economic community by solving important problems and providing transferable tools.

GERAD's research thrusts and fields of expertise are as follows:

- Development of mathematical analysis tools and techniques to solve the complex problems that arise in management sciences and engineering;

- Development of algorithms to resolve such problems efficiently;

- Application of these techniques and tools to problems posed in related disciplines, such as statistics, financial engineering, game theory and artificial intelligence;

- Application of advanced tools to optimization and planning of large technical and economic systems, such as energy systems, transportation/communication networks, and production systems;

- Integration of scientific findings into software, expert systems and decision-support systems that can be used by industry.

One of the marking events of the celebrations of the 25th anniversary of GERAD is the publication of ten volumes covering most of the Center's research areas of expertise. The list follows: **Essays and Surveys in Global Optimization**, edited by C. Audet, P. Hansen and G. Savard; **Graph Theory and Combinatorial Optimization**, edited by D. Avis, A. Hertz and O. Marcotte; **Numerical Methods in Finance**, edited by H. Ben-Ameur and M. Breton; **Analysis, Control and Optimization of Complex Dynamic Systems**, edited by E.K. Boukas and R. Malhamé; **Column Generation**, edited by G. Desaulniers, J. Desrosiers and M.M. Solomon; **Statistical Modeling and Analysis for Complex Data Problems**, edited by P. Duchesne and B. Rémillard; **Performance Evaluation and Planning Methods for the Next Generation Internet**, edited by A. Girard, B. Sansò and F. Vázquez-Abad; **Dynamic Games: Theory and Applications**, edited by A. Haurie and G. Zaccour; **Logistics Systems: Design and Optimization**, edited by A. Langevin and D. Riopel; **Energy and Environment**, edited by R. Loulou, J.-P. Waaub and G. Zaccour.

I would like to express my gratitude to the Editors of the ten volumes, to the authors who accepted with great enthusiasm to submit their work and to the reviewers for their benevolent work and timely response. I would also like to thank Mrs. Nicole Paradis, Francine Benoît and Louise Letendre and Mr. André Montpetit for their excellent editing work.

The GERAD group has earned its reputation as a worldwide leader in its field. This is certainly due to the enthusiasm and motivation of GERAD's researchers and students, but also to the funding and the infrastructures available. I would like to seize the opportunity to thank the organizations that, from the beginning, believed in the potential and the value of GERAD and have supported it over the years. These are HEC Montréal, École Polytechnique de Montréal, McGill University, Université du Québec à Montréal and, of course, the Natural Sciences and Engineering Research Council of Canada (NSERC) and the Fonds québécois de la recherche sur la nature et les technologies (FQRNT).

Georges Zaccour
Director of GERAD

Avant-propos

Le Groupe d'études et de recherche en analyse des décisions (GERAD) fête cette année son vingt-cinquième anniversaire. Fondé en 1980 par une poignée de professeurs et chercheurs de HEC Montréal engagés dans des recherches en équipe avec des collègues de l'Université McGill et de l'École Polytechnique de Montréal, le Centre comporte maintenant une cinquantaine de membres, plus d'une vingtaine de professionnels de recherche et stagiaires post-doctoraux et plus de 200 étudiants des cycles supérieurs. Les activités du GERAD ont pris suffisamment d'ampleur pour justifier en juin 1988 sa transformation en un Centre de recherche conjoint de HEC Montréal, de l'École Polytechnique de Montréal et de l'Université McGill. En 1996, l'Université du Québec à Montréal s'est jointe à ces institutions pour parrainer le GERAD.

Le GERAD est un regroupement de chercheurs autour de la discipline de la recherche opérationnelle. Sa mission s'articule autour des objectifs complémentaires suivants :

- la contribution originale et experte dans tous les axes de recherche de ses champs de compétence ;
- la diffusion des résultats dans les plus grandes revues du domaine ainsi qu'auprès des différents publics qui forment l'environnement du Centre ;
- la formation d'étudiants des cycles supérieurs et de stagiaires post-doctoraux ;
- la contribution à la communauté économique à travers la résolution de problèmes et le développement de coffres d'outils transférables.

Les principaux axes de recherche du GERAD, en allant du plus théorique au plus appliqué, sont les suivants :

- le développement d'outils et de techniques d'analyse mathématiques de la recherche opérationnelle pour la résolution de problèmes complexes qui se posent dans les sciences de la gestion et du génie ;
- la confection d'algorithmes permettant la résolution efficace de ces problèmes ;
- l'application de ces outils à des problèmes posés dans des disciplines connexes à la recherche opérationnelle telles que la statistique, l'ingénierie financière, la théorie des jeux et l'intelligence artificielle ;
- l'application de ces outils à l'optimisation et à la planification de grands systèmes technico-économiques comme les systèmes énergétiques, les réseaux de télécommunication et de transport, la logistique et la distributique dans les industries manufacturières et de service ;

- l'intégration des résultats scientifiques dans des logiciels, des systèmes experts et dans des systèmes d'aide à la décision transférables à l'industrie.

Le fait marquant des célébrations du 25ᵉ du GERAD est la publication de dix volumes couvrant les champs d'expertise du Centre. La liste suit : **Essays and Surveys in Global Optimization**, édité par C. Audet, P. Hansen et G. Savard ; **Graph Theory and Combinatorial Optimization**, édité par D. Avis, A. Hertz et O. Marcotte ; **Numerical Methods in Finance**, édité par H. Ben-Ameur et M. Breton ; **Analysis, Control and Optimization of Complex Dynamic Systems**, édité par E.K. Boukas et R. Malhamé ; **Column Generation**, édité par G. Desaulniers, J. Desrosiers et M.M. Solomon ; **Statistical Modeling and Analysis for Complex Data Problems**, édité par P. Duchesne et B. Rémillard ; **Performance Evaluation and Planning Methods for the Next Generation Internet**, édité par A. Girard, B. Sansò et F. Vázquez-Abad ; **Dynamic Games : Theory and Applications**, édité par A. Haurie et G. Zaccour ; **Logistics Systems : Design and Optimization**, édité par A. Langevin et D. Riopel ; **Energy and Environment**, édité par R. Loulou, J.-P. Waaub et G. Zaccour.

Je voudrais remercier très sincèrement les éditeurs de ces volumes, les nombreux auteurs qui ont très volontiers répondu à l'invitation des éditeurs à soumettre leurs travaux, et les évaluateurs pour leur bénévolat et ponctualité. Je voudrais aussi remercier Mmes Nicole Paradis, Francine Benoît et Louise Letendre ainsi que M. André Montpetit pour leur travail expert d'édition.

La place de premier plan qu'occupe le GERAD sur l'échiquier mondial est certes due à la passion qui anime ses chercheurs et ses étudiants, mais aussi au financement et à l'infrastructure disponibles. Je voudrais profiter de cette occasion pour remercier les organisations qui ont cru dès le départ au potentiel et à la valeur du GERAD et nous ont soutenus durant ces années. Il s'agit de HEC Montréal, l'École Polytechnique de Montréal, l'Université McGill, l'Université du Québec à Montréal et, bien sûr, le Conseil de recherche en sciences naturelles et en génie du Canada (CRSNG) et le Fonds québécois de la recherche sur la nature et les technologies (FQRNT).

Georges Zaccour
Directeur du GERAD

Contents

Contributing Authors

HATEM BEN AMOR
École Polytechnique and GERAD,
Canada
Hatem.Ben.Amor@gerad.ca

MARIELLE CHRISTIANSEN
Norwegian University of Science and
Technology, Norway
marielle.christiansen@iot.ntnu.no

EMILIE DANNA
Georgia Institute of Technology, USA
emilie.danna@isye.gatech.edu

JOSÉ VALÉRIO DE CARVALHO
Universidade do Minho, Portugal
vc@dps.uminho.pt

GUY DESAULNIERS
École Polytechnique and GERAD,
Canada
Guy.Desaulniers@gerad.ca

JACQUES DESROSIERS
HEC Montréal and GERAD, Canada
jacques.desrosiers@hec.ca

SYLVIE GÉLINAS
École Polytechnique and GERAD,
Canada
sylvieg@crt.umontreal.ca

HAN HOOGEVEEN
Utrecht University, The Netherlands
slam@cs.uu.nl

DENNIS HUISMAN
Erasmus University Rotterdam,
The Netherlands
huisman@few.eur.nl

STEFAN IRNICH
RWTH Aachen University, Germany
SIrnich@or.rwth-aachen.de

RAF JANS
Erasmus University Rotterdam,
The Netherlands
rjans@fbk.eur.nl

BRIAN KALLEHAUGE
Technical University of Denmark
bk@ctt.dtu.dk

DIEGO KLABJAN
University of Illinois at
Urbana-Champaign, USA
klabjan@uiuc.edu

JESPER LARSEN
Technical University of Denmark,
Denmark
jla@imm.dtu.dk

CLAUDE LE PAPE
ILOG S.A., France
clepape@ilog.fr

MARCO E. LÜBBECKE
Technische Universität Berlin, Germany
m.luebbecke@math.tu-berlin.de

OLI B.G. MADSEN
Technical University of Denmark,
Denmark
ogm@ctt.dtu.dk

BJØRN NYGREEN
Norwegian University of Science and
Technology, Norway
bjorn.nygreen@iot.ntnu.no

MARC PEETERS
Place de L'Université 16, Belgium
Marc.Peeters2@electrabel.com

DAVID M. RYAN
University of Auckland, New Zealand
d.ryan@auckland.ac.nz

FRANÇOIS SOUMIS
École Polytechnique and GERAD,
Canada
francois.soumis@gerad.ca

MIKKEL M. SIGURD
University of Copenhagen, Denmark
sigurd@diku.dk

MARIUS M. SOLOMON
Northeastern University, USA and
GERAD, Canada
m.solomon@neu.edu

NINA L. ULSTEIN
Norwegian University of Science and
Technology, Norway
nina.ulstein@iot.ntnu.no

FRANÇOIS VANDERBECK
Université Bordeaux 1, France
fv@math.u-bordeaux.fr

STEEF VAN DE VELDE
Erasmus University, The Netherlands
s.velde@fac.fbk.eur.nl

MARJAN VAN DEN AKKER
Utrecht University, The Netherlands
marjan@cs.uu.nl

ALBERT P.M. WAGELMANS
Erasmus University Rotterdam,
The Netherlands
wagelmans@few.eur.nl

Preface

GERAD is an Operations Research center founded in 1980 that brings together the top universities in Montréal: HEC Montréal, École Polytechnique de Montréal, McGill University, and Université du Québec à Montréal. It is organized across several research teams. The Gencol team is one of the oldest and best known. Led by Jacques Desrosiers and François Soumis, it also includes Jean-François Cordeau, Guy Desaulniers, Michel Gamache, Odile Marcotte, Gilles Savard, and Marius M. Solomon. The late Martin Desrochers was the first Ph.D. student. The group originally focused on the vehicle routing problem with time windows and then expanded their focus to more complex resource constrained vehicle routing and crew scheduling problems.

During these 25 years the team's efforts resulted in important academic, scientific, commercial, and industrial benefits. The academic spin-off consists of support offered to scores of Ph.D. and master students, analysts, and post-doctoral and visiting researchers. The scientific advances include numerous publications, many in premier journals and widely used survey papers. Overall, the Gencol group made significant advances in the integer programming column generation area. It is with the excitement of having participated in these developments and the modesty of being only a small part of the research community that we welcome the advancements described in the chapters of this book.

The book starts with *A Primer in Column Generation* by Jacques Desrosiers and Marco E. Lübbecke. It introduces the column generation technique in integer programming settings. The relevant theory and the more advanced ideas necessary to solve large-scale practical problems are illustrated with a variety of examples.

In the second chapter, *Shortest Path Problems with Resource Constraints*, Stefan Irnich and Guy Desaulniers offer a comprehensive survey of the problems used to cast the subproblem in most vehicle routing and crew scheduling applications solved by column generation. The following two chapters are dedicated to the *Vehicle Routing Problem with Time Windows*. Brian Kallehauge, Jesper Larsen, Oli B.G. Madsen, and Marius M. Solomon focus on the methodological evolution, including cutting planes, parallelism, acceleration strategies for the master problem and novel subproblem approaches. Emilie Danna and Claude Le Pape, *Branch-and-Price Heuristics: A Case Study on the Vehicle Routing Problem with Time Windows*, illustrate the benefits of using hybrid branch-and-price and heuristic solutions to rapidly produce good integer solutions. This technique along with stabilization methods proposed to improve the efficiency of the column generation process are revisited in the next chapter by Hatem Ben Amor and José M. Valério de Carvalho in the context of *Cutting Stock Problems*. The authors also explore the links between the extended Dantzig-Wolfe decomposition and the Gilmore-Gomory model.

The following three chapters deal with air and maritime transportation applications. *Large-scale Models in the Airline Industry* are presented by Diego Klabjan. He examines models involved in strategic business processes as well as operational processes. The former address schedule design and fleeting, aircraft routing, and crew scheduling, while the latter models cope with irregular operations. Then, Marielle Christiansen and Bjørn Nygreen describe *Robust Inventory Ship Routing by Column Generation*. They consider an actual integrated ship scheduling and inventory management problem where the transporter has the responsibility to keep the inventory level of a single product at all plants within predetermined limits without causing production stopages due to missed transportation opportunities or variability in sailing time. Next, Mikkel M. Sigurd, Nina L. Ulstein, Bjørn Nygreen, and David M. Ryan discuss the design of a sea-transport system for Norwegian companies who rely heavily on maritime transportation in *Ship Scheduling With Recurring Visits and Visit Separation Requirements*. The model determines the optimal fleet composition, including the potential investment in new ships, the ship routes and their visit-schedules when transport tonnage is pooled over all participating companies.

The next three chapters deal with production environments. First, Dennis Huisman, Raf Jans, Marc Peeters, and Albert P.M. Wagelmans propose *Combining Column Generation and Lagrangian Relaxation*. The authors focus on using Lagrangian relaxation to either directly solve the LP relaxation of the Dantzig-Wolfe master problem or to generate new columns. They illustrate their ideas with an application in lot-sizing and comment on applications in other areas. Second, Sylvie Gélinas and François Soumis propose *Dantzig-Wolfe Decomposition for Job Shop Scheduling*. They present a flexible formulation capable of handling several objectives. In this context, each subproblem is a single machine sequencing problem with time windows. Third, Marjan van den Akker, Han Hoogeveen, and Steef van de Velde survey *Applying Column Generation to Machine Scheduling*. In particular, they illustrate the success of column generation methods when the main objective is to divide the jobs across the machines.

The final chapter by François Vanderbeck is on *Implementing Mixed Integer Column Generation*. He reviews how to set-up the Dantzig-Wolfe reformulation, adapt standard MIP techniques to the column generation context (branching, preprocessing, primal heuristics), and deal with specific column generation issues (initialization, stabilization, column management strategies).

We think that this book offers an insightful overview of the state-of-the-art in integer programming column generation and its many applications, and we certainly hope that it will serve as a good reference for the novice as well as the experienced column generation user.

Finally, we would like to thank all the contributors for their fine work.

GUY DESAULNIERS
JACQUES DESROSIERS
MARIUS M. SOLOMON

Chapter 1

A PRIMER IN COLUMN GENERATION

Jacques Desrosiers
Marco E. Lübbecke

Abstract   We give a didactic introduction to the use of the column generation
technique in linear and in particular in integer programming. We touch
on both, the relevant basic theory and more advanced ideas which help
in solving large scale practical problems. Our discussion includes em-
bedding Dantzig-Wolfe decomposition and Lagrangian relaxation within
a branch-and-bound framework, deriving natural branching and cutting
rules by means of a so-called compact formulation, and understanding
and influencing the behavior of the dual variables during column generation. Most concepts are illustrated via a small example. We close with
a discussion of the classical cutting stock problem and some suggestions
for further reading.

1.   Hands-on experience

Let us start right away by solving a constrained shortest path problem.
Consider the network depicted in Figure 1.1. Besides a cost c_{ij} there is a
resource consumption t_{ij} attributed to each arc $(i, j) \in A$, say a traversal
time. Our goal is to find a shortest path from node 1 to node 6 such
that the total traversal time of the path does not exceed 14 time units.

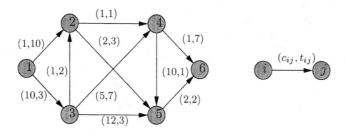

Figure 1.1.   Time constrained shortest path problem, (p. 599 Ahuja et al., 1993).

One way to state this particular network flow problem is as the integer program (1.1)–(1.6). One unit of flow has to leave the source (1.2) and has to enter the sink (1.4), while flow conservation (1.3) holds at all other nodes. The time resource constraint appears as (1.5).

$$z^\star := \min \sum_{(i,j)\in A} c_{ij} x_{ij} \tag{1.1}$$

$$\text{subject to} \sum_{j\,:\,(1,j)\in A} x_{1j} = 1 \tag{1.2}$$

$$\sum_{j\,:\,(i,j)\in A} x_{ij} - \sum_{j\,:\,(j,i)\in A} x_{ji} = 0 \quad i = 2,3,4,5 \tag{1.3}$$

$$\sum_{i\,:\,(i,6)\in A} x_{i6} = 1 \tag{1.4}$$

$$\sum_{(i,j)\in A} t_{ij} x_{ij} \le 14 \tag{1.5}$$

$$x_{ij} = 0 \text{ or } 1 \quad (i,j) \in A \tag{1.6}$$

An inspection shows that there are nine possible paths, three of which consume too much time. The optimal integer solution is path 13246 of cost 13 with a traversal time of 13. How would we find this out? First note that the resource constraint (1.5) prevents us from solving our problem with a classical shortest path algorithm. In fact, no polynomial time algorithm is likely to exist since the resource constrained shortest path problem is $\mathcal{NP}$-hard. However, since the problem is *almost* a shortest path problem, we would like to exploit this embedded well-studied structure algorithmically.

1.1   An equivalent reformulation: Arcs vs. paths

If we ignore the complicating constraint (1.5), the easily tractable remainder is $X = \{x_{ij} = 0 \text{ or } 1 \mid (1.2)\text{–}(1.4)\}$. It is a well-known result in network flow theory that an extreme point $\mathbf{x}_p = (x_{pij})$ of the polytope defined by the convex hull of X corresponds to a path $p \in P$ in the network. This enables us to express any arc flow as a convex combination of path flows:

$$x_{ij} = \sum_{p\in P} x_{pij}\lambda_p \quad (i,j) \in A \tag{1.7}$$

$$\sum_{p\in P} \lambda_p = 1 \tag{1.8}$$

$$\lambda_p \ge 0 \quad p \in P. \tag{1.9}$$

If we substitute for $\mathbf{x}$ in (1.1) and (1.5) we obtain the so-called *master problem*:

$$z^\star = \min \sum_{p \in P} \left(\sum_{(i,j) \in A} c_{ij} x_{pij} \right) \lambda_p \qquad (1.10)$$

$$\text{subject to} \sum_{p \in P} \left(\sum_{(i,j) \in A} t_{ij} x_{pij} \right) \lambda_p \leq 14 \qquad (1.11)$$

$$\sum_{p \in P} \lambda_p = 1 \qquad (1.12)$$

$$\lambda_p \geq 0 \quad p \in P \qquad (1.13)$$

$$\sum_{p \in P} x_{pij} \lambda_p = x_{ij} \quad (i,j) \in A \qquad (1.14)$$

$$x_{ij} = 0 \text{ or } 1 \quad (i,j) \in A. \qquad (1.15)$$

Loosely speaking, the structural information X that we are looking for a path is hidden in "$p \in P$." The cost coefficient of λ_p is the cost of path p and its coefficient in (1.11) is path p's duration. Via (1.14) and (1.15) we explicitly preserve the linking of variables (1.7) in the formulation, and we may recover a solution $\mathbf{x}$ to our *original* problem (1.1)–(1.6) from a master problem's solution. Always remember that integrality must hold for the original $\mathbf{x}$ variables.

1.2   The linear relaxation of the master problem

One starts with solving the linear programming (LP) relaxation of the master problem. If we relax (1.15), there is no longer a need to link the $\mathbf{x}$ and λ variables, and we may drop (1.14) as well. There remains a problem with nine path variables and two constraints. Associate with (1.11) and (1.12) dual variables π_1 and π_0, respectively. For large networks, the cardinality of P becomes prohibitive, and we cannot even explicitly state all the variables of the master problem. The appealing idea of *column generation* is to work only with a sufficiently meaningful subset of variables, forming the so-called *restricted master problem* (RMP). More variables are added only when needed: Like in the simplex method we have to find in every iteration a promising variable to enter the basis. In column generation an iteration consists

a) of optimizing the restricted master problem in order to determine the current optimal objective function value $\bar{z}$ and dual multipliers $\boldsymbol{\pi}$, and

Table 1.1. BB0: The linear programming relaxation of the master problem

Iteration	Master Solution	$\bar{z}$	π_0	π_1	$\bar{c}^\star$	p	c_p	t_p
BB0.1	$y_0 = 1$	100.0	100.00	0.00	-97.0	1246	3	18
BB0.2	$y_0 = 0.22, \lambda_{1246} = 0.78$	24.6	100.00	-5.39	-32.9	1356	24	8
BB0.3	$\lambda_{1246} = 0.6, \lambda_{1356} = 0.4$	11.4	40.80	-2.10	-4.8	13256	15	10
BB0.4	$\lambda_{1246} = \lambda_{13256} = 0.5$	9.0	30.00	-1.50	-2.5	1256	5	15
BB0.5	$\lambda_{13256} = 0.2, \lambda_{1256} = 0.8$	7.0	35.00	-2.00	0			
Arc flows:	$x_{12} = 0.8, x_{13} = x_{32} = 0.2, x_{25} = x_{56} = 1$							

b) of finding, if there still is one, a variable λ_p with *negative reduced cost*

$$\bar{c}_p = \sum_{(i,j) \in A} c_{ij} x_{pij} - \pi_1 \left(\sum_{(i,j) \in A} t_{ij} x_{pij} \right) - \pi_0 < 0. \qquad (1.16)$$

The implicit search for a minimum reduced cost variable amounts to optimizing a *subproblem*, precisely in our case: A shortest path problem in the network of Figure 1.1 with a modified cost structure:

$$\bar{c}^\star = \min_{(1.2)-(1.4),(1.6)} \sum_{(i,j) \in A} (c_{ij} - \pi_1 t_{ij}) x_{ij} - \pi_0. \qquad (1.17)$$

Clearly, if $\bar{c}^\star \geq 0$ there is no improving variable and we are done with the linear relaxation of the master problem. Otherwise, the variable found is added to the RMP and we repeat.

In order to obtain integer solutions to our original problem, we have to embed column generation within a branch-and-bound framework. We now give full numerical details of the solution of our particular instance. We denote by BB$n.i$ iteration number i at node number n ($n = 0$ represents the root node). The summary in Table 1.1 for the LP relaxation of the master problem also lists the cost c_p and duration t_p of path p, respectively, and the solution in terms of the value of the original variables x.

Since we have no feasible initial solution at iteration BB0.1, we adopt a big-M approach and introduce an artificial variable y_0 with a large cost, say 100, for the convexity constraint. We do not have any path variables yet and the RMP contains two constraints and the artificial variable. This problem is solved by inspection: $y_0 = 1, \bar{z} = 100$, and the dual variables are $\pi_0 = 100$ and $\pi_1 = 0$. The subproblem (1.17) returns path 1246 at reduced cost $\bar{c}^\star = -97$, cost 3 and duration 18. In iteration BB0.2, the RMP contains two variables: y_0 and λ_{1246}. An optimal solution with $\bar{z} = 24.6$ is $y_0 = 0.22$ and $\lambda_{1246} = 0.78$, which is still infeasible. The dual variables assume values $\pi_0 = 100$ and $\pi_1 = -5.39$. Solving the subproblem gives the feasible path 1356 of reduced cost -32.9, cost 24, and duration 8.

In total, four path variables are generated during the column generation process. In iteration BB0.5, we use 0.2 times the feasible path 13256 and 0.8 times the infeasible path 1256. The optimal objective function value is 7, with $\pi_0 = 35$ and $\pi_1 = -2$. The arc flow values provided at the bottom of Table 1.1 are identical to those found when solving the LP relaxation of the original problem.

1.3 Branch-and-bound: The reformulation repeats

Except for the integrality requirement (1.6) (or 1.15) all constraints of the original (and of the master) problem are satisfied, and a subsequent branch-and-bound process is used to compute an optimal integer solution. Even though it cannot happen for our example problem, in general the generated set of columns may not contain an integer feasible solution. To proceed, we have to start the reformulation and column generation again in each node.

Let us first explore some "standard" ways of branching on fractional variables, e.g., branching on $x_{12} = 0.8$. For $x_{12} = 0$, the impact on the RMP is that we have to remove path variables λ_{1246} and λ_{1256}, that is, those paths which contain arc $(1, 2)$. In the subproblem, this arc is removed from the network. When the RMP is re-optimized, the artificial variable assumes a positive value, and we would have to generate new $\boldsymbol{\lambda}$ variables. On branch $x_{12} = 1$, arcs $(1, 3)$ and $(3, 2)$ cannot be used. Generated paths which contain these arcs are discarded from the RMP, and both arcs are removed from the subproblem.

There are also many strategies involving more than a single arc flow variable. One is to branch on the sum of all flow variables which currently is 3.2. Since the solution is a path, an integer number of arcs has to be used, in fact, at least three and at most five in our example. Our freedom of making branching decisions is a powerful tool when properly applied.

Alternatively, we branch on $x_{13} + x_{32} = 0.4$. On branch $x_{13} + x_{32} = 0$, we simultaneously treat two flow variables; impacts on the RMP and the subproblem are similar to those described above. On branch $x_{13} + x_{32} \geq 1$, this constraint is first added to the original formulation. We exploit again the path substructure X, go through the reformulation process via (1.7), and obtain a new RMP to work with. Details of the search tree are summarized in Table 1.2.

Table 1.2.   Details of the branch-and-bound decisions

Iteration	Master Solution	$\bar{z}$	π_0	π_1	π_2	$\bar{c}^*$	p	c_p	t_p
BB1: BB0 and $x_{13} + x_{32} = 0$									
BB1.1	$y_0 = 0.067, \lambda_{1256} = 0.933$	11.3	100	−6.33	–	0			
BB1.2	Big-M increased to 1000								
	$y_0 = 0.067, \lambda_{1256} = 0.933$	71.3	1000	−66.33	–	−57.3	12456	14	14
BB1.3	$\lambda_{12456} = 1$	**14**	1000	−70.43	–	0			
BB2: BB0 and $x_{13} + x_{32} \geq 1$									
BB2.1	$\lambda_{1246} = \lambda_{13256} = 0.5$	9	15	−0.67	3.33	0			
Arc flows: $x_{12} = x_{13} = x_{24} = x_{25} = x_{32} = x_{46} = x_{56} = 0.5$									
BB3: BB2 and $x_{12} = 0$									
BB3.1	$\lambda_{13256} = 1$	**15**	15	0	0	-2	13246	13	13
BB3.2	$\lambda_{13246} = 1$	**13**	13	0	0	0			
BB4: BB2 and $x_{12} = 1$									
BB4.1	$y_0 = 0.067, \lambda_{1256} = 0.933$	111.3	100	−6.33	100	0			
Infeasible arc flows									

At node BB1, we set $x_{13} + x_{32} = 0$. In iteration BB1.1, paths 1356 and 13256 are discarded from the RMP, and arcs $(1,3)$ and $(3,2)$ are removed from the subproblem. The resulting RMP with $y_0 = 0.067$ and $\lambda_{1256} = 0.933$ is infeasible. The objective function assumes a value $\bar{z} = 11.3$, and $\pi_0 = 100$ and $\pi_1 = -6.33$. Given these dual multipliers, no column with negative reduced cost can be generated!

Here we face a drawback of the big-M approach. Path 12456 is feasible, its duration is 14, but its cost of 14 is larger than the current objective function value, computed as $0.067M + 0.933 \times 5$. The constant $M = 100$ is too small, and we have to increase it, say to 1000. (A different phase I approach, that is, minimizing the artificial variable y_0, would have easily prevented this.) Re-optimizing the RMP in iteration BB1.2 now results in $\bar{z} = 71.3$, $y_0 = 0.067$, $\lambda_{1256} = 0.933$, $\pi_0 = 1000$, and $\pi_1 = -66.33$. The subproblem returns path 12456 with a reduced cost of -57.3. In iteration BB1.3, the new RMP has an integer solution $\lambda_{12456} = 1$, with $\bar{z} = 14$, an upper bound on the optimal path cost. The dual multipliers are $\pi_0 = 1000$ and $\pi_1 = -70.43$, and no new variable is generated.

At node BB2, we impose $x_{13} + x_{32} \geq 1$ to the original formulation, and again, we reformulate these **x** variables in terms of the $\boldsymbol{\lambda}$ variables. The resulting new constraint (with dual multiplier π_2) in the RMP is $\sum_{p\in P}(x_{p13} + x_{p32})\lambda_p \geq 1$. From the value of $(x_{p13} + x_{p32})$ we learn how often arcs $(1,3)$ and $(3,2)$ are used in path p. The current problem at node BB2.1 is the following:

$$\begin{aligned}
\min \quad & 100y_0 + 3\lambda_{1246} + 24\lambda_{1356} + 15\lambda_{13256} + 5\lambda_{1256} \\
\text{subject to:} \quad & 18\lambda_{1246} + 8\lambda_{1356} + 10\lambda_{13256} + 15\lambda_{1256} \leq 14 \quad [\pi_1] \\
& \lambda_{1356} + 2\lambda_{13256} \geq 1 \quad [\pi_2] \\
y_0 + & \lambda_{1246} + \lambda_{1356} + \lambda_{13256} + \lambda_{1256} = 1 \quad [\pi_0] \\
& y_0, \ \lambda_{1246}, \ \lambda_{1356}, \ \lambda_{13256}, \ \lambda_{1256} \geq 0
\end{aligned}$$

From solving this linear program we obtain an increase in the objective function $\bar{z}$ from 7 to 9 with variables $\lambda_{1246} = \lambda_{13256} = 0.5$, and dual multipliers $\pi_0 = 15, \pi_1 = -0.67$, and $\pi_2 = 3.33$. The new subproblem is given by

$$\bar{c}^\star = \min_{(1.2)-(1.4), \ (1.6)} \sum_{(i,j)\in A} (c_{ij} - \pi_1 t_{ij})x_{ij} - \pi_0 - \pi_2(x_{13} + x_{32}). \quad (1.18)$$

For these multipliers no path of negative reduced cost exists. The solution of the flow variables is $x_{12} = x_{13} = x_{24} = x_{25} = x_{32} = x_{46} = x_{56} = 0.5$.

Next, we arbitrarily choose variable $x_{12} = 0.5$ to branch on. Two iterations are needed when x_{12} is set to zero. In iteration BB3.1, path variables λ_{1246} and λ_{1256} are discarded from the RMP and arc $(1,2)$ is removed from the subproblem. The RMP is integer feasible with $\lambda_{13256} = 1$ at cost 15. Dual multipliers are $\pi_0 = 15, \pi_1 = 0$, and $\pi_2 = 0$. Path 13246 of reduced cost -2, cost 13 and duration 13 is generated and used in the next iteration BB3.2. Again the RMP is integer feasible with path variable $\lambda_{13246} = 1$ and a new best integer solution at cost 13, with dual multipliers $\pi_0 = 15, \pi_1 = 0$, and $\pi_2 = 0$ for which no path of negative reduced cost exists.

On the alternative branch $x_{12} = 1$ the RMP is optimal after a single iteration. In iteration BB4.1, variable x_{13} can be set to zero and variables $\lambda_{1356}, \lambda_{13256}$, and λ_{13246} are discarded from the current RMP. After the introduction of an artificial variable y_2 in the second row, the RMP is infeasible since $y_0 > 0$ (as can be seen also from the large objective function value $\bar{z} = 111.3$). Given the dual multipliers, no columns of negative reduced cost can be generated, and the RMP remains infeasible. The optimal solution (found at node BB3) is path 13246 of cost 13 with a duration of 13 as well.

2. Some theoretical background

In the previous example we already saw all the necessary building blocks for a column generation based solution approach to integer programs: (1) an original formulation to solve which acts as the control

center to facilitate the design of natural branching rules and cutting
planes; (2) a master problem to determine the currently optimal dual
multipliers and to provide a lower bound at each node of the branch-
and-bound tree; (3) a pricing subproblem which explicitly reflects an
embedded structure we wish to exploit. In this section we detail the
underlying theory.

2.1   Column generation

Let us call the following linear program the *master problem* (MP).

$$z_{MP}^{\star} := \min \sum_{j \in J} c_j \lambda_j \tag{1.19}$$

$$\text{subject to } \sum_{j \in J} \mathbf{a}_j \lambda_j \geq \mathbf{b} \tag{1.20}$$

$$\lambda_j \geq 0, \quad j \in J. \tag{1.21}$$

In each iteration of the simplex method we look for a non-basic variable
to price out and enter the basis. That is, given the non-negative vector
$\boldsymbol{\pi}$ of dual variables we wish to find a $j \in J$ which minimizes $\bar{c}_j :=$
$c_j - \boldsymbol{\pi}^t \mathbf{a}_j$. This explicit pricing is a too costly operation when $|J|$ is huge.
Instead, we work with a reasonably small subset $J' \subseteq J$ of columns—
the restricted master problem (RMP)—and evaluate reduced costs only
by implicit enumeration. Let $\boldsymbol{\lambda}$ and $\boldsymbol{\pi}$ assume primal and dual optimal
solutions of the current RMP, respectively. When columns $\mathbf{a}_j, j \in J$, are
given as elements of a set $\mathcal{A}$, and the cost coefficient c_j can be computed
from $\mathbf{a}_j$ via a function c then the subproblem

$$\bar{c}^{\star} := \min\{c(\mathbf{a}) - \boldsymbol{\pi}^t \mathbf{a} \mid \mathbf{a} \in \mathcal{A}\} \tag{1.22}$$

performs the pricing. If $\bar{c}^{\star} \geq 0$, there is no negative $\bar{c}_j, j \in J$, and the
solution $\boldsymbol{\lambda}$ to the restricted master problem optimally solves the master
problem as well. Otherwise, we add to the RMP the column derived
from the optimal subproblem solution, and repeat with re-optimizing
the RMP. The process is initialized with an artificial, a heuristic, or a
previous ("warm start") solution. In what regards convergence, note that
each $\mathbf{a} \in \mathcal{A}$ is generated at most once since no variable in an optimal
RMP has negative reduced cost. When dealing with some finite set $\mathcal{A}$ (as
is practically always true), the column generation algorithm is exact. In
addition, we can make use of bounds. Let $\bar{z}$ denote the optimal objective
function value to the RMP. When an upper bound $\kappa \geq \sum_{j \in J} \lambda_j$ holds
for the optimal solution of the master problem, we have not only an
upper bound $\bar{z}$ on $z_{MP}^{\star}$ in each iteration, but also a lower bound: we

cannot reduce $\bar{z}$ by more than κ times the smallest reduced cost $\bar{c}^\star$:

$$\bar{z} + \kappa \bar{c}^\star \leq z^\star_{MP} \leq \bar{z}. \tag{1.23}$$

Thus, we may verify the solution quality at any time. In the optimum of (1.19), $\bar{c}^\star = 0$ for the basic variables, and $\bar{z} = z^\star_{MP}$.

2.2   Dantzig-Wolfe decomposition for integer programs

In many applications we are interested in optimizing over a discrete set X. For $X = \{\mathbf{x} \in \mathbb{Z}^n_+ \mid D\mathbf{x} \geq \mathbf{d}\} \neq \varnothing$ we have the special case of integer linear programming. Consider the following (*original* or *compact*) program:

$$z^\star := \min \mathbf{c}^t\mathbf{x} \tag{1.24}$$

$$\text{subject to } A\mathbf{x} \geq \mathbf{b} \tag{1.25}$$

$$\mathbf{x} \in X. \tag{1.26}$$

Replacing X by $\mathrm{conv}(X)$ in (1.24) does not change $z^\star$ which we assume to be finite. The Minkowski and Weyl theorems (see Schrijver, 1986) enable us to represent each $\mathbf{x} \in X$ as a convex combination of extreme points $\{\mathbf{x}_p\}_{p \in P}$ plus a non-negative combination of extreme rays $\{\mathbf{x}_r\}_{r \in R}$ of $\mathrm{conv}(X)$, i.e.,

$$\mathbf{x} = \sum_{p \in P} \mathbf{x}_p \lambda_p + \sum_{r \in R} \mathbf{x}_r \lambda_r, \quad \sum_{p \in P} \lambda_p = 1, \qquad \boldsymbol{\lambda} \in \mathbb{R}^{|P|+|R|}_+ \tag{1.27}$$

where the index sets P and R are finite. Substituting for $\mathbf{x}$ in (1.24) and applying the linear transformations $c_j = \mathbf{c}^t\mathbf{x}_j$ and $\mathbf{a}_j = A\mathbf{x}_j$, $j \in P \cup R$ we obtain an equivalent *extensive formulation*

$$z^\star := \min \sum_{p \in P} c_p \lambda_p + \sum_{r \in R} c_r \lambda_r \tag{1.28}$$

$$\text{subject to } \sum_{p \in P} \mathbf{a}_p \lambda_p + \sum_{r \in R} \mathbf{a}_r \lambda_r \geq \mathbf{b} \tag{1.29}$$

$$\sum_{p \in P} \lambda_p = 1 \tag{1.30}$$

$$\boldsymbol{\lambda} \geq \mathbf{0} \tag{1.31}$$

$$\sum_{p \in P} \mathbf{x}_p \lambda_p + \sum_{r \in R} \mathbf{x}_r \lambda_r = \mathbf{x} \tag{1.32}$$

$$\mathbf{x} \in \mathbb{Z}^n_+. \tag{1.33}$$

Equation (1.30) is referred to as the *convexity constraint*. When we relax the integrality of $\mathbf{x}$, there is no need to link $\mathbf{x}$ and $\boldsymbol{\lambda}$, and we may also relax (1.32). The columns of this special master problem are defined by the extreme points and extreme rays of $\mathrm{conv}(X)$. We solve the master by column generation to get its optimal objective function value $z_{MP}^\star$. Given an optimal dual solution $\boldsymbol{\pi}$ and π_0 to the current RMP, where variable π_0 corresponds to the convexity constraint, the subproblem is to determine $\min_{j\in P}\{c_j - \boldsymbol{\pi}^t\mathbf{a}_j - \pi_0\}$ and $\min_{j\in R}\{c_j - \boldsymbol{\pi}^t\mathbf{a}_j\}$. By our previous linear transformation and since π_0 is a constant, this results in

$$\bar{c}^\star := \min\{(\mathbf{c}^t - \boldsymbol{\pi}^t A)\mathbf{x} - \pi_0 \mid \mathbf{x} \in X\}. \tag{1.34}$$

This subproblem is an integer linear program. When $\bar{c}^\star \geq 0$, there is no negative reduced cost column, and the algorithm terminates. When $\bar{c}^\star < 0$ and finite, an optimal solution to (1.34) is an extreme point $\mathbf{x}_p$ of $\mathrm{conv}(X)$, and we add the column $[\mathbf{c}^t\mathbf{x}_p, (A\mathbf{x}_p)^t, 1]^t$ to the RMP. When $\bar{c}^\star = -\infty$ we identify an extreme ray $\mathbf{x}_r$ of $\mathrm{conv}(X)$ as a solution $\mathbf{x} \in X$ to $(\mathbf{c}^t - \boldsymbol{\pi}^t A)\mathbf{x} = 0$, and add the column $[\mathbf{c}^t\mathbf{x}_r, (A\mathbf{x}_r)^t, 0]^t$ to the RMP.

From (1.23) together with the convexity constraint we obtain in each iteration

$$\bar{z} + \bar{c}^\star \leq z_{MP}^\star \leq \bar{z}, \tag{1.35}$$

where $\bar{z} = \boldsymbol{\pi}^t\mathbf{b} + \pi_0$ is again the optimal objective function value of the RMP. Since $z_{MP}^\star \leq z^\star$, $\bar{z} + \bar{c}^\star$ is also a lower bound on $z^\star$. In general, $\bar{z}$ is not a valid upper bound on $z^\star$, except if the current $\mathbf{x}$ variables are integer. The algorithm is exact and finite as long as finiteness is ensured in optimizing the RMP.

The original formulation is the starting point to obtain integer solutions in the $\mathbf{x}$ variables. Branching and cutting constraints are added there, the reformulation as in Section 1.1 is re-applied, and the process continues with an updated master problem. It is important to see that it is our choice as to whether the additional constraints remain in the master problem (as in the previous section) or go into the subproblem (as we will see later).

Pricing out the original x variables.   Assume that in (1.24) we have a linear subproblem $X = \{\mathbf{x} \in \mathbb{R}_+^n \mid D\mathbf{x} \geq \mathbf{d}\} \neq \varnothing$. Column generation then essentially solves the linear program

$$\min \mathbf{c}^t\mathbf{x} \quad \text{subject to } A\mathbf{x} \geq \mathbf{b}, \quad D\mathbf{x} \geq \mathbf{d}, \qquad \mathbf{x} \geq 0.$$

We obtain an optimal primal solution $\mathbf{x}$ but only the dual multipliers $\boldsymbol{\pi}$ associated with the constraint set $A\mathbf{x} \geq \mathbf{b}$. However, following an idea of Walker (1969) we can also retrieve the dual variables $\boldsymbol{\sigma}$ associated with

$D\mathbf{x} \geq d$: It is the vector obtained from solving the linear subproblem in the last iteration of the column generation process. This full dual information allows for a pricing of the original variables, and therefore a possible elimination of some of them. Given an upper bound on the integer optimal objective function value of the original problem, one can eliminate an x variable if its reduced cost is larger than the optimality gap.

In the general case of a linear integer or even non-linear pricing subproblem, the above procedure does not work. Poggi de Aragão and Uchoa (2003) suggest to directly use the extensive formulation: If we keep the coupling constraint (1.32) in the master problem, it suffices to impose the constraint $\mathbf{x} \geq \epsilon$, for a small $\epsilon > 0$, at the end of the process. The shadow prices of these constraints are the reduced costs of the $\mathbf{x}$ vector of original variables. Note that there is no need to apply the additional constraints to already positive variables. Computational experiments underline the benefits of this procedure.

Block diagonal structure. For practical problems Dantzig-Wolfe decomposition can typically exploit a block diagonal structure of D, i.e.,

$$
D = \begin{pmatrix} D^1 & & & \\ & D^2 & & \\ & & \ddots & \\ & & & D^\kappa \end{pmatrix}, \quad \mathbf{d} = \begin{pmatrix} \mathbf{d}^1 \\ \mathbf{d}^2 \\ \vdots \\ \mathbf{d}^\kappa \end{pmatrix}. \tag{1.36}
$$

Each $X^k = \{D^k\mathbf{x}^k \geq \mathbf{d}^k, \mathbf{x}^k \geq 0 \text{ and integer}\}$, $k \in K := \{1,\ldots,\kappa\}$, gives rise to a representation as in (1.27). The decomposition yields κ subproblems, each with its own convexity constraint and associated dual variable:

$$
\bar{c}^{k\star} := \min\{(\mathbf{c}^{kT} - \boldsymbol{\pi}^t A^k)\mathbf{x}^k - \pi_0^k \mid \mathbf{x}^k \in X^k\}, \quad k \in K. \tag{1.37}
$$

The superscript k to all entities should be interpreted in the canonical way. The algorithm terminates when $\bar{c}^{k\star} \geq 0$, for all $k \in K$. Otherwise, extreme points and rays identified in (1.37) give rise to new columns to be added to the RMP. By linear programming duality, $\bar{z} = \boldsymbol{\pi}^t\mathbf{b} + \sum_{k=1}^{\kappa} \pi_0^k$, and we obtain the following bounds, see Lasdon (1970):

$$
\bar{z} + \sum_{k=1}^{\kappa} \bar{c}^{k\star} \leq z_{MP}^\star \leq \bar{z}. \tag{1.38}
$$

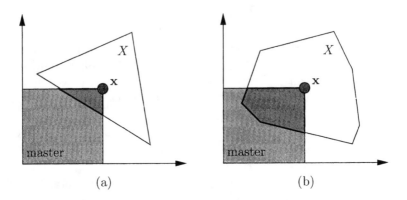

Figure 1.2.   Schematic illustration of the domains of master and subproblem X.

2.3   Useful working knowledge

When problems get larger and computationally much more difficult than our small constrained shortest path problem it is helpful to know more about mechanisms, their consequences, and how to exploit them.

Infeasible paths.   One may wonder why we kept infeasible paths in the RMP during column generation. Here, as for the whole process, we cannot overemphasize the fact that knowledge about the *integer* solution usually does not help us much in solving the *linear relaxation* program. Figure 1.2 illustrates the domain of the RMP (shaded) and the domain X of the subproblem. In part a), the optimal solution **x**, symbolized by the dot, is uniquely determined as a convex combination of the three extreme points of the triangle X, even though all of them are not feasible for the intersection of the master and subproblem. In our example, in iteration BB0.5, any convex combination of feasible paths which have been generated, namely 13256 and 1356, has cost larger than 7, i.e., is suboptimal for the linear relaxation of the master problem. Infeasible paths are removed only if needed during the search for an integer solution. In Figure 1.2 (a), **x** can be integer and no branch-and-bound search is needed.

In part b) there are many ways to express the optimal solution as a convex combination of three extreme points. This is a partial explanation of the slow convergence (*tailing off*) of linear programming column generation.

Lower and upper bounds.   Figure 1.3 gives the development of upper ($\bar{z}$) and lower ($\bar{z} + \bar{c}^*$) bounds on z_{MP}^* in the root node for our

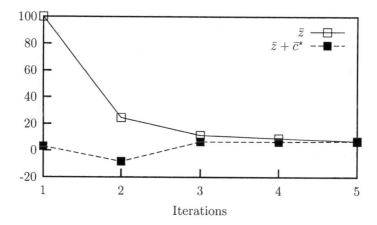

Figure 1.3.   Development of lower and upper bounds on z_{MP} in BB0.

Table 1.3.   Multipliers σ_i for the flow conservation constraints on nodes $i \in N$.

node i	1	2	3	4	5	6
σ_i	29	8	13	5	0	-6

small constrained shortest path example. The values for the lower bound are 3.0, -8.33, 6.6, 6.5, and finally 7. While the upper bound decreases monotonically (as expected when minimizing a linear program) there is no monotony for the lower bound. Still, we can use these bounds to evaluate the quality of the current solution by computing the optimality gap, and could stop the process when a preset quality is reached. Is there any use of the bounds beyond that?

Note first that $UB = \bar{z}$ is not an upper bound on $z^\star$. The currently (over all iterations) best lower bound LB, however, is a lower bound on $z^\star_{MP}$ and on $z^\star$. Even though there is no direct use of LB or UB in the master problem we can impose the additional constraints $LB \leq \mathbf{c}^t \mathbf{x} \leq UB$ to the subproblem structure X if the subproblem consists of a single block. Be aware that this cutting modifies the subproblem structure X, with all algorithmic consequences, that is, possible complications for a combinatorial algorithm. In our constrained shortest path example, two generated paths are feasible and provide upper bounds on the optimal integer solution $z^\star$. The best one is path 13256 of cost 15 and duration 10. Table 1.3 shows the multipliers σ_i, $i = 1, \ldots, 6$ for the flow conservation constraints of the path structure X at the last iteration of the column generation process.

Therefore, given the optimal multiplier $\pi_1 = -2$ for the resource constraint, the reduced cost of an arc is given by $\bar{c}_{ij} = c_{ij} - \sigma_i + \sigma_j - t_{ij}\pi_1$, $(i,j) \in A$. The reader can verify that $\bar{c}_{34} = 3 - 13 + 5 - (7)(-2) = 11$. This is the only reduced cost which exceeds the current optimality gap which equals to $15 - 7 = 8$. Arc $(3,4)$ can be permanently discarded from the network and paths 1346 and 13456 will never be generated.

Integrality property.   Solving the subproblem as an integer program usually helps in closing part of the integrality gap of the master problem (Geoffrion, 1974), except when the subproblem possesses the *integrality property*. This property means that solutions to the pricing problem are naturally integer when it is solved as a linear program. This is the case for our shortest path subproblem and this is why we obtained the value of the linear relaxation of the original problem as the value of the linear relaxation of the master problem.

When looking for an integer solution to the original problem, we need to impose new restrictions on (1.1)–(1.6). One way is to take advantage of a new X structure. However, if the new subproblem is still solved as a linear program, $z^\star_{MP}$ remains 7. Only solving the new X structure as an integer program may improve $z^\star_{MP}$.

Once we understand that we can modify the subproblem structure, we can devise other decomposition strategies. One is to define the X structure as

$$\sum_{(i,j)\in A} t_{ij}x_{ij} \le 14, \quad x_{ij} \text{ binary}, \qquad (i,j) \in A \qquad (1.39)$$

so that the subproblem becomes a knapsack problem which does not possess the integrality property. Unfortunately, in this example, $z^\star_{MP}$ remains 7. However, improvements can be obtained by imposing more and more constraints to the subproblem. An example is to additionally enforce the selection of one arc to leave the source (1.2) and another one to enter the sink (1.4), and impose constraint $3 \le \sum_{(i,j)\in A} x_{ij} \le 5$ on the minimum and maximum number of selected arcs. Richer subproblems, as long as they can be solved efficiently and do not possess the integrality property, may help in closing the integrality gap.

It is also our decision how much branching and cutting information (ranging from none to all) we put into the subproblem. This choice depends on where the additional constraints are more easily accommodated in terms of algorithms and computational tractability. Branching decisions imposed on the subproblem can reduce its solution space and may turn out to facilitate a solution as integer program. As an illus-

Table 1.4.   Lower bound cut added to the subproblem at the end of the root node.

Iteration	Master Solution	$\bar{z}$	π_0	π_1	$\bar{c}^*$	p	c_p	t_p	UB	LB
BB0.6	$\lambda_{13256} = 1$	**15**	15	0	-2	13246	13	13	15	13
BB0.7	$\lambda_{13246} = 1$	**13**	13	0					13	13

tration we describe adding the lower bound cut in the root node of our small example.

Imposing the lower bound cut $\mathbf{c^t x \geq 7}$.   Assume that we have solved the relaxed RMP in the root node and instead of branching, we impose the lower bound cut on the X structure, see Table 1.4. Note that this cut would not have helped in the RMP since $\bar{z} = \mathbf{c}^t\mathbf{x} = 7$ already holds. We start modifying the RMP by removing variables λ_{1246} and λ_{1256} as their cost is smaller than 7. In iteration BB0.6, for the re-optimized RMP $\lambda_{13256} = 1$ is optimal at cost 15; it corresponds to a feasible path of duration 10. UB is updated to 15 and the dual multipliers are $\pi_0 = 15$ and $\pi_1 = 0$. The X structure is modified by adding constraint $\mathbf{c}^t\mathbf{x} \geq 7$. Path 13246 is generated with reduced cost -2, cost 13, and duration 13. The new lower bound is $15 - 2 = 13$. On the downside of this significant improvement is the fact that we have destroyed the pure network structure of the subproblem which we have to solve as an integer program now. We may pass along with this circumstance if it pays back a better bound.

We re-optimize the RMP in iteration BB0.7 with the added variable λ_{13246}. This variable is optimal at value 1 with cost and duration equal to 13. Since this variable corresponds to a feasible path, it induces a better upper bound which is equal to the lower bound: Optimality is proven. There is no need to solve the subproblem.

Note that using the dynamically adapted lower bound cut right from the start has an impact on the solution process. For example, the first generated path 1246 would be eliminated in iteration BB0.3 since the lower bound reaches 6.6, and path 1256 is never generated. Additionally adding the upper bound cut has a similar effect.

Acceleration strategies.   Often acceleration techniques are key elements for the viability of the column generation approach. Without them, it would have been almost impossible to obtain quality solutions to various applications, in a reasonable amount of computation time. We sketch here only some strategies, see e.g., Desaulniers et al. (2001) for much more.

The most widely used strategy is to return to the RMP many negative reduced cost columns in each iteration. This generally decreases
the number of column generation iterations, and is particularly easy
when the subproblem is solved by dynamic programming. When the
number of variables in the RMP becomes too large, non-basic columns
with current reduced cost exceeding a given threshold may be removed.
Accelerating the pricing algorithm itself usually yields most significant
speed-ups. Instead of investing in a most negative reduced cost column,
any variable with negative reduced cost suffices. Often, such a column
can be obtained heuristically or from a pool of columns containing not
yet used columns from previous calls to the subproblem. In the case of
many subproblems, it is often beneficial to consider only few of them
each time the pricing problem is called. This is the well-known partial
pricing. Finally, in order to reduce the tailing off behavior of column generation, a heuristic rule can be devised to prematurely stop the linear
relaxation solution process, for example, when the value of the objective
function does not improve sufficiently in a given number of iterations.
In this case, the approximate LP solution does not necessarily provide a
lower bound but using the current dual multipliers, a lower bound can
still be computed.

With a careful use of these ideas one may confine oneself with a non-
optimal solution in favor of being able to solve much larger problems.
This turns column generation into optimization based heuristics which
may be used for comparison with other methods for a given class of
problems.

3.   A dual point of view

The dual program of the RMP is a relaxation of the dual of the master
problem, since constraints are omitted. Viewing column generation as
row generation in the dual, it is a special case of Kelley'scutting plane
method from 1961. Recently, this dual perspective attracted considerable attention and we will see that it provides us with several key
insights. Observe that the generation process as well as the stopping
criteria are driven entirely by the dual multipliers.

3.1   Lagrangian relaxation

A practically often used dual approach to solving (1.24) is *Lagrangian
relaxation*, see Geoffrion (1974). Penalizing the violation of $Ax \geq b$ via
Lagrangian multipliers $\pi \geq 0$ in the objective function results in the

Lagrangian subproblem relative to constraint set $Ax \geq b$

$$L(\boldsymbol{\pi}) := \min_{\mathbf{x} \in X} \mathbf{c}^t \mathbf{x} - \boldsymbol{\pi}^t(A\mathbf{x} - \mathbf{b}). \tag{1.40}$$

Since $L(\boldsymbol{\pi}) \leq \min\{\mathbf{c}^t \mathbf{x} - \boldsymbol{\pi}^t(A\mathbf{x} - \mathbf{b}) \mid A\mathbf{x} \geq \mathbf{b}, \mathbf{x} \in X\} \leq z^\star$, $L(\boldsymbol{\pi})$ is a lower bound on $z^\star$. The best such bound on $z^\star$ is provided by solving the *Lagrangian dual problem*

$$\mathcal{L} := \max_{\boldsymbol{\pi} \geq \mathbf{0}} L(\boldsymbol{\pi}). \tag{1.41}$$

Note that (1.41) is a problem in the dual space while (1.40) is a problem in $\mathbf{x}$. The Lagrangian function $L(\boldsymbol{\pi})$, $\boldsymbol{\pi} \geq \mathbf{0}$ is the lower envelope of a family of functions linear in $\boldsymbol{\pi}$, and therefore is a concave function of $\boldsymbol{\pi}$. It is piecewise linear with breakpoints where the optimal solution of $L(\boldsymbol{\pi})$ is not unique. In particular, $L(\boldsymbol{\pi})$ is not differentiable, but only sub-differentiable. The most popular, since very easy to implement, choice to obtain optimal or near optimal multipliers are subgradient algorithms. However, let us describe an alternative computation method, see Nemhauser and Wolsey (1988). We know that replacing X by $\text{conv}(X)$ in (1.24) does not change $z^\star$ and this will enable us to write (1.41) as a linear program.

When $X = \varnothing$, which may happen during branch-and-bound, then $\mathcal{L} = \infty$. Otherwise, given some multipliers $\boldsymbol{\pi}$, the Lagrangian bound is

$$L(\boldsymbol{\pi}) = \begin{cases} -\infty & \text{if } (\mathbf{c}^t - \boldsymbol{\pi}^t A)\mathbf{x}_r < 0 \text{ for some } r \in R \\ \mathbf{c}^t \mathbf{x}_p - \boldsymbol{\pi}^t(A\mathbf{x}_p - \mathbf{b}) & \text{for some } p \in P \text{ otherwise.} \end{cases}$$

Since we assumed $z^\star$ to be finite, we avoid unboundedness by writing (1.41) as

$$\max_{\boldsymbol{\pi} \geq \mathbf{0}} \min_{p \in P} \mathbf{c}^t \mathbf{x}_p - \boldsymbol{\pi}^t(A\mathbf{x}_p - \mathbf{b}) \text{ such that } (\mathbf{c}^t - \boldsymbol{\pi}^t A)\mathbf{x}_r \geq 0, \quad \forall r \in R,$$

or as a linear program with many constraints

$$\mathcal{L} = \max \pi_0$$
$$\text{subject to} \, \boldsymbol{\pi}^t(A\mathbf{x}_p - \mathbf{b}) + \pi_0 \leq \mathbf{c}^t \mathbf{x}_p, \quad p \in P$$
$$\boldsymbol{\pi}^t A\mathbf{x}_r \leq \mathbf{c}^t \mathbf{x}_r, \quad r \in R \tag{1.42}$$
$$\boldsymbol{\pi} \geq \mathbf{0}.$$

The dual of (1.42) reads as the linear relaxation of the master problem (1.28)–(1.33):

$$\mathcal{L} = \min \sum_{p \in P} \mathbf{c}^t \mathbf{x}_p \lambda_p + \sum_{r \in R} \mathbf{c}^t \mathbf{x}_r \lambda_r \tag{1.43}$$

$$\text{subject to } \sum_{p \in P} A\mathbf{x}_p \lambda_p + \sum_{r \in R} A\mathbf{x}_r \lambda_r \geq \mathbf{b} \sum_{p \in P} \lambda_p \tag{1.44}$$

$$\sum_{p \in P} \lambda_p = 1 \tag{1.45}$$

$$\boldsymbol{\lambda} \geq \mathbf{0}. \tag{1.46}$$

Observe that for a given vector $\boldsymbol{\pi}$ of multipliers and a constant π_0,

$$L(\boldsymbol{\pi}) = (\boldsymbol{\pi}^t \mathbf{b} + \pi_0) + \min_{\mathbf{x} \in \text{conv}(X)} (\mathbf{c}^t - \boldsymbol{\pi}^t A)\mathbf{x} - \pi_0 = \bar{z} + \bar{c}^{\star},$$

that is, each time the RMP is solved during the Dantzig-Wolfe decomposition, the computed lower bound in (1.35) is the same as the Lagrangian bound, that is, for optimal $\mathbf{x}$ and $\boldsymbol{\pi}$ we have $z^{\star} = \mathbf{c}^t \mathbf{x} = L(\boldsymbol{\pi})$.

When we apply Dantzig-Wolfe decomposition to (1.24) we satisfy complementary slackness conditions, we have $\mathbf{x} \in \text{conv}(X)$, and we satisfy $A\mathbf{x} \geq \mathbf{b}$. Therefore only the integrality of $\mathbf{x}$ remains to be checked. The situation is different for subgradient algorithms. Given optimal multipliers $\boldsymbol{\pi}$ for (1.41), we can solve (1.40) which ensures that the solution, denoted $\mathbf{x}_\pi$, is (integer) feasible for X and $\boldsymbol{\pi}^t(A\mathbf{x}_\pi - \mathbf{b}) = 0$. Still, we have to check whether the relaxed constraints are satisfied, that is, $A\mathbf{x}_\pi \geq \mathbf{b}$ to prove optimality. If this condition is violated, we have to recover optimality of a primal-dual pair $(\mathbf{x}, \boldsymbol{\pi})$ by branch-and-bound. For many applications, one is able to slightly modify infeasible solutions obtained from the Lagrangian subproblems with only a small degradation of the objective value. Of course these are only approximate solutions to the original problem. We only remark that there are more advanced (non-linear) alternatives to solve the Lagrangian dual like the *bundle method* (Hiriart–Urruty and Lemaréchal, 1993) based on quadratic programming, and the *analytic center cutting plane method* (Goffin and Vial, 1999), an interior point solution approach. However, the performance of these methods is still to be evaluated in the context of integer programming.

3.2   Dual restriction/Primal relaxation

Linear programming column generation remained "as is" for a long time. Recently, the dual point of view prepared the ground for technical advances.

Structural dual information.   Consider a master problem and its dual and assume both are feasible and bounded. In some situations we may have additional knowledge about an optimal dual solution which we may express as additional valid inequalities $F^t \pi \leq \mathbf{f}$ in the dual. To be more precise, we would like to add inequalities which are satisfied by at least one optimal dual solution. Such valid inequalities correspond to additional primal variables $\mathbf{y} \geq \mathbf{0}$ of cost $\mathbf{f}$ that are *not* present in the original master problem. From the primal perspective, we therefore obtain a relaxation. Devising such *dual-optimal inequalities* requires (but also exploits) a specific problem knowledge. This has been successfully applied to the one-dimensional cutting stock problem, see Valério de Carvalho (2003); Ben Amor et al. (2003).

Oscillation. It is an important observation that the dual variable values do not develop smoothly but they very much oscillate during column generation. In the first iterations, the RMP contains too few columns to provide any useful dual information, in turn resulting in non useful variables to be added. Initially, often the penalty cost of artificial variables guide the values of dual multipliers (calls this the *heading-in effect* Vanderbeck, 2004). One observes that the variables of an optimal master problem solution are generated in the last iterations of the process when dual variables finally come close to their respective optimal values. Understandably, dual oscillation has been identified as a major efficiency issue. One way to control this behavior is to impose lower and upper bounds, that is, we constrain the vector of dual variables to lie "in a box" around its current value. As a result, the RMP is modified by adding slack and surplus variables in the corresponding constraints. After re-optimization of the new RMP, if the new dual optimum is attained on the boundary of the box, we have a direction towards which the box should be relocated. Otherwise, the optimum is attained in the box's interior, producing the global optimum. This is the principle of the `Boxstep` method (Marsten, 1975; Marsten et al., 1975).

Stabilization. *Stabilized column generation* (see du Merle et al., 1999; Ben Amor and Desrosiers, 2003) offers more flexibility for controlling the duals. Again, the dual solution π is restricted to stay in a box, however, the box may be left at a certain penalty cost. This penalty may be a piecewise linear function. The size of the box and the penalty are updated dynamically so as to make greatest use of the latest available information. With intent to reduce the dual variables' variation, select a small box containing the (in the beginning estimated) current dual solution, and solve the modified master problem. Componentwise, if the new dual solution lies in the box, reduce its width and increase the penalty. Otherwise, enlarge the box and decrease the penalty. This allows for fresh dual solutions when the estimate was bad. The update

could be performed in each iteration, or alternatively, each time a dual
solution of currently best quality is obtained.

3.3   Dual aspects of our shortest path example

Optimal primal solutions.   Assume that we penalize the violation
of resource constraint (1.5) via the objective function with the single
multiplier $\pi_1 \leq 0$ which we determine using a subgradient method. Its
optimal value is $\pi_1 = -2$, as we know from solving the primal master
problem by column generation. The aim now is to find an optimal integer
solution **x** to our original problem. From the Lagrangian subproblem
with π_1 we get $L(-2) = 7$ and generate either the infeasible path 1256
of cost 5 and duration 15, or the feasible path 13256 of cost 15 and
duration 10. The important issue now, left out in textbooks, is how to
perform branch-and-bound in that context?

Assume that we generated path 1256. A possible strategy to start the
branch-and-bound search tree is to introduce cut $x_{12}+x_{25}+x_{56} \leq 2$ in the
original formulation (1.1)–(1.6), and then either incorporate it in X or
relax it (and penalize its violation) in the objective function via a second
multiplier. The first alternative prevents the generation of path 1256 for
any value of π_1. However, we need to re-compute its optimal value
according to the modified X structure, i.e., $\pi_1^\star = -1.5$. In this small
example, a simple way to get this value is to solve the linear relaxation
of the full master problem excluding the discarded path. Solving the
new subproblem results in an improved lower bound $L(-1.5) = 9$, and
the generated path 13256 of cost 15 and duration 10. This path is
feasible but suboptimal. In fact, this solution **x** is integer, satisfies the
path constraints but does not satisfy complementary slackness for the
resource constraint. That is, $\pi_1(\sum_{(i,j)\in A} t_{ij}x_{ij}-14) = -1.5(10-14) \neq 0$.
The second cut $x_{13} + x_{32} + x_{25} + x_{56} \leq 3$ in the X structure results in
$\pi_1^\star = -2$, an improved lower bound of $L(-2) = 11$, and the generated
path 1246 of cost 3 and duration 18. This path is *infeasible*, and adding
the third cut $x_{12} + x_{24} + x_{46} \leq 2$ in the subproblem X gives us the
optimal solution, that is, $\pi_1 = 0$, $L(0) = 13$ with the generated path
13245 of cost 13 and duration 13.

Alternatively, we could have penalized the cuts via the objective function which would *not* have destroyed the subproblem structure. We encourage the reader to find the optimal solution this way, making use of
any kind of branching and cutting decisions that can be defined on the
x variables.

A box method.   It is worthwhile to point out that a good deal of
the operations research literature is about Lagrangian relaxation. We

can steal ideas there about how to decompose problems and use them in column generation algorithms (see Guignard, 2004). In fact, the complementary availability of both, primal and dual ideas, brings us in a strong position which e.g., motivates the following. Given optimal multipliers π obtained by a subgradient algorithm, one can use very small boxes around these in order to rapidly derive an optimal primal solution x on which branching and cutting decisions are applied. The dual information is incorporated in the primal RMP in the form of initial columns together with the columns corresponding to the generated subgradients. This gives us the opportunity to initiate column generation with a solution which intuitively bears both, relevant primal and dual information.

Table 1.5. A box method in the root node with $-2.1 \le \pi_1 \le -1.9$.

Iteration	Master Solution	$\bar{z}$	π_0	π_1	$\bar{c}^*$	p	c_p	t_p	UB	LB
BoxBB0.1	$y_0 = 1, s_2 = 14$	73.4	100.0	-1.9	-66.5	1256	5	15	–	6.9
BoxBB0.2	$\lambda_{1256} = 1, s_1 = 1$	7.1	36.5	-2.1	-0.5	13256	15	10	7.1	6.6
BoxBB0.3	$\lambda_{13256} = 0.2, \lambda_{1256} = 0.8$	7.0	35.0	-2.0	0				7	7
Arc flows:	$x_{12} = 0.8, x_{13} = x_{32} = 0.2, x_{25} = x_{56} = 1$									

Alternatively, we have applied a box method to solve the primal master problem by column generation, c.f. Table 1.5. We impose the box constraint $-2.1 \le \pi_1 \le -1.9$. At start, the RMP contains the artificial variable y_0 in the convexity constraint, and surplus (s_1 with cost coefficient 1.9) and slack (s_2 with cost coefficient 2.1) variables in the resource constraint.

In the first iteration, denoted BoxBB0.1, the artificial variable $y_0 = 1$ and the slack variable $s_2 = 14$ constitute a solution. The current dual multipliers are $\pi_0 = 100$ and $\pi_1 = -1.9$. Path 1256 is generated (cost 5 and duration 15) and the lower bound already reaches 6.9. In the second iteration, $\lambda_{1256} = 1$ and surplus $s_1 = 1$ define an optimal solution to the RMP. This solution provides an upper bound of 7.1 and dual multipliers are $\pi_0 = 36.5$ and $\pi_1 = -2.1$. Experimentation reveals that a smaller box around $\pi = -2$ results in a smaller optimality gap. The subproblem generates path 13256 (cost 15 and duration 10) and the lower bound decreases to 6.6. Solving the RMP in BoxBB0.3 gives us an optimal solution of the linear relaxation of the master problem. This can be verified in two ways: The previous lower bound values 6.9 and 6.6, rounded up, equal the actual upper bound $\bar{z} = 7$; and the reduced cost of the subproblem is zero. Hence, the solution process is completed in only three iterations! The box constraint has to be relaxed when

Table 1.6. Hyperplanes (*lines*) defined by the extreme points of X, i.e., by the indicated paths.

p	1246	1256	12456	13246	13256	132456	1346	13456	1356
line	$3 - 4\pi$	$5 - \pi$	14	$13 + \pi$	$15 + 4\pi$	$24 + 5\pi$	$16 - 3\pi$	$27 + \pi$	$24 + 6\pi$

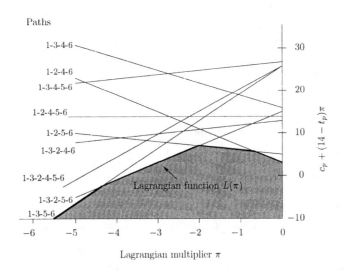

Figure 1.4. Lagrangian function $L(\pi)$.

branch-and-bound starts but this does not require any re-optimization iteration.

Geometric interpretation.   Let us draw the Lagrangian function $L(\pi)$, for $\pi \leq 0$, for our numerical example, where $\pi \equiv \pi_1$. Since the polytope X is bounded, there are no extreme rays and $L(\pi)$ can be written in terms of the nine possible extreme points (paths). That is, $L(\pi) = \min_{p \in P} c_p + (14 - t_p)\pi$, where c_p and t_p are the cost and duration of path p, respectively. Table 1.6 lists the lines (in general, hyperplanes) defined by $p \in P$, with an intercept of c_p and a slope of $14 - t_p$. We have plotted these lines in Figure 1.4.

Observe that for π given, the line of smallest cost defines the value of function $L(\pi)$. The Lagrangian function $L(\pi)$ is therefore the lower envelope of all lines and its topmost point corresponds to the value $\mathcal{L}$ of the Lagrangian dual problem.

If one starts at $\pi = 0$, the largest possible value is $L(0) = 3$, on the line defined by path 1246. At that point the slope is negative (the line is defined by $3 - 4\pi$) so that the next multiplier should be found on the left to the current point. In Dantzig-Wolfe decomposition, we found $\pi(\equiv \pi_1) = -97/18 \approx -5.4$. This result depends on big M: The exact value of the multiplier is $(3 - M)/18$. For any large M, path 1356 is returned, and here, $L(-97/18) = -25/3$, a lesser lower bound on $L(\pi)$.

The next multiplier is located where the two previous lines intersect, that is, where $3 - 4\pi = 24 + 6\pi$ for $\pi = -2.1$. $L(-2.1) = 6.6$ for path 13256 with an improvement on the lower bound. In the next iteration, the optimal multiplier value is at the intersection of the lines defined by paths 1246 and 13256, that is, $3 - 4\pi = 15 + 4\pi$ for $\pi = -1.5$. For that value, the Lagrangian function reaches 6.5 for path 1256. The final and optimal Lagrangian multiplier is at the intersection of the lines defined by paths 13256 and 1256, that is, $15 + 4\pi = 5 - \pi$ for $\pi = -2$ and therefore $L(-2) = 7$. We can now see why the lower bound is not strictly increasing: The point associated with the Lagrangian multiplier moves from left to right, and the value of the Lagrangian function is determined by the lowest line which is hit.

The Lagrangian point of view also teaches us why two of our methods are so successful: When we used the *box method* for solving the linear relaxation of the master problem by requiring that π has to lie in the small interval $[-2.1, -1.9]$ around the optimal value $\pi = -2$, only two paths are sufficient to describe the lower envelope of the Lagrangian function $L(\pi)$. This explains the very fast convergence of this stabilization approach.

Also, we previously added the cut $\mathbf{c}^t \mathbf{x} \geq 7$ in the subproblem, when we were looking for an integer solution to our resource constrained shortest path problem. In Figure 1.4 this corresponds to removing the two lines with an intercept smaller than 7, that is, for paths 1246 and 1256. The maximum value of function $L(\pi)$ is now attained for $\pi = 0$ and $L(0) = 13$.

4.   On finding a good formulation

Many vehicle routing and crew scheduling problems, but also many others, possess a multicommodity flow problem as an underlying basic structure (see Desaulniers et al., 1998). Interestingly, Ford and Fulkerson (1958), suggested to solve this "problem of some importance in applications" via a "specialized computing scheme that takes advantage of the structure": The birth of column generation which then inspired Dantzig and Wolfe (1960) to generalize the framework to a decomposition scheme for linear programs as presented in Section 2.1. Ford and Fulkerson had no idea "whether the method is practicable." In fact, at that time, it

was not. Not only because of the lack of powerful computers but mainly
because (only) linear programming was used to attack integer programs:
"That integers should result as the solution of the example is, of course,
fortuitous" (Gilmore and Gomory, 1961).

In this section we stress the importance (and the efforts) to find a
"good" formulation which is amenable to column generation. Our example is *the* classical column generation application, see Ben Amor and
Valério de Carvalho (2004). Given a set of rolls of width L and integer
demands n_i for items of length ℓ_i, $i \in I$ the aim of the *cutting stock*
problem is to find patterns to cut the rolls to fulfill the demand while
minimizing the number of used rolls. An item may appear more than
once in a cutting pattern and a cutting pattern may be used more than
once in a feasible solution.

4.1 Gilmore and Gomory (1961, 1963)

Let R be the set of all feasible cutting patterns. Let coefficient a_{ir}
denote how often item $i \in I$ is used in pattern $r \in R$. Feasibility of r
is expressed by the knapsack constraint $\sum_{i \in I} a_{ir} \ell_i \leq L$. The classical
formulation of Gilmore and Gomory (1961, 1963) makes use of non-
negative integer variables: λ_r reflects how often pattern r is cut in the
solution. We are first interested in solving the linear relaxation of that
formulation by column generation. Consider the following primal and
dual linear master problems P_{CS} and D_{CS}, respectively:

$$
\begin{array}{ll}
(P_{CS})\colon \sum_{r \in R} \lambda_r & \quad (D_{CS})\colon \max \sum_{i \in I} n_i \pi_i \\
\sum_{r \in R} a_{ir} \lambda_r \geq n_i, \quad i \in I & \quad \sum_{i \in I} a_{ir} \pi_i \leq 1, \quad r \in R \\
\lambda_r \geq 0, \quad r \in R & \quad \pi_i \geq 0, \quad i \in I \,.
\end{array}
$$

For $i \in I$, let π_i denote the associated dual multiplier, and let x_i count
the frequency item i is selected in a pattern. Negative reduced cost
patterns are generated by solving

$$
\min 1 - \sum_{i \in I} \pi_i x_i \equiv \max \sum_{i \in I} \pi_i x_i \text{ such that } \sum_{i \in I} x_i \ell_i \leq L, \ x_i \in \mathbb{Z}_+, \ i \in I.
$$

This pricing subproblem is a knapsack problem and the coefficients of
the generated columns are given by the value of variables x_i, $i \in I$.

Gilmore and Gomory (1961) showed that equality in the demand constraints can be replaced by greater than or equal. Column generation is
accelerated by this transformation: Dual variables π_i, $i \in I$ then assume
only non-negative values and it is easily shown by contradiction that
these *dual non-negativity constraints* are satisfied by all optimal solutions. Therefore they define a set of (simple) dual-optimal inequalities.

Although P_{CS} is known to provide a strong lower bound on the optimal number of rolls, its solution can be fractional and one has to resort to branch-and-bound. In the literature one finds several tailored branching strategies based on decisions made on the λ variables, see Barnhart et al. (1998); Vanderbeck and Wolsey (1996). However, we have seen that branching rules with a potential for exploiting more structural information can be devised when some compact formulation is available.

4.2   Kantorovich (1939, 1960)

From a technical point of view, the proposal by Gilmore and Gomory is a master problem and a pricing subproblem. For precisely this situation, Villeneuve et al. (2003) show that an equivalent compact formulation exists under the assumption that the sum of the variables of the master problem be bounded by an integer κ and that we have the possibility to also bound the domain of the subproblem. The variables and the domain of the subproblem are duplicated κ times, and the resulting formulation has a block diagonal structure with κ identical subproblems. Formally, when we start from Gilmore and Gomory's formulation, this yields the following formulation of the cutting stock problem.

Given the dual multipliers $\pi_i, i \in I$, the pricing subproblem can alternatively be written as

$$\min \left(x_0 - \sum_{i \in I} \pi_i x_i \right) \tag{1.47}$$

$$\text{subject to } \sum_{i \in I} \ell_i x_i \leq L x_0 \tag{1.48}$$

$$x_0 \in \{0, 1\} \tag{1.49}$$

$$x_i \in \mathbb{Z}_+ i \in I, \tag{1.50}$$

where x_0 is a binary variable assuming value 1 if a roll is used and 0 otherwise. When x_0 is set to 1, (1.50) is equivalent to solving a knapsack problem while if $x_0 = 0$, then $x_i = 0$ for all $i \in I$ and this null solution corresponds to an empty pattern, i.e., a roll that is not cut.

The constructive procedure to recover a compact formulation leads to the definition of a specific subproblem for each roll. Let $K := \{1, \ldots, \kappa\}$ be a set of rolls of width L such that $\sum_{r \in R} \lambda_r \leq \kappa$ for some feasible solution λ. Let $\mathbf{x}^k = (x_0^k, (x_i^k)_{i \in I}), k \in K$, be duplicates of the $\mathbf{x}$ variables, that is, x_0^k is a binary variable assuming value 1 if roll k is used and 0 otherwise, and x_i^k, $i \in I$ is a non-negative integer variable counting how often item i is cut from roll k. The compact formulation reads as follows:

$$\min \sum_{k \in K} x_0^k \tag{1.51}$$

$$\text{subject to } \sum_{k \in K} x_i^k \geq n_i i \in I \tag{1.52}$$

$$\sum_{i \in I} \ell_i x_i^k \leq L x_0^k k \in K \tag{1.53}$$

$$x_0^k \in \{0, 1\} k \in K \tag{1.54}$$

$$x_i^k \in \mathbb{Z}_+ k \in K, \quad i \in I, \tag{1.55}$$

which was proposed already by Kantorovich in 1939 (a translation of the Russian original report is Kantorovich (1960)). This formulation is known for the weakness of its linear relaxation. The value of the objective function is equal to $\sum_{i \in I} \ell_i / L$. Nevertheless, a Dantzig-Wolfe decomposition with (1.50) as an integer program pricing subproblem (in fact, κ identical subproblems, which allows for further simplification), yields an extensive formulation the linear programming relaxation of which is equivalent to that of P_{CS}. However, the variables of the compact formulation (1.55) are in a sense interchangeable, since the paper rolls are indistinguishable. One speaks of *problem symmetry* which may entail considerable difficulties in branch-and-bound because of many similar and thus redundant subtrees in the search.

4.3   Valério de Carvalho (2002)

Fortunately, the existence of a compact formulation in the "reversed" Dantzig-Wolfe decomposition process by Villeneuve et al. (2003) does not mean uniqueness. There may exist alternative compact formulations that give rise to the same linear relaxation of an extensive formulation, and we exploit this freedom of choice.

Valério de Carvalho (2002) suggests a very clever original network-based formulation for the cutting stock problem. Define the acyclic network $G = (N, A)$ where $N = \{0, 1, \ldots, L\}$ is the set of nodes and the set of arcs is given by $A = \{(u, v) \in N \times N \mid v - u = \ell_i, \forall i \in I\} \cup \{(u, v) \mid u \in N \backslash \{L\}\}$, see also Ben Amor and Valério de Carvalho (2004). Arcs link every pair of consecutive nodes from 0 to L without covering any item. An item $i \in I$ is represented several times in the network by arcs of length $v - u = \ell_i$. A path from the source 0 to the sink L encodes a feasible cutting pattern.

The proposed formulation of the cutting stock problem, which is pseudo-polynomial in size, reads as

$$\min z \tag{1.56}$$

$$\text{subject to} \sum_{(u,u+\ell_i)\in A} x_{u,u+\ell_i} \geq n_i \quad i \in I \tag{1.57}$$

$$\sum_{(0,v)\in A} x_{0,v} = z \tag{1.58}$$

$$\sum_{(u,v)\in A} x_{uv} - \sum_{(v,u)\in A} x_{vu} = 0 \quad v \in \{1,\ldots,L-1\} \tag{1.59}$$

$$\sum_{(u,L)\in A} x_{uL} = z \tag{1.60}$$

$$x_{uv} \in \mathbb{Z}_+ (u,v) \in A. \tag{1.61}$$

Keeping constraints (1.57) in the master problem, the subproblem is $X = \{(\mathbf{x}, z) \text{ satisfying (1.58)–(1.61)}\}$. This set X represents flow conservation constraints with an unknown supply of z from the source and a matching demand at the sink. Given dual multipliers π_i, $i \in I$ associated to the constraint (1.57), the subproblem is

$$\min_{(1.58)-(1.61)} z - \sum_{(u,u+\ell_i)\in A} \pi_i x_{u,u+\ell_i}. \tag{1.62}$$

Observe now that the solution $(\mathbf{x}, z) = (\mathbf{0}, 0)$ is the unique extreme point of X and that all other paths from the source to the sink are extreme rays. Such an extreme ray $r \in R$ is represented by a 0-1 flow which indicates whether an arc is used or not. An application of Dantzig-Wolfe decomposition to this formulation directly results in formulation P_{CS}, the linear relaxation of the extensive reformulation (this explains our choice of writing down P_{CS} in terms of extreme rays instead of extreme points). Formally, as the null vector is an extreme point, we should add one variable λ_0 associated to it in the master problem and the convexity constraint with only this variable. However, this empty pattern makes no difference in the optimal solution as its cost is 0.

The pricing subproblem (1.62) is a shortest path problem defined on a network of pseudo-polynomial size, the solution of which is also that of a knapsack problem. Still this subproblem suffers from some symmetries since the same cutting pattern can be generated using various paths. Note that this subproblem possesses the integrality property although the previously presented one (1.50) does not. Both subproblems construct the same columns and P_{CS} provides the same lower bound on the

value of an optimal integer solution. The point is that the integrality
property of a pricing subproblem, or the absence of this property, has
to be evaluated relative to its own compact formulation. In the present
case, the linear relaxation of Kantorovich's formulation provides a lower
bound that is weaker than that of P_{CS}, although it can be improved by
solving the integer knapsack pricing subproblem (1.50). On the other
hand, the linear relaxation of Valério de Carvalho's formulation already
provides the same lower bound as P_{CS}. Using this original formulation,
one can design branching and cutting decisions on the arc flow variables
of network G to get an optimal integer solution.

Let us mention that there are many important applications which
have a natural formulation as set covering or set partitioning problems,
without any decomposition. In such models it is usually the master
problem itself which has to be solved in integers (Barnhart et al., 1998).
Even though there is no explicit original formulation used, customized
branching rules can often be interpreted as branching on variables of
such a formulation.

5.   Further reading

Even though column generation originates from linear programming,
its strengths unfold in solving integer programming problems. The simultaneous use of two concurrent formulations, compact and extensive,
allows for a better understanding of the problem at hand and stimulates
our inventiveness in what concerns for example branching rules.

We have said only little about implementation issues, but there would
be plenty to discuss. Every ingredient of the process deserves its own
attention, see e.g., Desaulniers et al. (2001), who collect a wealth of
acceleration ideas and share their experience. Clearly, an implementation benefits from customization to a particular application. Still, it is
our impression that an off-the-shelf column generation software to solve
large scale integer programs is in reach reasonably soon; the necessary
building blocks are already available. A crucial part is to automatically
detect how to "best" decompose a given original formulation, see Vanderbeck (2004). This means in particular exploiting the absence of the
subproblem's integrality property, if applicable, since this may reduce
the integrality gap without negative consequences for the linear master program. Let us also remark that instead of a convexification of the
subproblem's domain X (when bounded), one can explicitly represent all
integer points in X via a *discretization* approach formulated by Vanderbeck (2000). The decomposition then leads to a master problem which
itself has to be solved in integer variables.

In what regards new and important technical developments, in addition to the stabilization of dual variables already mentioned, one can find a dynamic row aggregation technique for set partitioning master problems in Elhallaoui et al. (2003). This allows for a considerable reduction in size of the restricted master problem in each iteration. An interesting approach is also proposed by Valério de Carvalho (1999) where variables and rows of the original formulation are dynamically generated from the solutions of the subproblem. This technique exploits the fact that the subproblem possesses the integrality property. For a presentation of this idea in the context of a multicommodity network flow problem we refer to Mamer and McBride (2000).

This primer is based on our recent survey (Lübbecke and Desrosiers, 2002), and a much more detailed presentation and over one hundred references can be found there. For those interested in the many column generation applications in practice, the survey articles in this book will serve the reader as entry points to the large body of literature. Last, but not least, we recommend the book by Lasdon (1970), also in its recent second edition, as an indispensable source for alternative methods of decomposition.

Acknowledgments   We would like to thank Steffen Rebennack for cross-checking our numerical example, and Marcus Poggi de Aragão, Eduardo Uchoa, Geir Hasle, and José Manuel Valério de Carvalho, and the two referees for giving us useful feedback on an earlier draft of this chapter. This research was supported in part by an NSERC grant for the first author.

References

Ahuja, R.K., Magnanti, T.L., and Orlin, J.B. (1993). Network Flows: Theory, Algorithms and Applications. Prentice-Hall, Inc., Englewood Cliffs, New Jersey.

Barnhart, C., Johnson, E.L., Nemhauser, G.L., Savelsbergh, M.W.P., and Vance, P.H. (1998). Branch-and-price: Column generation for solving huge integer programs. *Operations Research*, 46(3):316–329.

Ben Amor, H. and Desrosiers, J. (2003). A proximal trust region algorithm for column generation stabilization. *Les Cahiers du GERAD* G-2003-43, HEC, Montreal, Canada. Forthcoming in: *Computers & Operations Research*.

Ben Amor, H., Desrosiers, J., and Valério de Carvalho, J.M. (2003). Dual-optimal inequalities for stabilized column generation. *Les*

Cahiers du GERAD G-2003-20, HEC, Montréal, Canada.

Ben Amor, H. and Valério de Carvalho, J.M. (2004). Cutting stock problems. In this book.

Dantzig, G.B. and Wolfe, P. (1960). Decomposition principle for linear programs. *Operations Research*, 8:101–111.

Desaulniers, G., Desrosiers, J., Ioachim, I., Solomon, M.M., Soumis, F., and Villeneuve, D. (1998). A unified framework for deterministic time constrained vehicle routing and crew scheduling problems. In: *Fleet Management and Logistics* (T.G. Crainic and G. Laporte, eds.), pp. 57–93, Kluwer, Norwell, MA.

Desaulniers, G., Desrosiers, J., and Solomon, M.M. (2001). Accelerating strategies in column generation methods for vehicle routing and crew scheduling problems. In: *Essays and Surveys in Metaheuristics* (C.C. Ribeiro and P. Hansen, eds.), pp. 309–324, Kluwer, Boston.

du Merle, O., Villeneuve, D., Desrosiers, J., and Hansen, P. (1999). Stabilized column generation. *Discrete Mathematics*, 194:229–237.

Elhallaoui, I., Villeneuve, D., Soumis, F., and Desaulniers, G. (2003). Dynamic aggregation of set partitioning constraints in column generation. *Les Cahiers du GERAD* G-2003-45, HEC, Montréal, Canada. Forthcoming in: *Operations Research*.

Ford, L.R. and Fulkerson, D.R. (1958). A suggested computation for maximal multicommodity network flows. *Management Science*, 5:97–101.

Geoffrion, A.M. (1974). Lagrangean relaxation for integer programming. *Mathematical Programming Study (Series)*, 2:82–114.

Gilmore, P.C. and Gomory, R.E. (1961). A linear programming approach to the cutting-stock problem. *Operations Research*, 9:849–859.

Gilmore, P.C. and Gomory, R.E. (1963). A linear programming approach to the cutting stock problem. Part II. *Operations Research*, 11:863–888.

Goffin, J.-L. and Vial, J.-Ph. (1999). Convex nondifferentiable optimization: A survey focused on the analytic center cutting plane method. *Technical Report* 99.02, Logilab, Université de Genève. Forthcoming in: *Optimization Methods & Software*.

Guignard, M. (2004). Lagrangean relaxation. In: *Handbook of Applied Optimization* (M. Resende and P. Pardalos, eds.), Oxford University Press.

Hiriart-Urruty, J.-B. and Lemaréchal, C. (1993). Convex Analysis and Minimization Algorithms, Part 2: Advanced Theory and Bundle Methods, volume 306, *Grundlehren der mathematischen Wissenschaften*, Springer, Berlin.

Kantorovich, L. (1960). Mathematical methods of organising and planning production. *Management Science*, 6:366–422.

Kelley Jr., J.E. (1961). The cutting-plane method for solving convex programs. *Journal of the Society for Industrial and Applied Mathematics*, 8(4):703–712.

Lasdon, L.S. (1970). Optimization Theory for Large Systems. Macmillan, London.

Lübbecke, M.E. and Desrosiers, J. (2002). Selected topics in column generation. *Les Cahiers du GERAD* G-2002-64, HEC, Montréal, Canada. Forthcoming in: *Operations Research*.

Nemhauser, G.L. and Wolsey, L.A. (1988). Integer and Combinatorial Optimization. John Wiley & Sons, Chichester.

Mamer, J.W. and McBride, R.D. (2000). A decomposition-based pricing procedure for large-scale linear programs—An application to the linear multicommodity flow problem. *Management Science*, 46(5):693–709.

Marsten, R.E. (1975). The use of the boxstep method in discrete optimization. *Mathematical Programming Study*, 3:127–144.

Marsten, R.E., Hogan, W.W., and Blankenship, J.W. (1975). The BOXSTEP method for large-scale optimization. *Operations Research*, 23:389–405.

Poggi de Aragão, M. and Uchoa, E. (2003). Integer program reformulation for robust branch-and-cut-and-price algorithms. In: *Proceedings of the Conference Mathematical Program in Rio: A Conference in Honor of Nelson Maculan*, pp. 56–61.

Schrijver, A. (1986). Theory of Linear and Integer Programming. John Wiley & Sons, Chichester.

Valério de Carvalho, J.M. (1999). Exact solution of bin-packing problems using column generation and branch-and-bound. *Annals of Operations Research*, 86:629–659.

Valério de Carvalho, J.M. (2002). LP models for bin-packing and cutting stock problems. *European Journal of Operational Research*, 141(2):253–273.

Valério de Carvalho, J.M. (2003). Using extra dual cuts to accelerate convergence in column generation. Forthcoming in: *INFORMS Journal of Computing*.

Vanderbeck, F. (2000). On Dantzig-Wolfe decomposition in integer programming and ways to perform branching in a branch-and-price algorithm. *Operations Research*, 48(1):111–128.

Vanderbeck, F. (2004). Implementing mixed integer column generation. In this book.

Vanderbeck, F. and Wolsey, L.A. (1996). An exact algorithm for IP column generation. *Operations Research Letters*, 19:151–159.

Villeneuve, D., Desrosiers, J., Lübbecke, M.E., and Soumis, F. (2003). On compact formulations for integer programs solved by column generation. *Les Cahiers du GERAD* G-2003-06, HEC, Montréal, Canada. Forthcoming in: *Annals of Operations Research*.

Walker, W.E. (1969). A method for obtaining the optimal dual solution to a linear program using the Dantzig-Wolfe decomposition. *Operations Research*, 17:368–370.

Chapter 2

SHORTEST PATH PROBLEMS WITH RESOURCE CONSTRAINTS

Stefan Irnich
Guy Desaulniers

Abstract   In most vehicle routing and crew scheduling applications solved by column generation, the subproblem corresponds to a shortest path problem with resource constraints (SPPRC) or one of its variants.

This chapter proposes a classification and a generic formulation for the SPPRCs, briefly discusses complex modeling issues involving resources, and presents the most commonly used SPPRC solution methods. First and foremost, it provides a comprehensive survey on the subject.

1.   Introduction

For more than two decades, column generation (also known as branch-and-price when embedded in a branch-and-bound framework) has been successful at solving a wide variety of vehicle routing and crew scheduling problems (see e.g. Desrosiers et al., 1995; Barnhart et al., 1998; Desaulniers et al., 1998), and most chapters in this book). In most of these applications, the master problem of the column generation method is a (possibly generalized) set partitioning or set covering problem with side constraints, where most of the variables, if not all, are associated with vehicle routes or crew schedules. These route and schedule variables are generated by one or several subproblems, each of them corresponding to a *shortest path problem with resource constraints* (SPPRC) or one of its variants. The SPPRC has contributed to the success of the column generation method for this class of problems for three main reasons. Firstly, through its resource constraints, it constitutes a flexible tool for modeling complex cost structures for an individual route or schedule, as well as a wide variety of rules that define the feasibility of a route or a

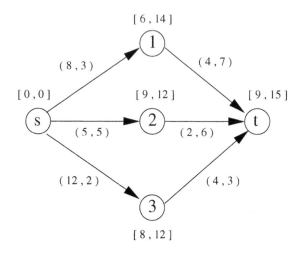

Figure 2.1. A small SPPRC example

schedule. Secondly, because it does not possess the integrality property
(i.e., there may be a positive gap between its optimal value and that
of its linear relaxation) as discussed in Desrosiers et al. (1984), the column generation approach can derive tighter bounds than those obtained
from the linear relaxation of arc-based formulations. Thirdly, there exist
efficient algorithms at least for some important variants of the SPPRC.

The SPPRC was introduced in the Ph.D dissertation of Desrochers
(1986) as a subproblem of a bus driver scheduling problem. It consists
of finding a shortest path among all paths that start from a source node,
end at a sink node, and satisfy a set of constraints defined over a set of
resources. A resource corresponds to a quantity, such as the time, the
load picked-up by a vehicle, or the duration of a break in a work shift,
that varies along a path according to functions, called resource extension functions (REFs). A REF is defined for every arc in the network
and every resource considered. It provides a lower bound on the value
that the corresponding resource can take at the head node of the corresponding arc, given the values taken by all the resources at its tail node.
The resource constraints are given as intervals, called resource windows,
which restrict the values that can be taken by the resources at every
node along a path. Such a constraint is defined for every node in the
network and every resource considered.

Figure 2.1 provides an SPPRC example that involves the resource
time. The source and sink nodes are denoted by s and t, respectively.
Each arc (i, j) bears a two-dimensional vector: The first component t_{ij}

provides the travel time (duration) of using the arc, while the second c_{ij} indicates the cost associated with it. Given a value T_i taken by the resource at a node i (T_i is said to be the visiting time at node i), the REF for an arc (i, j) is defined as $f_{ij}(T_i) = T_i + t_{ij}$, i.e., it computes the (*earliest*) *arrival time* at node j when starting at node i at time T_i. The resource window $[a_i, b_i]$ associated with each node i is specified in brackets beside it. It indicates at what time node i can be visited. If the arrival time of a path ending at a node i exceeds b_i, then this path is deemed infeasible. Otherwise, it is feasible even if its arrival time precedes a_i since waiting at a node is allowed, that is, the visiting time at node i can be greater than the arrival time at this node.

In the example of Figure 2.1, three paths link the source node s to the sink node t. The first path $P_1 = (s, 1, t)$, denoted by the sequence of nodes visited, is resource-feasible since it is possible to find visiting times along that path which satisfy all resource constraints. Indeed, setting $T_s = 0$ (the only feasible value at node s), it is easy to see that the arrival times ($T_1 = 8$ and $T_t = 12$) at nodes 1 and t provided by the appropriate REFs ($f_{s1}(T_s)$ and $f_{1t}(T_1)$) are all feasible with respect to the resource windows. The second path $P_2 = (s, 2, t)$ is also resource-feasible. However, waiting is needed at node 2 since the arrival time provided by $f_{s2}(0) = 5$ is smaller than $a_2 = 9$. In this case, the visiting time T_2 can be set at 9, and the subsequent visiting time T_t at 11, respectively. Finally, the third path $P_3 = (s, 3, t)$ is not resource-feasible since, along that path, $T_s = 0$, $T_3 \geq f_{s3}(0) = 12$, and the earliest arrival time at node t is $f_{3t}(12) = 16$. Hence, the resource window $[9, 15]$ at node t cannot be met. Since the cost of P_1 ($3 + 7 = 10$) is smaller than the cost of P_2 ($5 + 6 = 11$), the former path is optimal with respect to cost. However, path P_2 has a smaller earliest arrival time at node t. If the network in Figure 2.1 were only a sub-network within a bigger network, then extending path P_2 to a node could be feasible but extending P_1 could be infeasible.

This gives us a first glance at the core of SPPRC's difficulty. The SPPRC is very close to a multi-criteria problem. In the following we will consider both criteria, time and cost, as resources. Paths are *uncomparable* when one path is better than a second path in one criterion and worse in another criterion. Resource constraints make it necessary to consider all uncomparable paths that arrive at a node, since resource constraints might forbid extending any subset of these paths but allow an extension of the others.

The two-resource SPPRC, better known as the shortest path problem with time windows (SPPTW), was first studied in Desrosiers et al. (1983, 1984). The resource *cost* is unconstrained while the resource *time*

is restricted by corresponding time windows. Desrochers (1986) generalized the SPPTW to the case with several resources. Since then, several
variants of the SPPRC have appeared in the literature. For instance,
Ioachim et al. (1998) proposed the SPPTW with time dependent linear
costs at the nodes and Dumas et al. (1991) the SPPTW with pickups
and deliveries.

The contribution of this chapter is three-fold. Firstly, it presents a
classification of the SPPRC variants and provides a generic SPPRC formulation that includes all variants studied so far (Section 2). Secondly,
it discusses non-trivial modeling issues for the SPPRC (Section 3). Finally, it surveys the most important papers on this subject, namely,
those introducing a new variant of the SPPRC (Section 2) or proposing
an interesting methodological contribution (Section 4).

2.   Classification of the SPPRCs

The intention of this section is to provide a generic formulation for a
comprehensive class of shortest path problems with resource constraints
presented in the literature so far. Variants of the SPPRC, which we consider, are extensions of the classical shortest path problem, where the
cost is replaced by multi-dimensional resource vectors, which are accumulated along paths and constrained at intermediate nodes. Different
types of SPPRCs can be classified by

   (i)   the way in which resources are accumulated, leading to different
          definitions of resource feasible paths,

   (ii)   the existence of additional path-structural constraints excluding
          specific paths, e.g., non-elementary paths,

   (iii) the objective,

   (iv)   and the underlying network.

We state all SPPRCs on a digraph $G = (V, A)$, where V and A are
non-empty sets of nodes and arcs, respectively. A *path* $P = (e_1, \ldots, e_p)$
is a finite sequence of arcs (some arcs may occur more than once) where
the head node of $e_i \in A$ is identical to the tail node of $e_{i+1} \in A$ for
all $i = 1, \ldots, p - 1$. For the sake of convenience, we assume that G is
simple so that a path can be written as $P = (v_0, v_1, \ldots, v_p)$ with the
understanding that $(v_{i-1}, v_i) \in A$ holds for all $i \in \{1, \ldots, p\}$. The *length*
of this path is p.

2.1 Resource feasible paths

The description of *feasible paths* provides a basis for the generic definition of the SPPRC. In the following, we distinguish between feasibility *w.r.t. resources* and feasibility *w.r.t. path-structural constraints*. This section focuses on the first aspect while path-structural constraints are discussed in the next section.

Resource constraints can be formulated by means of (*minimal*) *resource consumptions* and *resource intervals* (e.g., the travel times t_{ij} and time windows $[a_i, b_i]$ in the SPPTW). Let R be the number of resources. A vector $T = (T^1, \ldots, T^R)^\top \in \mathbb{R}^R$ is called a *resource vector* and its components *resource variables* (remark: $x^\top$ denotes the transposed vector to the vector x). T is said to be not greater than (i.e., *dominates*) $S = (S^1, \ldots, S^R)^\top \in \mathbb{R}^R$ if the inequality $T^i \leq S^i$ holds for all components $i = 1, \ldots, R$. We denote this by $T \leq S$. For two resource vectors a and b the interval $[a, b]$ is defined as the set $\{T \in \mathbb{R}^R : a \leq T \leq b\}$.

Resource intervals, also called *resource windows*, associated with a node $i \in V$ are denoted by $[a_i, b_i]$ with $a_i, b_i \in \mathbb{R}^R$, $a_i \leq b_i$. The changes in the resource consumptions associated with an arc $(i, j) \in A$ are given by a vector $f_{ij} = (f_{ij}^r)_{r=1}^R$ of so-called *resource extension functions* (REFs). A REF $f_{ij}^r : \mathbb{R}^R \to \mathbb{R}$ depends on a resource vector $T_i \in \mathbb{R}^R$, which corresponds to the resource consumption accumulated along a path from s to i, i.e., up to the tail node i of arc (i, j). Hence, the result $f_{ij}(T_i) \in \mathbb{R}^R$ can be interpreted as a resource consumption accumulated along the path $(s, \ldots, i, j)$. "Classical" SPPRCs, like the SPPTW presented in the introduction, only consider REFs of the form

$$f_{ij}^r(T_i) = T_i^r + t_{ij}^r \tag{2.1}$$

where t_{ij}^r are constants associated with the arc (i, j). Classical REFs are separable by resources, i.e., there exist no interdependencies between different resources. The more general definition of REFs provides a powerful instrument for modeling practically relevant resource interdependencies.

Instead of giving an implicit MIP-formulation for the SPPRC, we state the resource constraints by considering individual paths. The reason for this is that node repetitions within a path (which are allowed in our path definition) prohibit to model resource consumptions by individual resource variables associated with a node. For a given path $P = (v_0, v_1, \ldots, v_p)$, one has to refer to the $p + 1$ different *positions* $i = 0, 1, \ldots, p$. A path P is *resource-feasible* if there exist resource vectors $T_i \in [a_{v_i}, b_{v_i}]$ for all positions $i = 0, 1, \ldots, p$ such that $f_{v_i, v_{i+1}}(T_i) \leq T_{i+1}$ holds for all $i = 0, \ldots, p - 1$. $\mathcal{T}(P)$ is defined as the set of all feasible resource vectors at the last node v_p of $P = (v_0, v_1, \ldots, v_p)$, i.e.,

$$\mathcal{T}(P) = \{T_p \in [a_{v_p}, b_{v_p}] : \exists T_i \in [a_{v_i}, b_{v_i}], f_{v_i, v_{i+1}}(T_i) \le T_{i+1}$$
$$\text{for all } i = 0, \ldots, p-1\}. \quad (2.2)$$

Let $\mathcal{F}(u, v)$ be the set of all resource-feasible paths from a node u to a node v. Note that $P \in \mathcal{F}(u, v)$ holds if and only if $\mathcal{T}(P) \ne \varnothing$.

2.2   Path-structural constraints

Path-structural constraints can model further requirements concerning the feasibility of paths, which are not covered by resources. Such additional requirements might either be an integral part of a feasible path's definition or be implied by branching rules, which come up in the context of branch-and-price and require modifications of the pricing problem. Sometimes, these modifications cannot be handled by simply removing some arcs or nodes of the underlying network. In order to specify those constraints, we need some definitions. An *elementary path* is a path in which all nodes are pairwise different. Contrarily, a *cycle* is a path $(v_0, v_1, \ldots, v_p)$ of length $p > 1$ having $v_0 = v_p$. We call any cycle of length less than or equal to k a k-*cycle*.

The following SPPRC variants have been proposed in the literature and defined according to path-structural constraints. Let $\mathcal{G}$ be the set of all paths feasible with respect to these constraints.

For the *elementary SPPRC* (ESPPRC), $\mathcal{G} = \{$elementary paths$\}$. On acyclic graphs, all paths are elementary so that SPPRC and ESPPRC coincide. In general (i.e., for networks with cycles), the ESPPRC has been identified to be $\mathcal{NP}$-hard in the strong sense (Dror, 1994) and has been first studied and solved by Beasley and Christofides (1989). In many vehicle routing applications the pricing problem is an ESPPRC. Feillet et al. (2004); Chabrier (2002); Rousseau et al. (2003) solved ESP-PRC pricing problems in the context of the vehicle routing problem with time windows (VRPTW). These approaches are known for their very tight lower bounds computed by the LP-relaxation of the VRPTW set-partitioning master program.

For the SPPRC, $\mathcal{G} = \{$all paths$\}$, that is, no path-structural constraints are imposed. The SPPRC occurs as a subproblem in numerous vehicle and crew scheduling problems which are most of the time formulated over acyclic time-space networks (see Desrosiers et al., 1984; Vance et al., 1997; Desaulniers et al., 1998; Gamache et al., 1999)).

Since the ESPPRC is very hard to solve (in some cases it is prohibitively hard), classical solution approaches for vehicle routing problems which are formulated over *cyclic graphs* are also based on the corresponding non-elementary SPPRC, because it can be solved using pseudo-polynomial algorithms (see Section 4.1). Influential contributions which rely on this idea were Desrosiers et al. (1986); Desrochers et al. (1992); Desrosiers et al. (1995). However, while solving the enclosing problem by branch-and-price, this subproblem relaxation sometimes leads to weak lower bounds and possibly impractical large branch-and-bound trees.

For the *SPPRC with k-cycle elimination* (SPPRC-k-cyc), $\mathcal{G} = \{k\text{-cycle-free paths}\}$. A compromise between solving the ESPPRC and the SPPRC is to forbid cycles of small length. Several examples of VRPTW instances, e.g., taken from the benchmark library of Solomon (1987), show that cycle elimination for small values of k can substantially improve the master program lower bounds. This justifies an additional effort to eliminate cycles (compared to solving a pure SPPRC) while the corresponding ESPPRC is practically impossible to solve. The case $k = 2$ was first analyzed by Houck et al. (1980) and used in the VRPTW context by Kolen et al. (1987); Desrochers et al. (1992). Irnich and Villeneuve (2003) recently proposed an algorithm for the general case of $k \geq 2$.

For the *SPPRC with forbidden paths* (SPPRCFP), $\mathcal{G} = \{\text{all paths}\} \setminus \mathcal{G}_{\text{forbidden}}$ where $\mathcal{G}_{\text{forbidden}}$ is a set of forbidden paths. This set is implicitly defined as the set of all paths that contain at least one element of a finite set of pre-specified sub-paths. Villeneuve and Desaulniers (2000) introduced this type of SPPRC which occurs two-fold in the context of branch-and-price. First, in some applications one wants to branch so that a route or schedule is excluded from the (restricted) master program (see Desaulniers et al., 2002b; Arunapuram et al., 2003). This makes it necessary to also exclude the corresponding path from being generated by the SPPRC pricing procedure. Second, some constraints might be impossible or very hard to model with resources. Instead of considering them directly, one iteratively solves relaxed SPPRCs to get tentative solutions, which are excluded from the SPPRC by means of forbidden paths as long as not all constraints are respected. Examples of hard-to-model constraints stem from aircrew scheduling applications, see e.g. Fahle et al. (2002).

Two additional types of constraints, *precedence constraints* and *pairing constraints*, are important in the pickup and delivery context. Given two nodes $i, j \in V$, a path P fulfills the (i, j)-pairing constraint if node i occurs as often as node j in P (possibly P contains none of them). A path P fulfills the (i, j)-precedence constraint if P contains no sub-path connecting j with i. The *SPPRC with pickups and deliveries* (SPPRCPD) is a subproblem of the vehicle routing problem with time windows, pickups and deliveries (see Dumas et al., 1991; Desaulniers et al.,

2002a). In this problem, transportation requests $i \in I$ must be satisfied
where a request requires a pickup at an origin i^+ and a delivery at a destination i^-. Consequently, the SPPRCPD contains an (i^+, i^-)-pairing
and an (i^+, i^-)-precedence constraint for each request $i \in I$.

In a branch-and-price context, each node and each arc represent a
(possibly empty) sequence of tasks, where a *task* (e.g., a flight leg, a
train segment, or a crew pairing) is associated with a set partitioning
constraint in the master problem. A task can be part of several sequences
and can therefore be represented by several nodes and arcs. For any path
$P = (v_0, v_1, \ldots, v_p)$ there is a (uniquely defined) *task sequence* $W(P)$
given by the concatenation of the sequences of tasks of v_0, (v_0, v_1), v_1,
$(v_1, v_2), \ldots, (v_{p-1}, v_p)$, v_p. All of the above path-structural constraints
might also be formulated w.r.t. the task sequences. For instance, the
task-ESPPRC considers only paths P for which $W(P)$ does not contain
task repetitions or the task-SPPRC-2-cyc does not allow paths having a
2-cycle in $W(P)$.

Several branching rules proposed in the literature impose additional
constraints on how two given tasks have to be covered by the paths. The
branching rules of Ryan and Foster (1981) decide whether two tasks i
and j are covered by the same path or by different paths. Hence, one
branch is simply an (i, j)-pairing constraint. The other branch is an
(i, j)-*anti-pairing constraint* which forbids tasks i and j to be together
in $W(P)$, i.e., $\mathcal{G} = \{P : i \notin W(P) \text{ or } j \notin W(P)\}$. Similarly, the inter-
task constraints (introduced in Desrochers and Soumis (1989)) decide
whether two given tasks i and j are performed consecutively or not. In
this case, an (i, j)-*follower constraint* guarantees on one branch that, for
each path $P \in \mathcal{G}$, $W(P)$ contains task i followed by task j or none of
these tasks. On the other branch, an (i, j)-*non-follower constraint* only
allows paths $P \in \mathcal{G}$ for which $W(P)$ does not contain task i followed by
task j.

Summing up the definitions of resource feasibility and path-structural
constraints, we know that the set $\mathcal{F} = \bigcup_{v \in V} (\mathcal{F}(s, v) \cap \mathcal{G})$ contains all
feasible paths to a one-to-all SPPRC problem.

2.3   Objectives and generic SPPRC formulation

The objective of the SPPRC is formulated by means of a resource
vector at the last nodes of feasible paths. Recall that in general, for
a single path $P \in \mathcal{F}$ there exist many feasible choices for the resource
vectors $T \in \mathcal{T}(P)$. Problems whose objective depends only on a single resource, called *cost* resource, are normally one-to-one shortest path
problems with a source node s and a sink node t. They can be formulated

as follows:

$$\min_{P \in \mathcal{F}(s,t) \cap \mathcal{G}} \left(\min_{T \in \mathcal{T}(P)} T^{\mathrm{cost}} \right). \tag{2.3}$$

Computing the minimum cost of a path $P = (v_0, \dots, v_p)$ requires the determination of feasible resource vectors $T_0, \dots, T_p$ along the path. Similarly to the feasibility problem $\mathcal{T}(P) \neq \varnothing$ discussed above, this can be a hard problem. In contexts with time windows, Dumas et al. (1990) optimized the cost of a given path for time-dependent convex inconvenience costs at all nodes.

A much more general formulation of the SPPRC is based on considering the set of Pareto-optimal resource vectors. For a given set $M \subset \mathbb{R}^R$, an element $m \in M$ is *Pareto-optimal* if $x \not\leq m$ holds for all $x \in M, x \neq m$. It means that none of the cones $x^\llcorner$ for $x \in M, x \neq m$ contain a Pareto-optimal point m, where a *cone* $T^\llcorner$ is defined as $\{S \in \mathbb{R}^R \colon S \geq T\}$. For $v \in V$, let $PO(v)$ be the set of Pareto-optimal vectors in $\bigcup_{P \in \mathcal{F}(s,v) \cap \mathcal{G}} \mathcal{T}(P)$. The SPPRC can be formulated as follows.

> **Generic SPPRC:** Find for each node $v \in V$ and for each Pareto-optimal resource vector $T \in PO(v)$ one feasible (representative) s-v-path $P \in \mathcal{F}(s, v) \cap \mathcal{G}$ having $T \in \mathcal{T}(P)$.

For the sake of convenience, we call the representative path P a *Pareto-optimal path*. Since all solutions to a problem $\min_{m \in M} \alpha^\top \cdot m$ for a non-negative weight vector $\alpha \in \mathbb{R}_+^R$, $\alpha \neq 0$ are Pareto-optimal points of M, the generic SPPRC formulation also solves all problems of the form

$$\min_{P \in \mathcal{F}(s,t) \cap \mathcal{G}} \left(\min_{T \in \mathcal{T}(P)} \alpha^\top T \right) \tag{2.4}$$

for any weight vector $\alpha \in \mathbb{R}_+^R$. Problem (2.3) is a special case of (2.4).

2.4   Properties of $\mathcal{T}(P)$

We will now study properties of the set $\mathcal{T}(P)$ for a fixed path $P = (v_0, v_1, \dots, v_p)$ under different assumptions concerning the REFs. Knowing $\mathcal{T}(P)$ and its structure is essential to (efficiently) resolve the following two basic tasks:

- Given a path P. Is P resource feasible, i.e., $P \in \mathcal{F}(v_0, v_p)$ or not?

- Given the prefix $P' = (v_0, \dots, v_{p-1})$ of $P = (v_0, \dots, v_{p-1}, v_p)$, compute $\mathcal{T}(P)$ using $\mathcal{T}(P')$.

Furthermore, compact implicit representations of $\mathcal{T}(P)$ are substantial for checking if a path P (or any of its extensions) is or might be a Pareto-optimal path. For instance, efficient dominance checks in the context of dynamic programming are based on representing $T(P)$ by either using a single Pareto-optimal point $T(P)$ or a function $g_P(\cdot)$ to describe the set of Pareto-optimal points in $T(P)$, see Section 4.1.

Before discussing different cases, we state the following universal property: If $T \in T(P)$ then $T^{\llcorner} \cap [a_{v_p}, b_{v_p}] \subseteq T(P)$, i.e., the set $T(P)$ contains the cone, restricted to the resource interval, generated by each point in this set.

Classical SPPRC and non-decreasing REFs.   In the classical SPPRC the set $T(P)$ has a simple representation as a cone restricted by $[a_{v_p}, b_{v_p}]$. Let $P_i = (v_0, \ldots, v_i)$, $i = 0, \ldots, p$ be the prefix of P of length i. Each set $T(P_i)$ has a unique cone-defining element $T(P_i) \in T(P_i)$ such that $T(P_i) = T(P_i)^{\llcorner} \cap [a_{v_i}, b_{v_i}]$ holds. The resource vector $T(P_i)$ can be recursively computed by

$$T(P_0) = a_{v_0} \quad \text{and}$$
$$T(P_i) = \max\{a_{v_i}, f_{v_{i-1}, v_i}(T(P_{i-1}))\} \quad \text{for all } i \in \{1, \ldots, p\}. \tag{2.5}$$

The same is true when all REFs are non-decreasing functions, meaning that each $f_{ij}^r(T_i^1, T_i^2, \ldots, T_i^R)$ is a non-decreasing function in one variable T_i^k, when the other $R - 1$ components are kept fixed. Under these assumptions $T(P)$ is still a cone. Formula (2.5) computes $T(P)$ with $T(P)^{\llcorner} \cap [a_{v_p}, b_{v_p}] = T(P)$ efficiently.

As a consequence, the generic SPPRC formulation can be simplified as follows.

> **Generic SPPRC with non-decreasing REFs:** Find for each node $v \in V$ one feasible representative s-v-path $P \in \mathcal{F}(s, v) \cap \mathcal{G}$ for which $T(P)$ is Pareto-optimal in $\{T(Q) : Q \in \mathcal{F}(s, v) \cap \mathcal{G}\}$.

Formulation (2.4) can then be re-written as $\min_{P \in \mathcal{F}(s,t) \cap \mathcal{G}} \alpha^{\top} T(P)$.

Linear REFs.   If the REFs are linear but not necessarily non-decreasing, it is easy to see that $T(P)$ is a bounded polyhedron. The description of the polyhedron $T(P)$ (e.g., by its extreme points) can get more and more complicated the longer the path P is (see Ioachim et al., 1998) and Section 4.1.2).

For instance, consider the path $P = (1, 2)$, $R = 2$ resources, resource intervals $[a_1, b_1] = [0, 1]^2$ and $[a_2, b_2] = [0, 1] \times [-1, 1]$ and the REF $f_{12}(T_1^1, T_1^2) = (T_1^1, T_1^2 - T_1^1)$. It is easy to see that $T(P)$ is $\{(T_2^1, T_2^2) \in [0, 1] \times [-1, 1] : T_2^2 \geq -T_2^1\}$. There exists no element $T \in T(P)$ such that $T(P) \subseteq T^{\llcorner}$ holds. Note that all vectors $T = (\lambda, -\lambda)$ for $\lambda \in [0, 1]$ are Pareto-optimal points of $T(P)$.

General REFs.   For arbitrary REFs, checking whether $P \in \mathcal{F}(u, v)$ or equivalently $\mathcal{T}(P) \neq \varnothing$ holds or not can be an $\mathcal{NP}$-hard problem. A known $\mathcal{NP}$-complete problem is the binary knapsack lower bound feasibility problem (KLBFP) (see Nemhauser and Wolsey, 1988): *Does there exist a feasible solution with profit at least lb for a given lower bound lb to the knapsack problem* $\max \sum_{i=1}^{n} p_i x_i$, $\sum_{i=1}^{n} w_i x_i \leq C$, $x \in \{0, 1\}^n$? One can easily transform this decision problem into an SP-PRC with three resources: Negative profit, weight, and decision. Let $G = (V, A)$ be a line graph with nodes $V = \{0, 1, \ldots, n\}$ and arcs $A = \{(0, 1), (1, 2), (2, 3), \ldots, (n{-}1, n)\}$. Let $[a_0, b_0] = [0, 0] \times [0, C] \times [0, 1]$, $[a_n, b_n] = [-\infty, -lb] \times [0, C] \times [0, 1]$, and $[a_i, b_i] = [-\infty, 0] \times [0, C] \times [0, 1]$ be the resource windows at all nodes $i \in V \setminus \{0, n\}$. Define the REFs to be $f_{i-1,i}(p, w, x) = (p, w, 0)$ for $x = 0$, and $f_{i-1,i}(p, w, x) = (p{-}p_i, w{+}w_i, 0)$ for $x \neq 0$. The answer to the KLBFP is "yes" if and only if $\mathcal{T}(P) \neq \varnothing$ for the path $P = (0, 1, \ldots, n)$.

2.5   Underlying network

The SPPRCs can also be differentiated according to whether or not their underlying network is acyclic or cyclic. The existence of cycles implies that there exist infinitely many different paths in G (not necessarily feasible w.r.t. resource and path-structural constraints). Thus, the SPPRC might be unbounded. In the following, we exclude these cases from our consideration.

The following discretization of $G = (V, A)$ formally makes the underlying network acyclic. If there exists at least one non-decreasing resource r (i.e., $f_{ij}^r(T_i) - T_i^r > 0$, or $t_{ij}^r > 0$ in the classical SPPRC with $f_{ij}^r(T_i) = T_i^r + t_{ij}^r$ for all $(i, j) \in A$, e.g., the resource time in many applications) it is possible to transform (V, A) into an acyclic time-space network. Each node $v \in V$ is replaced by several copies $\text{copy}^1(v), \ldots, \text{copy}^p(v)$ corresponding to a time discretization of the resource interval for r. Nevertheless, this transformation is only a formal device, e.g., used in the unified model of Desaulniers et al. (1998). Cycles of the original network correspond with paths visiting two or more copies of the same original node. Solving the ESPPRC in G is, therefore, equivalent to solving an SPPRC with task-cycle elimination in the discretized network.

3.   Modeling issues

The modeling of standard constraints like capacity constraints, path length restrictions and time windows is obvious from the introduction. Other simple examples can be found in Vance et al. (1997); Gamache et al. (1999); Desaulniers et al. (1999). This section will, therefore, focus

Table 2.1.   Resource intervals and REFs for task-related constraints.

Constraint	Type	Resource interval $[a_i^r, b_i^r]$ for all $i \in V$	REF $f_{ij}^r(T_i)$ for all $(i,j) \in A$
(k,ℓ)-pairing	$R^=$	$[0,0]$ for $i = s, t$ $[-M, M]$ for $i \in V \setminus \{s, t\}$	$T_i^r + \delta_{ik} - \delta_{i\ell}$
(k,ℓ)-anti-pairing	$R^=$	$[0,0]$ for $i = s$ $[0, M]$ for $i = k$, $[-M, 0]$ for $i = \ell$ $[-M, M]$ for $i \in V \setminus \{s, k, \ell\}$	$T_i^r + \delta_{ik} - \delta_{i\ell}$
(k,ℓ)-precedence	$R^\leq$	$[0, 1 - \delta_{ik}]$	$T_i^r + \delta_{i\ell}$
(k,ℓ)-pairing and precedence	$R^=$	$[0,0]$ for $i = s, k, t$ $[-1, -1]$ for $i = \ell$ $[-1, 1]$ for all $i \in V \setminus \{s, t, k, \ell\}$	$T_i^r + \delta_{i\ell} - \delta_{ik}$
(k,ℓ)-follower and (k,ℓ)-non-follower	$R^=$	$\left[l\big(W(s)\big), l\big(W(s)\big)\right]$ for $i = s$ $[0, N]$ for $i \in V \setminus \{s\}$	(see equation (2.6))

on non-trivial modeling issues, provide examples and give references to some relevant literature.

In some applications, one wants to model *exact resource consumptions* instead of *minimal resource consumptions*. For the SPPTW it means that waiting is not allowed so that the arrival time at each node is always identical to the visiting time. In general, the inequalities in (2.2) defining a resource-feasible path $P = (v_0, v_1, \ldots, v_p)$ have to be replaced by $T_{i+1}^r = f_{v_i, v_{i+1}}^r(T_i)$. By $R^=$ (resp. $R^\leq$) we denote the resources which force an equality (resp. inequality) in (2.2). However, as suggested in Gamache et al. (1998), a resource $r \in R^=$ might equivalently be replaced by two resources $r_1, r_2 \in R^\leq$ where the resource intervals and REFs for r_1 are identical to those for r while those for r_2 are $[a_i^{r_2}, b_i^{r_2}] = [-b_i^r, -a_i^r]$ and $\tilde{f}_{ij}^{r_2}(\tilde{T}_i) = -f_{ij}^r(\tilde{T}_i^1, \ldots, \tilde{T}_i^{r-1}, -\tilde{T}_i^{r_2}, \tilde{T}_i^{r+1}, \ldots, \tilde{T}_i^R)$ (the $\tilde{}$ symbol refers to the case with the r_1 and r_2 resources).

Section 2.2 has provided several examples of path-structural constraints. Most of them can be modeled with additional resources (one for each constraint) in a standard SPPRC. For the ESPPRC, Beasley and Christofides (1989) proposed to add to $R^\leq$ an additional resource r_v for each node $v \in V$. (For a compact notation, we use the Kronecker-symbol with $\delta_{ij} = 1$ if $i = j$, and $\delta_{ij} = 0$, otherwise.) The resource intervals are defined as $[a_i^{r_v}, b_i^{r_v}] = [0, 1 - \delta_{si}]$ for all $i \in V$ and the REFs by $f_{ij}^{r_v}(T_i) = T_i^{r_v} + \delta_{iv}$ for all $(i, j) \in A$.

Table 2.1 gives an overview of how (anti-)pairing constraints, precedence constraints, and (non-)follower constraints can be modeled by means of resources. In this table, M is a sufficiently large positive integer. For the first group (pairing, anti-pairing, and precedence) we assume that a single task is associated with each node. Note that the modeling proposed for the (k, ℓ)-pairing and precedence constraints is equivalent to the set component proposed by Dumas et al. (1991) for the SPPRCPD.

If a single task is associated with each node, follower and non-follower constraints simply imply the removal of some of the arcs (see e.g. Desrochers and Soumis (1989)). Therefore, we present these constraints for the case that sequences of tasks are associated with arcs and nodes. We assume that tasks are numbered from 1 to N, the last task of any non-empty task sequence $W(\cdot)$ is denoted by $l(W(\cdot))$. For empty task sequences one defines $l(W(\varnothing)) = 0$.

All follower and non-follower constraints can be modeled with a single resource r, where $T_i^r \in \{1, \ldots, N\}$ means that the last task of the task sequence of the current path $(s, \ldots, v_i)$ was the one with number T_i^r. $T_i^r = 0$ means that the current path has an empty task sequence. The definition of the corresponding REFs is:

$$
f_{ij}^r(T_i) = \begin{cases}
T_i^r & \text{if } W\big((i,j), j\big) = \varnothing \\
l\big(W((i,j), j)\big) & \text{if } T_i^r \neq 0, \, W\big((i,j), j\big) \neq \varnothing, \\
& \text{and } \big(T_i^r, W((i,j), j)\big) \text{ feasible} \\
l\big(W((i,j), j)\big) & \text{if } T_i^r = 0, \, W\big((i,j), j\big) \neq \varnothing, \\
& \text{and } W\big((i,j), j\big) \text{ feasible} \\
-1 & \text{otherwise.}
\end{cases} \tag{2.6}
$$

The strength of the non-classical REF concept is that it allows multiple resources to depend on each other. In several applications such as the aircrew pairing problem Vance et al. (1997), the cost of a path depends on several resources. A second example of non-trivial dependent REFs stems from the capacity constraints of the *VRPTW with simultaneous pickups and deliveries*, see Min (1989); Desaulniers et al. (1998). Here, each customer $i \in V \setminus \{s, t\}$ has demanded for delivery q_i^d and for pickup q_i^p. A vehicle of capacity Q starts at the depot s with the entire delivery demand of the tour loaded. It services each customer (pickup after delivery) so that the vehicle reaches the final depot t having the entire pickup demand on board. A feasible path (route) is one in which the pickups of already visited nodes plus the deliveries of the following customers do not exceed the vehicle capacity on any arc traveled. The feasibility problem is modeled with two dependent resources $r_p, r_{\max} \in R^{\leq}$, where the resource variable $T_i^{r_p}$ is *demand already picked* (directly after node i) and $T_i^{r_{\max}}$ is the *maximum load* in the vehicle on the path from s

to i. Obviously, one has $[a_i^{r_p}, b_i^{r_p}] = [a_i^{r_{max}}, b_i^{r_{max}}] = [0, Q]$ for all $i \in V$ and $f_{ij}^{r_p}(T_i^{r_p}, T_i^{r_{max}}) = T_i^{r_p} + q_j^p$ for all $(i, j) \in A$. For the maximum load, one has non-linear but non-decreasing REFs $f_{ij}^{r_{max}}(T_i^{r_p}, T_i^{r_{max}}) = \max\{T_i^{r_p} + q_j^p, T_i^{r_{max}} + q_j^d\}$. It means that the maximum load at node j (following node i) is either the entire pickup demand at the end of the path, computed by $T_i^{r_p} + q_j^p$, or results from the maximum load on the sub-path $(0, \ldots, i)$ to which the delivery of j has to be added.

The modeling of other non-linear resource consumptions is straight-forward, e.g., *soft time windows* (see Dumas et al., 1990), *load-dependent travel costs* or *time-dependent travel times* (connections (i, j) with different travel durations depending on the time of the day). Complex schedule regulations and their modeling can be found in Desaulniers et al. (1997); Vance et al. (1997).

Another non-trivial example of dependent resources is the computation of the *minimal waiting time* for an SPPTW path. With the notation for the SPPTW given in the introduction, the total waiting time along path $P = (v_0, v_1, \ldots, v_p)$ is given by $T_p - T_0 - \sum_{i=1}^{p} t_{i-1,i}$. Desaulniers and Villeneuve (2000) showed that three resources with non-decreasing REFs are enough to compute both the earliest arrival time and the minimal waiting time (or equivalently, an associated waiting cost).

4.   Solution methods

This section describes different methodologies developed for solving the SPPRCs, namely, dynamic programming which has been used extensively, Lagrangean relaxation, constraint programming, and heuristics. It also presents a graph modification approach for the SPPRCFP.

4.1   Dynamic programming and labeling algorithms

Dynamic programming solution approaches for the SPPRC systematically build new paths, starting from the trivial path $P = (s)$, by extending paths one-by-one into all feasible directions. Their efficiency depends on the ability to identify and discard paths which are not useful either to build a Pareto-optimal set of paths or to be extended into Pareto-optimal paths. Discarding non-useful paths is achieved by a dominance sub-algorithm based on dominance rules, which strongly depend on the path-structural constraints and the properties of the REFs.

For the sake of efficiency, paths in the dynamic programming algorithms are encoded by *labels*. Paths sharing a common prefix are represented by using a single chain of labels for their common prefix. This is implemented with the help of a tree data structure in which a label corresponding to path $P = (v_0, \ldots, v_{p-1}, v_p)$ is directly linked back to the label of the prefix path $(v_0, \ldots, v_{p-1})$ (see e.g. Ahuja et al., 1993, for an introduction to labeling algorithms). Beside encoding the path itself, the label typically stores a representation of $\mathcal{T}(P)$, e.g., given by the unique resource vector $T(P)$ in case of non-decreasing REFs. In Ioachim et al. (1998) a more complex representation of $\mathcal{T}(P)$ is stored in the labels, while Irnich and Villeneuve (2003) store additional (compressed) information to accelerate the dominance algorithm.

In order to formalize the above ideas, we need some definitions. For a given path $P = (v_0, v_1, \ldots, v_p)$ we call $v(P) = v_p$ the *resident node* of P. A path $P = (v_0, v_1, \ldots, v_p)$ is a *feasible extension* of path $Q = (w_0, w_1, \ldots, w_q)$ if $(Q, P) = (w_0, \ldots, w_q, v_0, \ldots, v_p) \in \mathcal{F}(w_0, v_p) \cap \mathcal{G}$. The set of all feasible extensions is $\mathcal{E}(Q) = \{P \colon (Q, P) \in \mathcal{F}(w_0, v(P)) \cap \mathcal{G}\}$.

Labeling algorithms rely on the manipulation of two sets. The first set $\mathcal{U}$ is the set of *unprocessed paths*, which have not yet been extended. The second set $\mathcal{P}$ is the set of *useful paths*. Useful paths $P \in \mathcal{P}$ have already been processed. They have been identified to be Pareto-optimal or might be prefixes of Pareto-optimal paths (note that Pareto-optimal paths might have prefixes which are not Pareto-optimal, see Section 4.1.2). Both sets, $\mathcal{U}$ and $\mathcal{P}$, change dynamically in the course of the labeling algorithm.

One can identify two basic procedures invoked by the labeling algorithm (see the pseudo-code below). In the *path extension step* an unprocessed path $Q \in \mathcal{U}$ is chosen, all feasible extensions (Q, v) with $v \in V$ are constructed and added to $\mathcal{U}$, while Q itself is removed from $\mathcal{U}$. Thus, the extension step replaces one element of $\mathcal{U}$ by all of its feasible one-node extensions. Once processed, an element is transferred to the set $\mathcal{P}$. If possible, the *dominance algorithm* reduces the sets $\mathcal{U}$ and $\mathcal{P}$. Its goal is to accelerate the overall labeling procedure by limiting the number of necessary extension steps.

The path extension step and the dominance algorithm maintain the following invariant: *The useful paths $\mathcal{P}$ and all extensions of unprocessed paths $\mathcal{U}$ together contain a solution of the SPPRC.* Recall from Section 2.3 that an SPPRC solution is not necessarily unique since it contains representatives taken from a set of desired solutions, e.g., one path for each Pareto-optimal resource vector. Therefore, let Σ be the set of all different solutions of an SPPRC, where each element $\mathcal{S} \in \Sigma$ is a set of paths, e.g., Pareto-optimal paths. The above invariant is

$$\exists \mathcal{S} \in \Sigma \colon \mathcal{S} \subseteq \{(Q, P) \colon Q \in \mathcal{U}, P \in \mathcal{E}(Q)\} \cup \mathcal{P}. \qquad (2.7)$$

The algorithm is initialized with $\mathcal{U} = \{P_0\}$ and $\mathcal{P} = \varnothing$ where $P_0 = (s)$ is the trivial path. Each path $P = (v_0, v_1, \ldots, v_p) \in \mathcal{F}$ results from an extension of P_0, i.e., $(v_1, \ldots, v_p) \in \mathcal{E}(P_0)$. Hence, condition (2.7) holds for the initialization. Obviously, the path extension step also maintains the invariant. The crucial point is to define dominance rules in such a way that the dominance algorithm also respects (2.7). We focus on that aspect in Section 4.1.2. By doing so, the algorithm finally terminates with an $\mathcal{S} \subseteq \mathcal{P}$ for some $\mathcal{S} \in \Sigma$. In a post-processing *filtering step* Pareto-optimal solutions can be extracted from $\mathcal{P}$.

Generic Dynamic Programming SPPRC Algorithm {
   (* Initialize *)
   SET $\mathcal{U} = \{(s)\}$ and $\mathcal{P} = \varnothing$
   WHILE $\mathcal{U} \neq \varnothing$ DO
      (* Path extension step *)
      CHOOSE a path $Q \in \mathcal{U}$ and REMOVE Q from $\mathcal{U}$
      FORALL arcs $(v(Q), w) \in A$ of the forward star of $v(Q)$ DO
         IF $(Q, w) \in \mathcal{F}(s, w) \cap \mathcal{G}$ THEN ADD (Q, w) to $\mathcal{U}$
      ADD Q to $\mathcal{P}$
      (* Dominance step *)
      IF (* any condition *)
         APPLY dominance algorithm to paths from $\mathcal{U} \cup \mathcal{P}$ ending
            at some node v
   (* Filtering step *)
   FILTER $\mathcal{P}$, i.e., identify a solution $\mathcal{S} \subseteq \mathcal{P}$
}

Several remarks should be made.

1 If one performs path extension steps only, but no dominance steps, the result is $\mathcal{P} = \mathcal{F}$, i.e., the algorithm computes all feasible paths.

2 The path extension step leaves the freedom to choose paths $Q \in \mathcal{U}$ according to different processing strategies. These path selection strategies can lead to *label setting* or *label correcting* algorithms depending on the underlying network and the REFs. These issues will be discussed in Section 4.1.1.

3 The dominance algorithm can be applied at any time in the course of the algorithm. In order to keep the effort small, it makes sense to delay the dominance algorithm to a point when there is a chance to remove several of the paths at the same time, before they are processed in the path extension step.

The dominance rules strongly depend on the problem at hand. Section 4.1.2 discusses the impact of different path-structural constraints and classical, non-decreasing, special or general REFs.

4 There exist efficient algorithms for the filtering step to identify, e.g., Pareto-optimal paths (see Bentley, 1980; Kung et al., 1975).

4.1.1   Label setting and label correcting algorithms.   The defining property of a *label setting algorithm* is that those labels chosen to be extended (in the path extension step) are kept until the end of the labeling process. They will not be identified as discardable in subsequent calls of the dominance algorithm. Labeling algorithms that do not guarantee this behavior are called *label correcting algorithms*. The general ideas of label setting as well as label correcting algorithms in the context of the one-dimensional shortest path problem (SPP) are, for instance, explained in the book of Ahuja et al. (1993).

An acyclic network $G = (V, A)$ naturally gives rise to label setting algorithms if paths are treated (that is, chosen and extended) according to a topological order of their resident nodes. More precisely, the above generic algorithm loops over the topologically sorted nodes $v = s, v_2, \ldots, v_{|V|}$, applies the dominance algorithm to the paths $\{P \in \mathcal{U} \cup \mathcal{P} : v(P) = v\}$ resident at the current node v, and extends those paths who survive the dominance process into all feasible directions.

It is possible to mimic an acyclic network for the treatment of labels if the resource consumptions for at least one resource r are strictly positive, i.e., $f_{ij}^r(T_i) - T_i^r > 0$ holds for all $(i, j) \in A$ and all $T_i \in [a_i, b_i]$. In this case, the labeling algorithm chooses unprocessed paths $Q \in \mathcal{U}$ with minimum (or "small") $T(Q)^r$ for extension first. It is guaranteed that paths Q already treated only produce extensions (Q, P) with $T(Q, P)^r > T(Q)^r$. Hence, newly generated paths cannot enforce the elimination of already treated paths. Desrochers and Soumis (1988) used the concept of generalized buckets to identify paths with small value $T(Q)^r$.

Label correcting algorithms solve shortest path instances with negative arc lengths. The existence of negative resource consumptions $f_{ij}^r(T_i) - T_i^r$ for an arc (i, j) and all resources r (i.e., negative t_{ij}^r for the classical SPPRC) means that the strategy of treating paths in a strictly increasing ord er of their resource vectors has to be replaced by a more flexible processing strategy. The well-known Ford-Bellman label correcting algorithm for the SPP adds newly generated labels to the end of a queue and extends labels one-by-one starting with the label currently at the top of the queue. Powell and Chen (1998) have presented a more sophisticated generalized label correcting strategy for the SPPRC, which is directly applicable to the general SPPRC case.

4.1.2   Dominance rules and dominance algorithms.   Efficient dominance rules have been described for the SPPRC, ESPPRC and SPPRC-k-cyc with non-decreasing REFs. Recall that in these cases each path $P \in \mathcal{F}(s, v)$ has a unique resource vector $T(P) \in \mathcal{T}(P)$, which is the only Pareto-optimal point of $\mathcal{T}(P)$.

Dominance rules identify paths Q to be non-useful in the following sense: Q is neither necessary to describe the set of Pareto-optimal solutions $PO(v(Q))$, nor feasible extensions $Q' \in \mathcal{E}(Q)$ lead to paths (Q, Q') necessary to construct $PO(v(Q'))$. Such a path Q can be discarded. Typically, dominance rules identify non-useful paths by comparing $T(Q)$ and $\mathcal{E}(Q)$ with the corresponding values $T(P)$ and $\mathcal{E}(P)$ of paths P resident at node $v(P) = v(Q)$. We discuss the cases SPPRC, ESPPRC, SPPRC-2-cyc, and SPPRC-k-cyc with non-decreasing REFs in detail.

SPPRC.   Given two different paths $P, Q \in \mathcal{U} \cup \mathcal{P}, v(P) = v(Q)$ with $T(P) \leq T(Q)$, the dominance algorithm can discard path Q while keeping P, which results from the following two arguments. First, $T(P) \leq T(Q)$ means that $T(Q)$ is not necessary to represent Pareto-optimal paths ending at $v(Q)$. Second, one has to investigate possible extensions of Q. The fact $T(P) \leq T(Q)$, the absence of any path-structural constraints and the non-decreasing REF imply $\mathcal{E}(P) \supseteq \mathcal{E}(Q)$. Therefore, any $Q' \in \mathcal{E}(Q)$ fulfills $(P, Q') \in \mathcal{F}$ and $T(P, Q') \leq T(Q, Q')$. There do not result any Pareto-optimal resource consumptions from extensions of Q which could not have been built using extensions of P. Hence Q can be discarded.

Note that dominance rules are sensitive to the occurrence of paths with identical resource vectors. Therefore, one has to distinguish between *dominance* and *discarding dominated paths*. Two paths $P, Q \in \mathcal{F}(s, v)$ with $T(P) = T(Q)$ dominate each other but only one of these two can be eliminated (while the other one is kept). (Irnich and Villeneuve, 2003) propose techniques to resolve ambiguity and analyze them for the SPPRC and SPPRC-k-cyc cases.

ESPPRC.   In presence of path-structural constraints, the relation $T(P) \leq T(Q)$ does not necessarily imply the relation $\mathcal{E}(P) \supseteq \mathcal{E}(Q)$. For the ESPPRC, the reason is that paths $P \in \mathcal{G}$ can only be extended to nodes not already visited. We denote the set of visited nodes by $V(P)$. A restricted dominance rule for the ESPPRC allows to discard path Q if $T(P) \leq T(Q)$ and $V(P) \subseteq V(Q)$ since both conditions together imply $\mathcal{E}(P) \supseteq \mathcal{E}(Q)$. Beasley and Christofides (1989) modeled the sets $V(P)$ for paths $P \in \mathcal{F}$ by one additional resource for each node of V.

Feillet et al. (2004) improved the idea of Beasley and Christofides. They interpreted the set $V(P)$ differently as the *"set of nodes which cannot be visited any more"*. By analyzing the resource vector $T(P)$

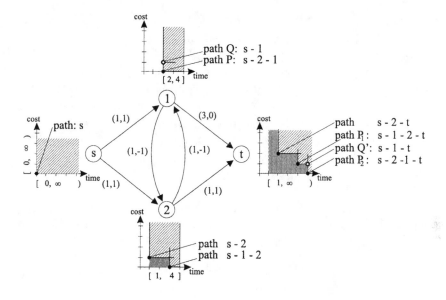

Figure 2.2. Example of an SPPTW with 2-cycle elimination.

they identified additional unvisited nodes which are impossible to reach
(e.g., because of current time, time window constraints and non-negative
travel times). These nodes are added to the set $V(P)$ to form the set
$\overline{V}(P)$. As a result, the above dominance rule based on the "extended"
sets $\overline{V}(P)$ can eliminate more paths.

SPPRC-2-cyc.   An informal description of a dominance rule for
the 2-cycle elimination case is the following: *Keep only a Pareto-best
path P_1 and a second-best path P_2 which is extended from a different
predecessor node.* For any path $P = (v_0, \ldots, v_{p-1}, v_p)$ with $p \geq 1$, the
node v_{p-1} is called the *predecessor node* and denoted $\mathrm{pred}(P)$. It is
easy to see that the SPPRC dominance rule applies to paths P, Q,
$v(P) = v(Q)$, $T(P) \leq T(Q)$ having identical predecessor nodes. Kohl
(1995); Larsen (1999) showed that if $\mathcal{E}(P)$ does not contain the one-
node path $(\mathrm{pred}(P))$, i.e., the dominating path P cannot be extended
to its predecessor node, the SPPRC dominance rule also remains valid.
Contrarily, given three different paths P_1, P_2, Q, $v(P_1) = v(P_2) = v(Q)$,
$T(P_1), T(P_2) \leq T(Q)$ with different predecessors $\mathrm{pred}(P_1) \neq \mathrm{pred}(P_2)$,
one can discard path Q while keeping P_1 and P_2. The proof of this rule
is based on the fact that $\mathrm{pred}(P_1) \neq \mathrm{pred}(P_2)$ implies $\mathcal{E}(P_1) \cup \mathcal{E}(P_2) \supseteq
\mathcal{E}(Q)$.

An example of an SPPTW with 2-cycle elimination is shown in Figure 2.2 and illustrates the two above-mentioned dominance rules.

First, at node 1 the paths P and Q fulfill $T(P) \leq T(Q)$. Since $\text{pred}(P) \neq \text{pred}(Q)$ it is not allowed to eliminate Q. This is substantial because the dominated path $Q = (s, 1)$ is a prefix of the Pareto-optimal path $P_1 = (s, 1, 2, t)$ at the sink t. The path-structural constraints imply that some dominated paths, like Q, are still useful paths. Second, path Q' at node t can be discarded because the two dominating paths P_1 and P_2 have different predecessor nodes (alternatively, because P_2 and Q' have the same predecessor node).

SPPRC-k-cyc.   Handling the k-cycle elimination case for $k \geq 3$ needs sophisticated data structures (see Irnich and Villeneuve, 2003). In essence, the dominance rule efficiently checks whether

$$\mathcal{E}(Q) \subseteq \bigcup_{P \in \mathcal{P} \cup \mathcal{U}:\, T(P) \leq T(Q), v(P) = v(Q)} \mathcal{E}(P) \tag{2.8}$$

holds, i.e., all extensions of dominating paths cover the extensions of Q. A path Q for which (2.8) holds can be discarded. There exists a finite representation of the right hand side of (2.8), which uses up to $(k-1)!^2$ vectors (so-called *set forms*) with $\binom{k}{2}$ entries. Moreover, these set forms can be used to efficiently encode and update the relation (2.8) so that the evaluation of (2.8) can be performed in constant time. From a complexity point of view, the main result of this dominance rule is that the maximum number of paths stored in $\mathcal{P} \cup \mathcal{U}$ grows by a factor $\alpha(k)$ compared to the classical SPPRC. The factor $\alpha(k)$ is independent of the size of the underlying network and bounded by $\alpha(k) \leq k(k-1)!^2$.

SPPTWTC.   Another case where efficient dominance rules have been described is the shortest path problem with time windows and time costs (SPPTWTC)(see Ioachim et al., 1998). An SPPTWTC instance is uniquely defined by the SPPTW data, i.e., travel costs c_{ij}, travel times t_{ij}, and time windows $[a_j, b_j]$, together with arbitrary node costs $w_j \in \mathbb{R}$ (positive as well as negative) for the nodes $j \in V$. Visiting the node j at time T_j^{time} causes additional *time costs or profits* of $w_j T_j^{\text{time}}$. Hence, depending on the sign of w_j it is advantageous to visit node j as early or as late as possible. When negative and positive time costs occur together at the nodes of a path, the determination of feasible visiting times T_j^{time} with minimum overall cost is an optimization problem in itself.

Formally, the SPPTWTC is a two-resource problem with a resource *time* and a time-dependent resource *cost*. The REFs for time are given

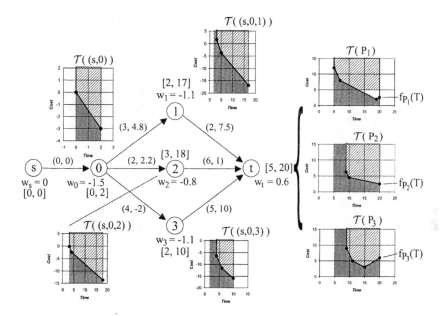

Figure 2.3. Example of an SPPTWTC: Travel time and travel cost are given as pairs (t_{ij}, c_{ij}) for each arc (i, j), time windows $[a_i, b_i]$ and linear node costs w_i are given for each node i, paths ending at node t are $P_1 = (s, 0, 1, t)$, $P_2 = (s, 0, 2, t)$, and $P_3 = (s, 0, 3, t)$.

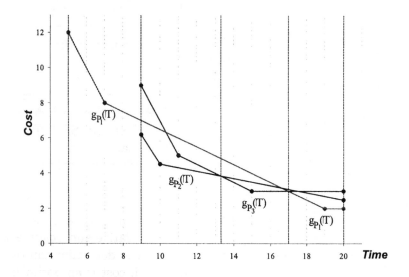

Figure 2.4. Piecewise linear cost functions representing $\mathcal{T}(P)$ for the paths $P_1 = (s, 1, t)$, $P_2 = (s, 2, t)$, and $P_3 = (s, 3, t)$.

by $f_{ij}^{\mathrm{time}}(T_i^{\mathrm{time}}) = T_i^{\mathrm{time}} + t_{ij}$ and for cost by $f_{ij}^{\mathrm{cost}}(T_j^{\mathrm{time}}, T_i^{\mathrm{cost}}) = T_i^{\mathrm{cost}} + c_{ij} + w_j T_j^{\mathrm{time}}$. This is a (minor) extension of the REF concept of Section 2 because f_{ij}^{cost} depends on a resource variable T_j^{time} of the node j (and not only on resource variables T_i at node i). Figure 2.3 shows a small SPPTWTC example. Next to each node, the resource space (a time-cost diagram) shows the set $\mathcal{T}(P)$ for each of the feasible paths P. Obviously, $\mathcal{T}(P)$ is a bounded polyhedron.

Dominance rules for the SPPTWTC were proposed by Ioachim et al. (1998). Although presented differently, their main ideas are the following. The set $\mathcal{T}(P)$ is determined by its lower envelope, which is a piecewise linear (cost) function $f_P(T^{\mathrm{time}})$ with a maximum of $p + 1$ pieces when P has length p (in the following we use $T = T^{\mathrm{time}}$). The function $f_P(T)$ is convex and only its first strictly decreasing part is relevant for dominance (since the objective is to find the minimum-cost path, the nonnegative slope segments are useless (see Ioachim et al., 1998, p. 196). Hence, the relevant piecewise linear cost function is

$$g_P(T) = \begin{cases} f_P(T), & \text{for } T \le \arg\min_T f_P(T) \\ f_*, & \text{for } T \ge \arg\min_T f_P(T) \end{cases}$$

with the minimum $f_* = \min_T f_P(T)$. Simple update formulas allow to compute $g_P(T)$ from $g_{P'}(T)$ when P' is the prefix path of $P = (P', v)$. A path Q can be discarded if there exists paths $P_1, \ldots, P_k$ ending at $v = v(Q) = v(P_1) = \cdots = v(P_k)$ with $\mathcal{T}(Q)^{\llcorner} \subseteq \bigcup_{i=1}^k \mathcal{T}(P_i)^{\llcorner}$ (for a set X the symbol $X^{\llcorner}$ denotes the set $\bigcup_{x \in X} x^{\llcorner}$). This dominance rule can be implemented by computing the minimum cost function $G_v(T) = \min_P g_P(T)$ over all paths P ending at node v. Each path Q with $v(Q) = v$ which does not contribute to the minimum cost function $G_v(T)$ can be discarded. Figure 2.4 shows the situation for the three paths P_1, P_2, and P_3 ending at node t from the above example. All paths P_1, P_2, P_3 contribute to $G_t(T)$, which is composed of four pieces imposed by $g_{P_1}(T)$ for $T \in [5, 9)$, $g_{P_2}(T)$ for $T \in [9, 13.\bar{3}]$, $g_{P_3}(T)$ for $T \in [13.\bar{3}, 17]$, and $g_{P_1}(T)$ for $T \in [17, 20]$. None of the paths are dominated by the other paths.

4.2   Lagrangean relaxation

The constrained shortest path problem (CSPP) is a specialized s-t-SPPRC with independent additive resource consumptions along arcs. The resource consumption is constrained only as a whole and not by individual resource intervals. The objective is to find a least-cost s-t-path with resource consumptions within a pre-specified interval. Among others, Beasley and Christofides (1989); Borndörfer et al. (2001) proposed to solve the CSPP with Lagrangean relaxation for computing lower bounds and a tree search procedure exploiting these computed lower bounds. For the remainder of this section we assume that the underlying network $G = (V, A)$ is acyclic.

For a formal description of the CSPP, consider R different resources including *cost* as the last resource R with cost matrix $C = (c_{ij}) = (t_{ij}^R)$. For the remaining resources, let $\mathcal{R} = (t_{ij}^r) \in \mathbb{R}^{(R-1)\times|A|}$ be the *resource consumption matrix* with non-negative consumptions t_{ij}^r for $r = 1, \ldots, R - 1$. The REFs are $f_{ij}^r(T_i) = T_i + t_{ij}^r$ for all $r = 1, \ldots, R$ and $(i, j) \in A$, and the resource accumulation is $T_j = f_{ij}(T_i)$ whenever arc (i, j) is traversed. Lower bounds $l \in \mathbb{R}^{R-1}$ and upper bounds $u \in \mathbb{R}^{R-1}$ on the overall accumulated resource consumptions are implied by defining $[a_s, b_s] = [0, 0,$ $[a_t, b_t] = [l, u]$, and $[a_i, b_i] = [0, u]$ at all other node $i \in V \setminus \{s, t\}$. For a given path P and its incidence vector $x \in \{0, 1\}^{|A|}$, the resource consumption is $\mathcal{R}x$ and the cost is $c^\top x$. P is feasible if $l \leq \mathcal{R}x \leq u$ holds. Borndörfer et al. (2001) have added a *goal value* $g \in [l, u]$ for the resource consumption $\mathcal{R}x$ to the formulation of Beasley and Christofides (1989). Slack and surplus variables z_+, z_- measure the deviation of $\mathcal{R}x$ from g, which is penalized by $p_+, p_- \in \mathbb{R}_{\geq 0}^{R-1}$. The CSPP can be stated as follows:

$$z_{\text{CSPP}} = \min c^\top x + p_-^\top z_- + p_+^\top z_+ \qquad (2.9a)$$

$$\text{subject to } Ix = e_s - e_t \qquad (2.9b)$$

$$\mathcal{R}x + z_+ - z_- = g \qquad (2.9c)$$

$$(z_-, z_+) \leq (u - g, g - l) \qquad (2.9d)$$

$$x \in \{0, 1\}^{|A|}, \quad z_-, z_+ \in \mathbb{R}_{\geq 0}^R \qquad (2.9e)$$

Cost (2.9a) is a combination of accumulated travel costs and the penalty for the deviation of $\mathcal{R}x$ from g. Flow conservation constraints (2.9b) are given by means of the arc-node incidence matrix $I \in \{-1, 0, 1\}^{|V|\times|A|}$ and unit vectors $e_s, e_t \in \{0, 1\}^{|V|}$. They guarantee that $\{(i, j) \colon x_{ij} = 1\}$ forms a path in the acyclic network G. Constraints (2.9d) bounds the slack and surplus variables so that $l \leq \mathcal{R}x \leq u$ is ensured.

A Lagrangean relaxation of (2.9) can be obtained by relaxing the resource consumption constraints (2.9c). Let $\pi \in \mathbb{R}^{R-1}$ be an associated dual price vector. The Lagrangean dual of (2.9) is $\max_{\pi \in \mathbb{R}^{R-1}} z_{\text{DCSPP}}(\pi)$ where the Lagrangean subproblem decomposes into the following two parts:

$$z_{\text{DCSPP}}(\pi) = P(\pi) + B(\pi) + \pi^\top g \qquad (2.10a)$$

$$\text{with } P(\pi) = \min(c^\top - \pi^\top \mathcal{R})x,$$

$$Ix = e_s - e_t, \quad x \in \{0,1\}^{|A|} \tag{2.10b}$$

$$\text{and} \quad B(\pi) = \min(p_+^\top - \pi^\top)z_+ + (p_-^\top + \pi^\top)z_-,$$
$$0 \leq z_- \leq u - g, \quad 0 \leq z_+ \leq g - l. \tag{2.10c}$$

The first part (2.10b) is an SPP, which can be solved with a label setting algorithm (see Ahuja et al., 1993). The second part (2.10c) is a minimization problem defined over a box, which is trivial to solve by inspection of the signs of the components of $(p_+^\top - \pi^\top)$ and $(p_-^\top + \pi^\top)$.

High quality solutions for the above Lagrangean dual formulation can be computed with any subgradient optimization method, e.g., a coordinate ascent method as in Borndörfer et al. (2001). The same authors proposed to use such a dual solution π^* and the dual solution of (2.10b) obtained for $\pi = \pi^*$ (i.e., a distance vector $(h_v(\pi^*))_{v \in V}$) to compute so-called *Lagrangean distance labels*:

$$g_v(\pi^*) = h_v(\pi^*) - h_s(\pi^*) + B(\pi^*) + g^\top \pi^*, \quad \text{for all } v \in V.$$

These labels are very useful to prune the search tree because of the following property. Let $x^1 \in \{0,1\}^{|A|}$ and $x^2 \in \{0,1\}^{|A|}$ be path incidence vectors of an s-v-path and a v-t-path, respectively. If $x = x^1 + x^2$ is a feasible CSPP path and $\pi^* \in \mathbb{R}^{R-1}$ a Lagrangean multiplier vector, then

$$z_{\text{CSPP}}(x) \geq g_v(\pi^*) + (c^\top - \pi^{*\top}\mathcal{R})x^2 \tag{2.11}$$

where $z_{\text{CSPP}}(x)$ denotes the cost of path x. The inequality means that if the right-hand side is non-negative then there exists no prefix path x^1 such that $x^1 + x^2$ has a negative (reduced) cost. Consequently, one should implement a tree search for finding negative (reduced) cost CSPP paths in G by systematically building v-t-paths x^2 starting at the sink node t. A tentative path x^2 can be discarded if the right-hand side of (2.11) becomes non-negative. Note that additional constraints that could not be considered in (2.9) can always be taken into account in the search phase.

4.3   Constraint programming

Constraint programming (CP) relies on a model which is defined by a set of variables, each with an initial domain, and a set of constraints. A CP approach is composed of a *search mechanism* to explore the solution space, a *domain reduction algorithm* for each constraint that tries to remove inconsistent values from the domains of the variables involved in that constraint, and a *propagation algorithm* that propagates these domain changes among the constraints. It allows to consider a wide spectrum of constraints (algebraic and non-algebraic), including some that cannot be modeled using resources or simple path-structural constraints: For instance, an employee cannot work more than 8 hours in every 24-hour period. Within column generation approaches, CP has recently been used to tackle the SPPRC on an acyclic network (de Silva, 2001; Fahle et al., 2002) and the ESPPRC (Rousseau et al., 2003). In both cases, the goal is solely to find at least one feasible path with a negative (reduced) cost. This goal is modeled as a constraint (the cost of a feasible path must be negative), yielding a constraint satisfaction problem.

Fahle et al. (2002) considered an SPPRC on an acyclic network where a task, defined by a starting and an ending time, is associated with each node. They proposed a model where a boolean variable is associated with each node. Such a variable is set to *true* if the corresponding node is part of the path currently built. In this case, we will say that the node is *selected*. Additional variables are also used to specify, for instance, the minimal amount of rest to assign after each task. Their model includes simple constraints such as the boolean variables associated with two nodes whose tasks must be performed concurrently cannot be set at *true* simultaneously, or the total duration of the tasks associated with the selected nodes cannot exceed the maximum worked time in a schedule. Given a set of selected nodes, these two types of constraints can be used to fix some boolean variables to *false*.

In de Silva (2001), a different CP model is used. It involves variables to indicate the successor node $next[t]$ of each node t and variables to specify the amount of accumulated resource consumptions at each node. Nodes with $next[t] = t$ are not included in the current path. Path constraints model resource consumptions along the selected (partial) path, e.g., for the reduced cost, total working time, etc. Each time that a successor node is selected, the propagation algorithm is invoked, i.e., constraints are verified by solving an SPP for every unselected node of the underlying network. For instance, one can exclude a node (i.e., set $next[t] = t$) if the value of the path with the shortest worked time and passing through that node t and all selected nodes exceeds the maximum total worked time. A similar decision propagation based on the (reduced) cost of a path can also be executed. So-called *goals*, e.g., based on reduced cost shortest path computations, control how new tasks are added to the current partial path. The search tree is usually explored until a prespecified number of negative cost paths are found or until a time limit is reached.

For the ESPPRC, Rousseau et al. (2003) used a similar model with variables for the successor node and variables for the accumulated resource consumptions. Some of the constraints they consider are: All successor nodes must be different, no subtours are allowed, lower bounds provided by the resource REFs must be respected, the (reduced) cost of a feasible path must be negative. For verifying this last constraint, the authors compute a lower bound by solving an assignment problem. The choice of the next variable to branch on in the search tree is made in such a way to construct a path from the source to the sink node.

4.4   Heuristics

Even with sophisticated solution methods, solving an SPPRC instance might still be very time-consuming. In the column generation context, solving SPPRCs to proven optimality is only necessary to show that no negative reduced cost paths exist in the last pricing step. In preceding iterations it is sufficient to approximately solve the SPPRC, i.e., to compute any negative reduced cost feasible paths. That is the point where heuristics for the SPPRC come into play. In addition, they might be applied when the entire column generation problem is treated heuristically. In the following, we distinguish between three major areas of application for heuristics: *Pre-processing*, *dynamic programming*, and *direct search*.

Classical *pre-processing* techniques eliminate arcs and reduce the resource intervals (see e.g. Desrochers et al., 1992). The heuristic version of this idea is to solve a given SPPRC instance on a hierarchy of restricted networks, where each of the restricted networks contains only a limited number of arcs, e.g., defined by the $p > 0$ "nearest neighbors" of each node. Starting with the smallest p-nearest-neighbor network, one solves the associated SPPRC, and if no solution is found, one continues with the next p. This idea has been used in many implementations (e.g. Dumas et al. (1991); Savelsbergh and Sol (1998); Larsen (1999); Irnich and Villeneuve (2003)). Another idea is to replace some of the resources by less accurate resources to get an easier-to-solve SPPRC network. Gamache et al. (1999) gave the example where a restricted network measures time rounded up to the nearest hour while the exact global network uses minutes.

Dynamic programming heuristics are based on the techniques of Section 4.1 but heuristically accelerate the computation. For the VRPTW, Larsen (1999) used a so-called *forced early stop* rule to quit from the dynamic program when an adequate number of negative reduced cost columns has been found and a pre-defined number of labels has been generated. Chabrier (2002) tried to solve the ESPPRC by using the standard path extension step (i.e., not extending a path to a node already visited) with the stronger SPPRC dominance rule (i.e., only the resource vectors are compared but not the visited nodes). Clearly, this procedure is quick but might fail to detect any negative reduced cost path. Therefore, he proposed to iteratively apply a dynamic programming procedure which combines the ESPPRC extension step with a gradually parametrizable dominance rule. A parameter *DomLevel* (between 0 and ∞) defines the length of a path after which the ESPPRC dominance rule is applied. If the partial path is shorter, the heuristic SPPRC dominance rule is applied. Larger values of *DomLevel* make the modified dynamic programming procedure substantially faster. The case *DomLevel* = 0 corresponds with the exact ESPPRC and is expected to be quite slow (especially for non-adjusted dual variables). Hence, starting with a large value for *DomLevel*, the dynamic programming algorithm with the modified dominance rule is iteratively applied with decreasing values of *DomLevel* until a negative reduced cost path is found (or the ESPPRC is solved exactly).

Finally, *direct search heuristics* are mainly based on local search. Such improvement procedures start from a given feasible path P and delete, insert, or replace nodes or exchange arcs in order to find an improving feasible path P' with smaller reduced cost. Note that after solving the restricted master program, the basic variables provide a set of paths with reduced cost 0 from which an improvement algorithm might start. Successful column generation applications which use these techniques can be found in Savelsbergh and Sol (1998); Xu et al. (2003).

4.5   A graph modification approach for the SPPRCFP

The graph modification approach for the SPPRCFP defined on a given network $G = (V, A)$ is not a solution method in itself but a method that manipulates G to obtain a new network $G' = (V', A')$ from which all forbidden paths are removed while the other paths of G are still feasible. One can then apply any of the proposed methods for the SPPRC to the network G' to solve the given SPPRCFP. Formally, let $\mathcal{H}$ be the set of forbidden sub-paths and let $\mathcal{G}_{\text{forbidden}} = \{(P, Q, P'): P, Q, P' \text{ paths}, Q \in \mathcal{H}\}$ so that $\mathcal{G} = \{ \text{all paths}\} \setminus \mathcal{G}_{\text{forbidden}}$ is the set of all feasible paths for the SPPRCFP. The approach of Villeneuve and Desaulniers (2000) merges the original graph G with the state graph of a finite automaton, which identifies the infeasible sub-paths in $\mathcal{H}$. We illustrate the procedure by an example in which G is given in Figure 2.5(a) and $\mathcal{H} = \{(1, 2, 4), (2, 1), (2, 3, 1)\}$.

The approach works in two stages. First, the algorithm of Aho and Corasick (1975) is used to construct the state graph $S = (V_S, A_S)$ of a

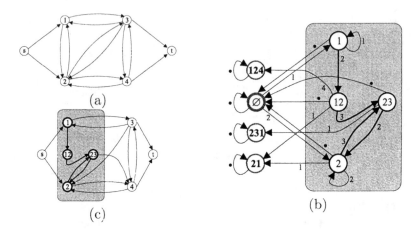

Figure 2.5. (a) Example network $G = (V, A)$ for the SPPRCFP. (b) State graph $S = (V_S, A_S)$ of a finite automaton which identifies the sub-paths of $\mathcal{H} = \{(1, 2, 4), (2, 1), (2, 3, 1)\}$. "•" stands for any label $v \in V$ except those corresponding to the other out-arcs of the same node. (c) Resulting SPPRCFP network $G' = (V', A')$, node 12 corresponds with node 2 of the original network and node 23 with node 3.

finite automaton, which processes the nodes of a path P to detect the first sub-path in $\mathcal{H}$ it contains. The nodes (states) in V_S correspond to the prefixes of the sub-paths in $\mathcal{H}$, i.e., $V_S = \{\varnothing, 1, 12, 124, 2, 21, 23, 231\}$ in the example (see Figure 2.5(b)). Each time that a node of P is processed, the automaton performs a state transition. Possible transitions are represented by labeled arcs. There is an arc $(z_1, z_2) \in A_S$ labeled with $v \in V$ (v represents a possible node of P) that connects two different states z_1 and z_2 if $z_1 \notin \mathcal{H}$ and $z_2 = (\sigma(z_1), v)$, where $\sigma(z_1)$ is the longest (possibly empty) suffix of z_1 for which $(\sigma(z_1), v) \in V_S$. Further loops, i.e., $(124, 124)$, $(231, 231)$, and $(21, 21)$ guarantee that once a forbidden sequence has been detected, the automaton stays in the corresponding state. The remaining transitions connect a state z_1 back to the initial state $\varnothing$. The states $z \in \mathcal{H}$ indicate that a forbidden path has been detected.

Second, the original graph G has to be merged with the state graph S to produce a new graph $G' = (V', A')$. Prefixes $z \in V_S$ of length 1 are identified with the nodes of V. The new node set V' consists of all original nodes V and all nodes of V_S except the state $\varnothing$ and the states $z \in \mathcal{H}$. In the example, the new node set V' is $\{s, 1, 2, 3, 4, 12, 23, t\}$. In order to get the new arc set A', one has to join the sets A and $\{(z, z') \in A_S : z, z' \in V'\}$ and to perform three operations:

(i) remove all loops of the arc set A_S;

(ii) remove from the original arc set A the first arc of each sub-path in $\mathcal{H}$;

(iii) replace each transition $(z, \varnothing)$ of the finite automaton by a set of arcs (z, v) with $v \in V$ such that $(z, v) \notin V_S$ but $(\lambda(z), v) \in A$ where $\lambda(z)$ denotes the last node of the prefix z.

In the example, all loops and the arcs $(1, 2)$, $(2, 1)$ and $(2, 3)$ are removed while the arc $(23, 4)$ replaces the transition $(23, \varnothing)$. The new digraph $G' = (V', A')$ is depicted in Figure 2.5(c). A node z in V' represents the node $\lambda(z)$ in V so that paths in G' are in correspondence with paths in G. For instance, the path $(s, 1, 12, 23, 2, 4, t)$ corresponds with the feasible path $(s, 1, 2, 3, 2, 4, t)$ of the original network.

5.   Conclusions

This survey has highlighted the richness of the SPPRC. In particular, it showed its great flexibility to incorporate a wide variety of constraints, yielding numerous SPPRC variants as well as diversified solution methods. We have given a new classification scheme and a generic formulation, which integrates the special purpose SPPRC formulations presented in the literature so far. Future research on the SPPRC will focus on developing more efficient exact and heuristic algorithms for some of the most difficult SPPRCs such as the ESPPRC or the SPPRC with general REFs. Additionally, with the application of column generation to a wider class of vehicle routing and crew scheduling problems, one should expect new variants of the SPPRC that will require the adaptation of existing solution methods or the development of new ones.

References

Aho, A. and Corasick, M. (1975). Efficient string matching: An aid to bibliographic search. *Journal of the ACM*, 18(6):333–340.

Ahuja, R., Magnanti, T., and Orlin, J.(1993). Network Flows: Theory, Algorithms, and Applications. Prentice Hall, Englewood Cliffs, New Jersey.

Arunapuram, S., Mathur, K., and Solow, D.(2003). Vehicle routing and scheduling with full truck loads. *Transportation Science*, 37(2):170–182.

Barnhart, C., Johnson, E., Nemhauser, G., Savelsbergh, M., and Vance, P. (1998). Branch-and-price: Column generation for solving huge integer programs. *Operations Research*, 46(3):316–329.

Beasley, J. and Christofides, N.(1989). An algorithm for the resource constrained shortest path problem. *Networks*, 19:379–394.

Bentley, J. (1980). Multidimensional divide-and-conquer. *Communications of the ACM*, 23(4):214–229.

Borndörfer, R., Grötschel, M., and Löbel, A. (2001). Scheduling duties by adaptive column generation. *Technischer Bericht* (ZIB-Report) 01-02, Konrad-Zuse-Zentrum für Informationstechnik Berlin (ZIB), Berlin.

Chabrier, A. (2002). Vehicle routing problem with elementary shortest path based column generation. *Technical Report*, ILOG, Madrid.

Desaulniers, G., Desrosiers, J., Dumas, Y., Marc, S., Rioux, B., Solomon, M. M., and Soumis, F. (1997). Crew pairing at Air France. *European Journal of Operational Research*, 97:245–259.

Desaulniers, G., Desrosiers, J., Erdmann, A., Solomon, M., and Soumis, F. (2002a). VRP with pickup and delivery. In: *The Vehicle Routing Problem* (P. Toth and D. Vigo, D., eds.), Chapter 9, pp. 225–242. Siam, Philadelphia.

Desaulniers, G., Desrosiers, J., Ioachim, I., Solomon, M., Soumis, F., and Villeneuve, D. (1998). A unified framework for deterministic time constrained vehicle routing and crew scheduling problems. In: *Fleet Management and Logistics* (T. Crainic and G. Laporte, eds.), Chapter 3, pp. 57–93. Kluwer Academic Publisher, Boston, Dordrecht, London.

Desaulniers, G., Desrosiers, J., Lasry, A., and Solomon, M. M. (1999). Crew pairing for a regional carrier. In: *Computer-Aided Transit Scheduling* (N. Wilson, ed.), Lecture Notes in Computer Science, Volume 471, pp. 19–41. Springer, Berlin.

Desaulniers, G., Langevin, A., Riopel, D., and Villeneuve, B. (2002b). Dispatching and conflict-free routing of automated guided vehicles: An exact approach. *Les Cahiers du GERAD* G-2002-31, HEC, Montréal, Canada. Forthcoming in: *International Journal of Flexible Manufacturing Systems*.

Desaulniers, G. and Villeneuve, D. (2000). The shortest path problem with time windows and linear waiting costs. *Transportation Science*, 34(3):312–319.

Desrochers, M., Desrosiers, J., and Solomon, M. (1992). A new optimization algorithm for the vehicle routing problem with time windows. *Operations Research*, 40(2):342–354.

Desrochers, M. and Soumis, F. (1988). A generalized permanent labelling algorithm for the shortest path problem with time windows. *Information Systems and Operations Research*, 26(3):191–212.

Desrochers, M. and Soumis, F. (1989). A column generation approach to the urban transit crew scheduling problem. *Transportation Science*, 23(1):1–13.

Desrochers, M. (1986). La Fabrication d'horaires de travail pour les conducteurs d'autobus par une méthode de génération de colonnes. *Ph.D Thesis*, Centre de recherche sur les Transports, Publication #470, Université de Montréal, Canada.

Desrosiers, J., Dumas, Y., Solomon, M., and Soumis, F. (1995). Time constrained routing and scheduling. In: *Handbooks in Operations Research and Management Science* (M. Ball, T. Magnanti, C. Monma, and G. Nemhauser, eds.), Volume 8, *Network Routing*, Chapter 2, pp. 35–139. Elsevier, Amsterdam.

Desrosiers, J., Pelletier, P., and Soumis, F. (1983). Plus court chemin avec contraintes d'horaires. *RAIRO*, 17:357–377.

Desrosiers, J., Soumis, F., Desrochers, M., and Sauve, M. (1986). Methods for routing with time windows. *European Journal of Operational Research*, 23:236–245.

Desrosiers, J., Soumis, F., and Desrochers, M. (1984). Routing with time windows by column generation. *Networks*, 14:545–565.

de Silva, A. (2001). Combining constraint programming and linear programming on an example of bus driver scheduling. *Annals of Operations Research*, 108:277–291.

Dror, M. (1994). Note on the complexity of the shortest path models for column generation in VRPTW. *Operations Research*, 42(5):977–978.

Dumas, Y., Desrosiers, J., and Soumis, F. (1991). The pick-up and delivery problem with time windows. *European Journal of Operational Research*, 54:7–22.

Dumas, Y., Soumis, F., and Desrosiers, J. (1990). Optimizing the schedule for a fixed vehicle path with convex inconvenience costs. *Transportation Science*, 24(2):145–152.

Fahle, T., Junker, U., Karisch, S., Kohl, N., Sellmann, M., and Vaaben, B. (2002). Constraint programming based column generation for crew assignment. *Journal of Heuristics*, 8(1):59–81.

Feillet, D., Dejax, P., Gendreau, M., and Gueguen, C. (2004). An exact algorithm for the elementary shortest path problem with resource constraints: Application to some vehicle routing problems. *Networks*, 44(3):216–229.

Gamache, M., Soumis, F., and Marquis, G. (1999). A column generation approach for large-scale aircrew rostering problems. *Operations Research*, 47(2):247–263.

Gamache, M., Soumis, F., Villeneuve, D., Desrosiers, J., and Gélinas, E. (1998). The preferential bidding system at Air Canada. *Transportation Science*, 32(3):246–255.

Houck, D., Picard, J., Queyranne, M., and Vemuganti, R. (1980). The travelling salesman problem as a constrained shortest path problem: Theory and computational experience. *Opsearch*, 17:93–109.

Ioachim, I., Gélinas, S., Desrosiers, J., and Soumis, F. (1998). A dynamic programming algorithm for the shortest path problem with time windows and linear node costs. *Networks*, 31:193–204.

Irnich, S. and Villeneuve, D. (2003). The shortest path problem with resource constraints and k-cycle elimination for $k \geq 3$. *Les Cahiers du GERAD* G-2003-55, HEC, Montréal, Canada.

Kohl, N. (1995). Exact methods for time constrained routing and related scheduling problems. *Ph.D Thesis*, Institute of Mathematical Modelling, Technical University of Denmark, Lyngby, Denmark.

Kolen, A., Rinnooy-Kan, A., and Trienekens, H. (1987). Vehicle routing with time windows. *Operations Research*, 35(2):266–274.

Kung, H., Luccio, F., and Preparata, F. (1975). On finding maxima of a set of vectors. *Journal of the ACM*, 22(4):469–476.

Larsen, J. (1999). Parallelization of the gehicle routing problem with time windows. *Ph.D Thesis*, Department of Mathematical Modelling, *Technical Report*, University of Denmark.

Min, H. (1989). The multiple vehicle routing problem with simultaneous delivery and pick-up points. *Transportation Research*, 23:377–386.

Nemhauser, G. and Wolsey, L. (1988). Integer and Combinatorial Optimization. Wiley, New York.

Powell, W. and Chen, Z. (1998). A generalized threshold algorithm for the shortest path problem with time windows. In: *DIMACS Series in*

Discrete Mathematics and Theoretical Computer Science (P. Pardalos and D. Du, eds.), pp. 303–318. American Mathematical Society.

Rousseau, L.-M., Focacci, F., Gendreau, M., and Pesant, G. (2003). Solving VRPTWs with constraint programming based column generation. *Publication CRT*-2003-10, Center for Research on Transportation, Université de Montréal, Canada.

Ryan, D. and Foster, B. (1981). An integer programming approach to scheduling. In: *Computer Scheduling of Public Transport Urban Passenger Vehicle and Crew Scheduling* (A. Wren, ed.), pp. 269–280. North-Holland, Amsterdam.

Savelsbergh, M. and Sol, M. (1998). Drive: Dynamic routing of independent vehicles. *Operations Research*, 46(4):474–490.

Solomon, M. (1987). Algorithms for the vehicle routing and scheduling problem with time window constraints. *Operations Research*, 35(2):254–265.

Vance, P., Barnhart, C., Johnson, E., and Nemhauser, G. (1997). Airline crew scheduling: A new formulation and decomposition algorithm. *Operations Research*, 45(2):188–200.

Villeneuve, D. and Desaulniers, G. (2000). The shortest path problem with forbidden paths. *Les Cahiers du GERAD* G-2000-41, HEC, Montréal, Canada. Forthcoming in: *European Journal of Operational Research*.

Xu, H., Chen, Z., Rajagopal, S., and Arunapuram, S. (2003). Solving a practical pickup and delivery problem. *Transportation Science*, 37(3):347–364.

Chapter 3

VEHICLE ROUTING PROBLEM WITH TIME WINDOWS

Brian Kallehauge
Jesper Larsen
Oli B.G. Madsen
Marius M. Solomon

Abstract   In this chapter we discuss the Vehicle Routing Problem with Time Windows in terms of its mathematical modeling, its structure and decomposition alternatives. We then present the master problem and the subproblem for the column generation approach, respectively. Next, we illustrate a branch-and-bound framework and address acceleration strategies used to increase the efficiency of branch-and-price methods. Then, we describe generalizations of the problem and report computational results for the classic Solomon test sets. Finally, we present our conclusions and discuss some open problems.

1.   Introduction

The vehicle routing problem (VRP) involves finding a set of routes, starting and ending at a depot, that together cover a set of customers. Each customer has a given demand, and no vehicle can service more customers than its capacity permits. The objective is to minimize the total distance traveled or the number of vehicles used, or a combination of these. In this chapter, we consider the vehicle routing problem with time windows (VRPTW), which is a generalization of the VRP where the service at any customer starts within a given time interval, called a time window. Time windows are called soft when they can be considered nonbiding for a penalty cost. They are hard when they cannot be violated, i.e., if a vehicle arrives too early at a customer, it must wait until the time window opens; and it is not allowed to arrive late. This is the case we consider here.

The remarkable advances in information technology have enabled companies to focus on efficiency and timeliness throughout the supply chain. In turn, the VRPTW has increasingly become an invaluable tool in modeling a variety of aspects of supply chain design and operation. Important VRPTW applications include deliveries to supermarkets, bank and postal deliveries, industrial refuse collection, school bus routing, security patrol service, and urban newspaper distribution. Its increased practical visibility has evolved in parallel with the development of broader and deeper research directed at its solution. Significant progress has been made in both the design of heuristics and the development of optimal approaches.

In this chapter we will concentrate on exact methods for the VRPTW based on column generation. These date back to Desrochers, Desrosiers and Solomon (1992) who used column generation in a Dantzig-Wolfe decomposition framework and Halse (1992) who implemented a decomposition based on variable splitting (also known as Lagrangean decomposition). Later, Kohl and Madsen (1997) developed an algorithm exploiting Lagrangean relaxation. Then, Kohl, Desrosiers, Madsen, Solomon and Soumis (1999); Larsen (1999); Cook and Rich (1999) extended the previous approaches by developing Dantzig-Wolfe based decomposition algorithms involving cutting planes and/or parallel platforms. Kallehauge (2000) suggested a hybrid algorithm based on a combination of Lagrangean relaxation and Dantzig-Wolfe decomposition. Recently, Chabrier (2005); Chabrier, Danna and Le Pape (2002); Feillet, Dejax, Gendreau and Gueguen (2004); Irnich and Villeneuve (2005); Rousseau, Gendreau and Pesant (2004) have proposed algorithms based on enhanced subproblem methodology. Advancements in master problem approaches have been made by Danna and Le Pape (2005); Larsen (2004).

This chapter has the following organization. In Section 2 we describe the mathematical model of the VRPTW and in Section 3 we discuss the structure of the problem and decomposition alternatives. Next, Sections 4 and 5 present the master problem and the subproblem for the column generation approach, respectively. Section 6 illustrates the branch-and-bound framework, while Section 7 addresses acceleration strategies used to increase the efficiency of branch-and-price methods. Then, we describe generalizations of the VRPTW in Section 8 and report computational results for the classic Solomon test sets in Section 9. Finally we present our conclusions and discuss some open problems in 10.

2.   The model

The VRPTW is defined by a fleet of vehicles, $\mathcal{V}$, a set of customers, $\mathcal{C}$, and a directed graph $\mathcal{G}$. Typically the fleet is considered to be homogeneous, that is, all vehicles are identical. The graph consists of $|\mathcal{C}| + 2$ vertices, where the customers are denoted $1, 2, \ldots, n$ and the depot is represented by the vertices 0 ("the starting depot") and $n + 1$ ("the returning depot"). The set of all vertices, that is, $0, 1, \ldots, n+1$ is denoted $\mathcal{N}$. The set of arcs, $\mathcal{A}$, represents direct connections between the depot and the customers and among the customers. There are no arcs ending at vertex 0 or originating from vertex $n + 1$. With each arc (i, j), where $i \neq j$, we associate a *cost* c_{ij} and a *time* t_{ij}, which may include service time at customer i.

Each vehicle has a capacity q and each customer i a demand d_i. Each customer i has a *time window* $[a_i, b_i]$ and a vehicle must arrive at the customer before b_i. If it arrives before the time window opens, it has to wait until a_i to service the customer. The time windows for both depots are assumed to be identical to $[a_0, b_0]$ which represents the *scheduling horizon*. The vehicles may not leave the depot before a_0 and must return at the latest at time b_{n+1}.

It is assumed that q, a_i, b_i, d_i, c_{ij} are non-negative integers and t_{ij} are positive integers. Note that this assumption is necessary to develop an algorithm for the shortest path with resource constraints used in the column generation approach presented later. Furthermore it is assumed that the triangle inequality is satisfied for both c_{ij} and t_{ij}.

The model contains two sets of decision variables x and s. For each arc (i, j), where $i \neq j$, $i \neq n + 1$, $j \neq 0$, and each vehicle k we define x_{ijk} as

$$
x_{ijk} = \begin{cases} 1, & \text{if vehicle } k \text{ drives directly from vertex } i \text{ to vertex } j, \\ 0, & \text{otherwise.} \end{cases}
$$

The decision variable s_{ik} is defined for each vertex i and each vehicle k and denotes the time vehicle k starts to service customer i. In case vehicle k does not service customer i, s_{ik} has no meaning and consequently it's value is considered irrelevant. We assume $a_0 = 0$ and therefore $s_{0k} = 0$, for all k.

The goal is to design a set of routes that minimizes total cost, such that

- each customer is serviced exactly once,

- every route originates at vertex 0 and ends at vertex $n + 1$, and

- the time windows of the customers and capacity constraints of the vehicles are observed.

This informal VRPTW description can be stated mathematically as a multicommodity network flow problem with time windows and capacity constraints:

$$\min \sum_{k \in \mathcal{V}} \sum_{i \in \mathcal{N}} \sum_{j \in \mathcal{N}} c_{ij} x_{ijk} \ s.t., \tag{3.1}$$

$$\sum_{k \in \mathcal{V}} \sum_{j \in \mathcal{N}} x_{ijk} = 1 \quad \forall i \in \mathcal{C}, \tag{3.2}$$

$$\sum_{i \in \mathcal{C}} d_i \sum_{j \in \mathcal{N}} x_{ijk} \le q \quad \forall k \in \mathcal{V}, \tag{3.3}$$

$$\sum_{j \in \mathcal{N}} x_{0jk} = 1 \quad \forall k \in \mathcal{V}, \tag{3.4}$$

$$\sum_{i \in \mathcal{N}} x_{ihk} - \sum_{j \in \mathcal{N}} x_{hjk} = 0 \quad \forall h \in \mathcal{C}, \ \forall k \in \mathcal{V}, \tag{3.5}$$

$$\sum_{i \in \mathcal{N}} x_{i,n+1,k} = 1 \quad \forall k \in \mathcal{V}, \tag{3.6}$$

$$x_{ijk}(s_{ik} + t_{ij} - s_{jk}) \le 0 \quad \forall i, j \in \mathcal{N}, \ \forall k \in \mathcal{V}, \tag{3.7}$$

$$a_i \le s_{ik} \le b_i \quad \forall i \in \mathcal{N}, \ \forall k \in \mathcal{V}, \tag{3.8}$$

$$x_{ijk} \in \{0, 1\} \quad \forall i, \ j \in \mathcal{N}, \ \forall k \in \mathcal{V}. \tag{3.9}$$

The objective function (3.1) minimizes the total travel cost. The constraints (3.2) ensure that each customer is visited exactly once, and (3.3) state that a vehicle can only be loaded up to it's capacity. Next, equations (3.4), (3.5) and (3.6) indicate that each vehicle must leave the depot 0; after a vehicle arrives at a customer it has to leave for another destination; and finally, all vehicles must arrive at the depot $n + 1$. The inequalities (3.7) establish the relationship between the vehicle departure time from a customer and its immediate successor. Finally constraints (3.8) affirm that the time windows are observed, and (3.9) are the integrality constraints. Note that an unused vehicle is modeled by driving the "empty" route $(0, n + 1)$.

The model can also incorporate a constraint giving an upper bound on the number of vehicles, as is the case in Desrosiers, Dumas, Solomon and Soumis (1995):

$$\sum_{k \in \mathcal{V}} \sum_{j \in \mathcal{N}} x_{0jk} \le |V| \quad \forall k \in \mathcal{V}, \ \forall j \in \mathcal{N} \tag{3.10}$$

Note also that the nonlinear restrictions (3.7) can be linearized as:

$$s_{ik} + t_{ij} - M_{ij}(1 - x_{ijk}) \leq s_{jk} \quad \forall i, \; j \in \mathcal{N}, \; \forall k \in \mathcal{V}. \qquad (3.11)$$

The large constants M_{ij} can be decreased to $\max\{b_i + t_{ij} - a_j\}$, $(i, j) \in A$.

For each vehicle, the service start variables impose a unique route direction thereby eliminating any subtours. Hence, the classical VRP subtour elimination constraints become redundant. Finally, the objective function (3.1) has been universally used when solving the VRPTW to optimality. In the research on heuristics it has been common to minimize the number of vehicles which may lead to additional travel cost.

The VRPTW is a generalization of both the traveling salesman problem (TSP) and the VRP. When the time constraints (3.7) and (3.8)) are not binding the problem relaxes to a VRP. This can be modeled by setting $a_i = 0$ and $b_i = M$, where M is a large scalar, for all customers i. If only one vehicle is available the problem becomes a TSP. If several vehicles are available and the cost structure is: $c_{0j} = 1$, $j \in \mathcal{C}$ and $c_{ij} = 0$, otherwise, we obtain the bin-packing problem. Since trips between customers are "free", the order in which these are visited becomes unimportant and the objective turns to "squeezing" as much demand as possible into as few vehicles (bins) as possible. In case the capacity constraints (3.2) are not binding the problem becomes a m-TSPTW, or, if only one vehicle is available, a TSPTW.

3.   Structure and decomposition

A closer look at the above model reveals that only the assignment constraints (3.2) are coupling the vehicles while the remaining constraints are dealing with each vehicle separately. This strongly suggests the use of Lagrangean relaxation (LR) or decomposition, for example Dantzig-Wolfe (DWD), to break up the overall problem into a subproblem for each vehicle and a master problem. To date, the most successful decomposition approaches for the VRPTW cast the subproblem as a constrained shortest path structure. The master problem is an integer program whose solution cannot be obtained directly, so its LP relaxation is solved. The column generation process alternates between solving this linear master problem and the subproblem. The former finds new multipliers to send to the latter which uses this information to find new columns to send back. A lower bound on the optimal integer solution of the VRPTW model is obtained at the end of this back and forth process. This is then used within a branch-and-bound framework to obtain the optimal VRPTW solution. If the vehicles are identical, as we have assumed here, all subproblems will be equivalent and therefore it is necessary to only solve one. The master problem and the subproblem will be discussed in more detail in Sections 4 and 5, respectively. The complete column generation process is described in Section 1, while the subproblem forms the subject of Section 2.

In addition, other LRs are possible but not promissing. One may consider relaxing the time and capacity constraints (3.3), (3.7) and (3.8). This yields a linear network flow problem which possesses the integrality property. The corresponding bound can be calculated very fast, but is not likely to be very strong unless capacity is not binding and time windows are very narrow (see Desrosiers, Dumas, Solomon and Soumis, 1995). Relaxing only the capacity or time window constraints also does not seem sensible since the relaxed problem is not generally easier to solve than the original.

Desrochers, Desrosiers and Solomon (1992) were the first to apply DWD with a free number of vehicles. The assignment constraints were considered the coupling constraints, while the subproblem was a shortest path problem with resource constraints. Relaxing the same constraint set and applying LR was first proposed by Kohl and Madsen (1997). Desrosiers, Sauvé and Soumis (1988) have used a similar relaxation to calculate a lower bound for the minimum fleet size for the m-TSPTW.

Jörnsten, Madsen and Sørensen (1986) suggested solving the VRPTW by variable splitting (later called Lagrangean decomposition, or LD). In follow-up work, Halse (1992) described three different variable splitting methods where $\sum_j x_{ijk}$ was replaced by y_{ik} in constraint set (3.2) and possibly (3.3). In turn, the constraint $y_{ik} = \sum_j x_{ijk}$ was introduced and Lagrangeanly relaxed. The problem decomposes into two problems, one in the x- and s-variables and the other in the y-variables. The former problem is further decomposed by vehicle and it is a shortest path problem with resource constraints. The latter is an assignment-type problem. Specifically, the approaches are:

- VS1: Keep constraints (3.2) and (3.3) in the y-problem. This represents a generalized assignment problem (GAP) and the x/s-problem becomes a shortest path problem with time windows (SPPTW). The GAP has the special structure where all right hand sides in (3.3) are identical and d_i does not depend on k.

- VS2: Keep constraints (3.2) in the y-problem. The y-problem becomes a "Semi assignment" problem (SAP) consisting of constraints (3.2) only. The x/s-problem is equivalent to a shortest path problem with time windows and capacity constraints (SPPTWCC). The SAP is easily solvable and possesses the integrality property.

- VS3: Keep constraints (3.2) in the y-problem and constraints (3.3) in both the y- and the x/s-problem. The y-problem is a GAP and the x/s-problem constitutes a SPPTWCC.

In the LD master problem, whose role is to find multipliers to the relaxed equation relating x and y, the number of multipliers is larger than in the LR considered above. This clearly makes the master problem more difficult. Also the subproblems are no longer identical since the LD multipliers depend on both customer and vehicle. Note that only VS1 and VS2 have been implemented.

We now define LB(VS1), LB(VS2) and LB(VS3) as the best lower bounds obtainable from the three variable splitting approaches, respectively. It can be shown that the previous LR and the DWD yield the same lower bound LB(LR/DWD). Provided that the vehicles are identical, Kohl (1995) has derived the following results:

$$LB(VS3) \geq LB(VS1),$$
$$LB(VS3) \geq LB(VS2),$$
$$LB(LR/DWD) = LB(VS2).$$

There exist instances for which LB(VS3) > LB(VS1). He further showed that LB(VS2) = LB(VS3) under some weak supplementary conditions. This is surprising because it implies there is no additional gain to be derived from solving two hard integer problems (the SPPTWCC and GAP) instead of just one (the SPPTWCC). However, in the more general case where vehicles have different capacities it might be possible that the VS3 model yields a better bound than VS2.

To conclude, in VRPTW case, the variable splitting methods mentioned above generally provide similar lower bounds to those obtained from the ordinary LR or DWD.

4.   The master problem

The column generation methodology has been successfully applied to the VRPTW by numerous researchers. It represents a generalization of the linear DWD since the master problem and the subproblem are integer and mixed-integer programs, respectively. Often the master problem is simply stated as a set partitioning problem on which column generation is applied, thereby avoiding the description of the DWD on which it is based. To gain an appreciation for different cutting and branching opportunities compatible with column generation, here we present the master problem by going through the steps of the DWD based on the multicommodity network flow formulation (3.1)–(3.9).

The column generation approach exploits the fact that only constraint set (3.2) links the vehicles together. Hence, the integer master problem is defined through (3.1)–(3.2) and (3.9), that is, it contains the objective function, the assignment of customers to exactly one vehicle and the binary requirement on the flow variables. The rest of the constraints and (3.9) are part of the subproblem which has a modified objective function that decomposes into $|V|$ independent subproblems, one for each vehicle. In the rest of this section we will focus on the linear master problem (3.1)–(3.2). Branching, necessary to solve the integer master problem, will be discussed in Section 6.

Let $\mathcal{P}^k$ be the set of feasible paths for vehicle k, $k \in \mathcal{V}$. Hence, $p \in \mathcal{P}^k$ corresponds to an elementary path which can also be described by using the binary values x_{ijp}^k, where $x_{ijp}^k = 1$, if vehicle k goes directly from vertex i to vertex j on path p, and $x_{ijp}^k = 0$, otherwise. Any solution x_{ij}^k to the master problem (3.1)–(3.2) can be written as a non-negative convex combination of a finite number of elementary paths, i.e.,

$$x_{ij}^k = \sum_{p \in \mathcal{P}^k} x_{ijp}^k y_p^k \quad \forall k \in \mathcal{V}, \ \forall (i,j) \in \mathcal{A}, \tag{3.12}$$

$$\sum_{p \in \mathcal{P}^k} y_p^k = 1 \quad \forall k \in \mathcal{V}, \tag{3.13}$$

$$y_p^k \geq 0 \quad \forall k \in \mathcal{V}, \ \forall p \in \mathcal{P}^k. \tag{3.14}$$

Using x_{ijp}^k we can define the cost of a path, c_p^k, and the number of times a customer i is visited by vehicle k, a_i^k, as:

$$c_p^k = \sum_{(i,j) \in \mathcal{A}} c_{ij}^k x_{ijp}^k \quad \forall k \in \mathcal{V}, \ \forall p \in \mathcal{P}^k,$$

$$a_{ip}^k = \sum_{j \in \mathcal{N} \cup \{n+1\}} x_{ijp}^k \quad \forall k \in \mathcal{V}, \ \forall i \in \mathcal{N}, \ \forall p \in \mathcal{P}^k.$$

Now we can substitute these values into (3.1)–(3.2) and arrive at the revised formulation of the master problem:

$$\min \sum_{k \in \mathcal{V}} \sum_{p \in \mathcal{P}^k} c_p^k y_p^k \ s.t., \tag{3.15}$$

$$\sum_{k \in \mathcal{V}} \sum_{p \in \mathcal{P}^k} a_{ip}^k y_p^k = 1 \quad \forall i \in \mathcal{C}, \tag{3.16}$$

$$\sum_{p \in \mathcal{P}^k} y_p^k = 1 \quad \forall k \in \mathcal{V}, \tag{3.17}$$

$$y_p^k \geq 0 \quad \forall k \in \mathcal{V}, \ \forall p \in \mathcal{P}^k. \tag{3.18}$$

The mathematical formulation (3.15)–(3.18) is then the linear relaxation of a set partitioning type problem with an additional constraint on the total number of vehicles and a set of convex combination constraints.

In the usual case of a single depot and a homogeneous fleet of vehicles with the same initial conditions for all vehicles, all $\mathcal{P}^k$ are identical, that is, $\mathcal{P}^k = \mathcal{P}$, $k \in \mathcal{V}$. Furthermore, the networks for the subproblems are also identical. Therefore constraints (3.17) can be aggregated. By letting $y_p = \sum_{k \in \mathcal{V}} y_p^k$, the index k can be eliminated from the formulation (3.15)–(3.18). The resulting model given below is the classical linear relaxation of the set partitioning formulation:

$$\min \sum_{p \in \mathcal{P}} c_p y_p \ s.t., \tag{3.19}$$

$$\sum_{p \in \mathcal{P}} a_{ip} y_p = 1 \quad \forall i \in \mathcal{C}, \tag{3.20}$$

$$y_p \geq 0 \quad \forall p \in \mathcal{P}. \tag{3.21}$$

In the column generation methodology, the set of columns in the linear master problem is limited to only those that have already been generated, hence the term *restricted* master problem. It consists of finding a set of minimum cost paths among all paths presently in the master problem. The restricted master problem can mathematically be stated as:

$$\min \sum_{p \in \mathcal{P}'} c_p y_p \ s.t., \tag{3.22}$$

$$\sum_{p \in \mathcal{P}'} a_{ip} y_p = 1 \quad \forall i \in \mathcal{C}, \tag{3.23}$$

$$y_p \geq 0 \quad \forall p \in \mathcal{P}'. \tag{3.24}$$

Each decision variable y_p counts the number of times path p is used. This is not necessarily integer, but can be any real number in the interval $[0; 1]$. The set $\mathcal{P}'$ contains all the paths generated, a_{ip} denotes the number of times customer i is serviced on path p, and, c_p is the cost of the path. The parameter a_{ip} should in principle be either 0 or 1, but since the subproblem is relaxed (see Section 5) it can take larger integer values.

Solving the restricted master yields a solution $y = (y_1, y_2, \ldots, y_{|\mathcal{P}'|})$ which might be integer but this is not guaranteed. If it is integer, a feasible but not necessarily optimal solution to the VRPTW has been found. In addition to the primal solution, a dual solution $\phi = (\phi_1, \phi_2, \ldots, \phi_{|\mathcal{C}|})$ is also obtained.

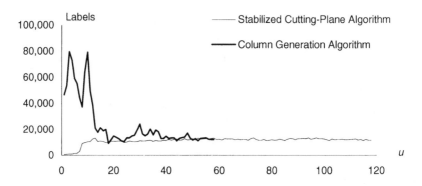

Figure 3.1. Number of labels generated in the subproblem wrt. the iteration number for the Dantzig-Wolfe method and the bundle method on the Solomon instance R104 with 100 customers (from Kallehauge, 2000).

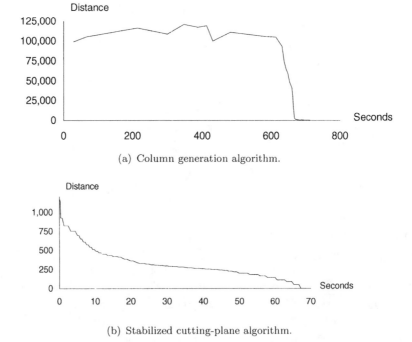

Figure 3.2. The Euclidian distance between the current dual variables and the optimum dual variables. Observe the different scales.

An initial start for the restricted master problem is often the set of routes visiting a single customer, that is, routes of the type depot-i-depot (cf. Section 8). When the optimal solution to the restricted master problem is found, the simplex algorithm asks for a new variable (i.e. a column/path $p \in \mathcal{P} \setminus \mathcal{P}'$) with negative reduced cost. Such a column is found by solving a subproblem, sometimes called the pricing problem. For the VRPTW, the subproblem should solve the problem "Find the path with minimal reduced cost." Solving the subproblem is in fact an implicit enumeration of all feasible paths, and the process terminates when the optimal objective of the subproblem is non-negative (it will actually be 0).

It is not surprising that the behavior of the dual variables plays a pivotal role in the overall performance of the column generation principle for the VRPTW. It has been observed by Kallehauge (2000) that dual variables do not converge smoothly to their respective optima. Assume that the paths $(0, i, n+1)$ are used to initialize the algorithm. Figure 3.1 illustrates the instability of the column generation algorithm compared to the stabilized cutting-plane algorithm presented in the above paper. Furthermore, Figure 3.2 illustrates the effect of the size of the multipliers on the computational difficulty of the SPPTWCC subproblems. Whereas the multipliers are large in the Dantzig-Wolfe process, they are small in the cutting-plane approach. This problem originates in the coordination between the master problem and the subproblem.

Finally, in many routing problems the optimal solution remains unchanged even if overcovering rather than exact covering of customers is allowed. Due to the triangle inequality in the VRPTW, overcovering will always be more expensive than just covering and therefore an optimal solution will always be one where each customer is visited exactly once. The advantage of allowing overcovering is that the linear relaxation of the Set Covering Problem is easier to solve than that of the Set Partitioning Problem, and this will in turn lead to the computation of good estimates of the dual variables.

5.   The subproblem

In the column generation approach for the VRPTW, the subproblem decomposes into $|\mathcal{V}|$ identical problems, each one being a shortest path problem with resource constraints (time windows and vehicle capacity). More specifically, the subproblem is an Elementary Shortest Path Problem with Time Windows and Capacity Constraints (ESPPTWCC), where elementary means that each customer can appear at most once in the shortest path. It can be formulated as:

$$\min \sum_{i \in \mathcal{N}} \sum_{j \in \mathcal{N}} \hat{c}_{ij} x_{ij}, \ s.t., \tag{3.25}$$

$$\sum_{i \in \mathcal{C}} d_i \sum_{j \in \mathcal{N}} x_{ij} \leq q, \tag{3.26}$$

$$\sum_{j \in \mathcal{N}} x_{0j} = 1, \tag{3.27}$$

$$\sum_{i \in \mathcal{N}} x_{ih} - \sum_{j \in \mathcal{N}} x_{hj} = 0 \quad \forall h \in \mathcal{C}, \tag{3.28}$$

$$\sum_{i \in \mathcal{N}} x_{i,n+1} = 1, \tag{3.29}$$

$$s_i + t_{ij} - M_{ij}(1 - x_{ij}) \leq s_j \quad \forall i, \ j \in \mathcal{N}, \tag{3.30}$$

$$a_i \leq s_i \leq b_i \quad \forall i \in \mathcal{N}, \tag{3.31}$$

$$x_{ij} \in \{0, 1\} \quad \forall i, \ j \in \mathcal{N} \tag{3.32}$$

Constraint (3.26) is the capacity constraint, constraints (3.30) and (3.31) are time constraints, while constraint (3.32) ensures integrality. The constraints (3.27), (3.28) and (3.29) are flow constraints resulting in a path from the depot 0 to the depot $n + 1$. When solving the ESPPTWCC as the subproblem in the VRPTW, $\hat{c}_{ij}$ is the *modified cost* of using arc (i, j), where $\hat{c}_{ij} = c_{ij} - \pi_i$. Note that while c_{ij} is a non-negative integer, $\hat{c}_{ij}$ can be any real number.

This subproblem does not posses the integrality property, and therefore solving it as a linear mixed-integer programming problem will potentially result in a reduction of the integrality gap between the optimal solution of the LP-relaxed version of the VRPTW and the optimal integer solution to the problem.

Since the ESPPTWCC is NP-hard in the strong sense (see Dror, 1994; Kohl, 1995), the usual approach has been to slightly alter the problem by relaxing some of the constraints. In particular, allowing cycles changes the problem to the Shortest Path Problem with Time Windows and Capacity Constraints (SPPTWCC). Since arcs can now be used more than once (and customers may therefore be visited more than once), the decision variables x_{ij} and s_i are replaced by x_{ij}^l and s_i^l. The variable x_{ij}^l is set to 1 if the arc (i, j) is used as the l'th arc on the shortest path, and 0 otherwise, and the variable s_i^l is set to the start of service at customer i as customer number l, where $l \in \mathcal{L} = \{1, 2, \ldots, |\mathcal{L}|\}$, $|\mathcal{L}| = \lfloor b_{n+1}/\min t_{ij} \rfloor$. The SPPTWCC can now be described by the following mathematical model:

$$\min \sum_{l\in\mathcal{L}} \sum_{i\in\mathcal{N}} \sum_{j\in\mathcal{N}} \hat{c}_{ij} x_{ij}^l, \ s.t. \tag{3.33}$$

$$\sum_{i\in\mathcal{N}} \sum_{j\in\mathcal{N}} x_{ij}^1 = 1, \tag{3.34}$$

$$\sum_{i\in\mathcal{N}} \sum_{j\in\mathcal{N}} x_{ij}^l - \sum_{i\in\mathcal{N}} \sum_{j\in\mathcal{N}} x_{ij}^{l-1} \le 0 \quad \forall l \in \mathcal{L} - \{1\}, \tag{3.35}$$

$$\sum_{i\in\mathcal{C}} d_i \sum_{l\in\mathcal{L}} \sum_{j\in\mathcal{N}} x_{ij}^l \le q, \tag{3.36}$$

$$\sum_{j\in\mathcal{N}} x_{0j}^1 = 1, \tag{3.37}$$

$$\sum_{i\in\mathcal{N}} x_{ih}^{l-1} - \sum_{j\in\mathcal{N}} x_{hj}^l = 0 \quad \forall h \in \mathcal{C} \ \forall l \in \mathcal{L} - \{1\}, \tag{3.38}$$

$$\sum_{l\in\mathcal{L}} \sum_{i\in\mathcal{N}} x_{i,n+1}^l = 1, \tag{3.39}$$

$$s_i^l + t_{ij} - K(1 - x_{ij}^l) \le s_j^l \quad \forall i, \ j \in \mathcal{N} \ \forall l \in \mathcal{L} - \{1\}, \tag{3.40}$$

$$a_i \le s_i^l \le b_i \quad \forall i \in \mathcal{N}, \tag{3.41}$$

$$x_{ij}^l \in \{0,1\} \quad \forall i, j \in \mathcal{N}. \tag{3.42}$$

In this formulation, (3.34) forces the first arc to be used only once, while (3.35) states that arc l can only be used provided that arc $l-1$ is used. The remaining constraints are the original constraints (3.3) to (3.9) extended to include the additional superscript l and the changes related to its inclusion. Note that (3.34) is redundant as it is covered by (3.37), but it has been kept in the model as to indicate the origin node.

This problem can be solved by a pseudo-polynomial algorithm described in Desrochers, Desrosiers and Solomon (1992). This and all other current approaches are based on dynamic programming. Even though negative cycles are possible, the time windows and the capacity constraints prohibits infinite cycling. Note that capacity is accumulated every time a customer is serviced in a cycle. If the distance used to compute the cost of routes satisfies the triangle inequality, the optimal solution contains only elementary routes. Solving the SPPTWCC instead of the ESPPTWCC augments the size of the set of admissible columns generated for the master problem. Consequently the lower bound on the master problem may decrease. A slight improvement can be obtained by implementing 2-cycle elimination in the solution process which dates back to Kolen, Rinnooy Kan and Trienekens (1987).

While the SPPTWCC relaxation was at the time a computational necessity, the ESPPTWCC has recently been tackled directly. Work on this problem and k-cycle elimination, where $k \geq 3$, proved very successful in expanding the scope of the VRPTW problems solved. Even though the ESPPTWCC continues to be regarded as difficult to solve when time windows are wide, two research groups have recently used it directly in VRPTW optimal algorithms. Chabrier (2005); Chabrier, Danna and Le Pape (2002), and independently Feillet, Dejax, Gendreau and Gueguen (2004) have extended the dynamic programming approach of Desrochers, Desrosiers and Solomon (1992) to the ESPPTWCC by adapting the path dominance rule. They then incorporated several heuristic modifications to make the algorithm much faster. Chabrier (2005) and Chabrier, Danna and Le Pape (2002) obtained lower bounds superior to those based on the SPPTWCC resulting in excellent computational results to be described in Section 9. A different approach that has not yet been tried on the VRPTW is presented in Dumitrescu and Boland (2003). The authors compare three scaling techniques and a standard label-setting method. They show that integrating preprocessing information within the label-setting method can be very beneficial in terms of both memory and run time. Further improvements of the label-setting method can be obtained by using Lagrangean relaxation.

Instead of dealing with the computational burden of the ESPPTWCC or the weakened lower bound provided by the SPPTWCC, one could consider a middle of the road approach. That is, disallow cycles of small length. As discussed above, cycle elimination corresponding to $k = 2$ has been a common technique. In the SPPTWCC-k-cyc, paths with cycles of length of at most k are eliminated. The case $k \geq 3$ has been considered by Irnich and Villeneuve (2005) with encouraging results presented in Section 9. Recently Rousseau, Gendreau and Pesant (2004) have presented results where Constraint Programming is used to solve the subproblem. Taking into account the difference in computer power, the authors conclude that their approach is not any faster than that of Desrochers, Desrosiers and Solomon (1992).

6.   Branch-and-bound

The column generation approach does not automatically guarantee integer solutions and often the solutions obtained will indeed be fractional. Therefore a branch-and-bound framework has to be established. The calculations are organised in a branching tree. For the VRPTW only binary strategies have been proposed in the literature although it should be noted that it is generally not difficult to come up with non-binary branching trees for the problem. The branching decisions are generally based on considerations related to the original 3-index flow formulation (3.2)–(3.9). The column generation process is then repeated at each node in the branch-and-bound tree.

6.1   Branching on the number of vehicles

Branching on the number of vehicles was originally proposed by Desrochers, Desrosiers and Solomon (1992). If the number of vehicles is fractional we introduce a bound on the number of vehicles. Note that this branching strategy does not require that the flow and time variables of the original model be computed.

This branching rule can be implemented fairly easily and only concerns the master problem. We denote the flow over an arc by f_{ij} and this is the sum of all flows over that arc, that is $f_{ij} = \sum_{k \in \mathcal{V}} x_{ijk}$. The f_{ij} values can easily be derived from the solution of the master problem. When we branch on the number of vehicles, two child nodes are created, one imposing on the master problem parent node the additional constraint $\sum_{j \in C} f_{0j} \geq \lceil l \rceil$ while the other forcing $\sum_{j \in C} f_{0j} \leq \lfloor l \rfloor$, where l is the fractional sum of all variables in the master problem.

Note that branching on the number of vehicles is not necessarily enough to obtain an integer solution as it is possible to derive solutions where the sum of the vehicles is integer, but yet there are fractional vehicles driving around the network.

6.2   Branching on flow variables

Branching on a single variable x_{ijk} is possible only if each vehicle can be distinguished. In column generation this can be achieved by solving the subproblem for *each* vehicle individually and by introducing an additional constraint in the master problem

$$\sum_{p \in P_k} y_p = 1 \quad \forall k \in \mathcal{V}$$

where P_k is the set of routes generated for each vehicle k and y_p is the binary variable indicating whether route p is used.

Since most cases described in the literature assume a homogeneous fleet, it doesn't make sense to branch on individual vehicles. Instead, branching can be done on sums of flows, that is either on $\sum_j x_{ijk}$ or on $\sum_k x_{ijk}$ (equivalent to f_{ij}). Branching on $\sum_j x_{ijk}$ results in a different subproblem for each vehicle, even though the vehicles are identical. That is because imposing $\sum_j x_{ijk} = 1$ forces customer i to be visited by vehicle

k, while $\sum_j x_{ijk} = 0$ implies that customer i is assigned to any vehicle but k.

The standard practice has been to branch on $\sum_k x_{ijk}$ since the branching decision can easily be transfered to the master problem and subproblem. This was proposed independently by Halse (1992); Desrochers, Desrosiers and Solomon (1992). When $\sum_k x_{ijk} = 1$, customer j succeeds customer i on the same route, while if $\sum_k x_{ijk} = 0$, customer i does not immediately precede j. If there is more than one candidate for branching, that is, there are several fractional variables, we would generally like to choose a candidate that is not close to either 0 or 1 in order to make an impact. When selecting among the nodes to branch on, a common heuristic is to branch on the variable maximizing $c_{ij}(\min\{x_{ijk}, 1 - x_{ijk}\})$ using a best-first strategy In order to create more complex strategies the branching schemes can be applied hierarchically, such as first branching on the number of vehicles and then on $\sum_k x_{ijk}$, or mixed.

6.3   Branching on resource windows

Branching on resource windows was first proposed by Gélinas, Desrochers, Desrosiers and Solomon (1995) and is presently the only alternative to branching on flow variables. In the VRPTW model resource windows can be interpreted as either the time windows or the capacity constraints. We will only discuss branching on time windows, as capacity is significantly less constraining in many cases. In Gélinas, Desrochers, Desrosiers and Solomon (1995) only branching on time windows was used.

Branching on time windows results in splitting a time window into two smaller ones. Branching has to be done in such a way that at least one route is infeasible in each of the two sub-windows.

In order to branch on time windows three decisions have to be taken:

1) How should the node for branching be chosen?

2) Which time window should be divided?

3) Where should the partition point be?

In order to decide on the above issues, we define *feasibility intervals* $[l_i^r, u_i^r]$ for all vertices $i \in \mathcal{N}$ and all routes r with fractional flow. l_i^r is the earliest time that service can start at vertex i on route r, and u_i^r is the latest time that service can start, that is, $[l_i^r, u_i^r]$ is the time interval during which route r must visit vertex i to remain feasible.

The intervals can easily be computed by a recursive formula. Additionally we define

$$L_i = \max_{\text{fractional routes } r} \{l_i^r\}, \quad i \in \mathcal{N}, \tag{3.43}$$

$$U_i = \min_{\text{fractional routes } r} \{u_i^r\}, \quad i \in \mathcal{N}. \tag{3.44}$$

If $L_i > U_i$ at least two routes (or two visits by the same route) have disjoint feasibility intervals, i.e., the vertex is a candidate for branching on time windows. We can branch on a candidate vertex i by dividing the time windows $[a_i, b_i]$ at any integer value in the open interval $[U_i, L_i[$. It should be noted that situations can arise where there are no candidates for branching on time windows, but the solution is not feasible.

Three different strategies were proposed by Gélinas, Desrochers, Desrosiers and Solomon (1995) aiming at the elimination of cycles, the minimization of the number of visits to a customer i and the balancing of flow in the two branch-and-bound nodes.

After having chosen the candidate vertex i for branching, an integer $t \in [U_i, L_i[$ has to be selected in order to determine the division. Here t is chosen in order to divide the time window of the customer such that 1) the flow is balanced and 2) the time window is divided as evenly as possible.

7. Acceleration strategies

7.1 Preprocessing

The aim of preprocessing is to narrow the solution space by tightening the formulation before the actual optimization is started. This can be done by fixing some variables, reducing the interval of values a variable can take and so on. In the VRPTW, the time windows can be narrowed if the triangle inequality holds. Accordingly, Kontoravdis and Bard (1995) propose the following scheme. The earliest time a vehicle can arrive at a customer is by arriving straight from the depot and the latest time it can leave is by going directly back to the depot. Hence, for each customer i, its time window can be strengthened from $[a_i, b_i]$ to $[\max\{a_0 + t_{0i}, a_i\}, \min\{b_{n+1} - t_{i,n+1}, b_i\}]$.

A further reduction of the time windows can be achieved by the method developed by Desrochers, Desrosiers and Solomon (1992). The time windows are reduced by applying the following four rules in a cyclic manner. The process is stopped when one whole cycle is performed without changing any of the time windows. The four rules are:

1) Minimal arrival time from predecessors:

$$a_l = \max \left\{ a_l, \min \left\{ b_l, \min_{(i,l)}\{a_i + t_{il}\} \right\} \right\}.$$

2) Minimal arrival time to successors:

$$a_l = \max \left\{ a_l, \min \left\{ b_l, \min_{(l,j)}\{a_j - t_{lj}\} \right\} \right\}.$$

3) Maximal departure time from predecessors:

$$b_l = \min \left\{ b_l, \max \left\{ a_l, \max_{(i,l)}\{b_i + t_{il}\} \right\} \right\}.$$

4) Maximal departure time to successors:

$$b_l = \min \left\{ b_l, \max \left\{ a_l, \max_{(l,j)}\{b_j - t_{lj}\} \right\} \right\}.$$

The first rule adjusts the start of the time window to the earliest time a vehicle can arrive coming straight from any possible predecessor. In a similar fashion, the second rule modifies the start of the time window in order to minimize the excess time spent before the time windows of all possible successors open if the vehicle continues to a successor as quickly as possible. The two remaining rules use the same principles to adjust the closing of the time window. With respect to capacity, an arc (i, j) can obviously be removed if $d_i + d_j > q$.

7.2   Subproblem strategies

A well known strategy for accelerating column generation is to return many negative marginal cost columns to the master problem. Even though in principle only one needs to be returned, several can be if they are available. Computational tests conducted by Kohl (1995); Larsen (1999) confirm the benefits of this approach.

7.3   Master problem strategies

Along with the novel perspectives on the subproblem solution described in 5, master problem acceleration strategies have been key to the evolution of VRPTW approaches over the last few years. One approach is to accelerate the solution at the root node of the branch-and-bound tree by using a local search method to generate a set of initial columns. This helps the column generation process get a fast increase in the quality of the dual variables. It has been implemented by numerous researchers and has finally been discussed in the literature by Danna and Le Pape (2005). The authors use a local search method based on the savings algorithm incorporating time windows which produces a set of routes better than the trivial depot-customer-depot ones. Furthermore, local search is used along with a MIP solver throughout the branch-and-price process to generate good integer solutions fast. Two different heuristics, a local search method based on large neighborhood search and a guided tabu search, were tested and proved beneficial, especially on Solomon's R1 and RC1 problem classes.

Two new approaches have been suggested by Larsen (2004, 1999). First, during the execution of the branch-and-price a large number of columns are generated and many of these only participate in a few computations and will not be used afterwards. If kept, each column will increase computing time when solving the relaxed set partitioning problem and when adjusting the upper bounds on variables due to branching decisions. Therefore Larsen (2004) suggests to keep track of how long a column is part of a basis. If it does not participate in a basis for a given number of branch-and-bound nodes it is removed from the model. This was also suggested by Desaulniers, Desrosiers and Solomon (2002) where it was also noted that a certain number of nonbasic columns should remain in the problem. Larsen (2004) reports that deleting columns that have not been part of the basis for the last 20 branch-and-bound nodes outperforms the code without column deletion by a factor of 2.5 aggregated over 27 instances.

The second acceleration approach is to stop the algorithm for the SPPTWCC before it completes. Computations can be stopped as soon as at least one route with negative cost has been generated. This approach is denoted "forced early stop" in Larsen (1999) and results in dramatic running time reductions, especially for problems with large time windows. For these, the values of the dual variables at the beginning of the procedure will however be of poor quality. Only when the subproblem proves optimality it cannot be stopped prematurely.

7.4   Cutting planes

The barebone column generation methodology for solving the VRPTW is part of the popular approach for solving difficult integer programming problems by relaxing the integrality constraints of the original problem. Typically, the optimal solution to the relaxed problem is not feasible for the original problem and branch-and-bound is used in order to get integer solutions.

Cutting planes has been proposed to improve the polyhedral description of the relaxed problem in order to get an integer solution or at least narrow the integrality gap. Kohl, Desrosiers, Madsen, Solomon and Soumis (1999) suggested three cuts in order to tighten the LP formulation of the VRPTW problem. As these cuts are only introduced at the root node, this is not a branch-and-cut approach, where cuts can be introduced at any node of the search tree.

The method is based on subtour elimination constraints and comb inequalities transferred from the TSP, and 2-path cuts. To detect subtour elimination constraints, a separation algorithm by Crowder and Padberg (1980) was implemented. With respect to the comb inequalities, only combs with 3 teeth and 2 nodes were detected. The separation algorithm was a primitive enumeration scheme. Neither of these constraints had a large impact on tightening the bound.

A new idea introduced by Kohl, Desrosiers, Madsen, Solomon and Soumis (1999) was the inclusion of 2-path cuts. The basis of this set of cuts is the subtour elimination inequality in the strong form: $x(S) \geq k(S), \forall S \subseteq C$, where $x(S)$ is the flow leaving the set S, and $k(S)$ is the minimum number of vehicles needed to service the customers in S. Determining $k(S)$ is not an easy task, but using the triangle inequality on the travel times we have that $S_1 \subset S_2 \Rightarrow k(S_1) \leq k(S_2)$. Sets S that satisfy $x(S) < 2$ and $k(S) > 1$ must now be found. As $k(S)$ is an integer, $k(S) > 1$ implies $k(S) \geq 2$. So we need to identify sets S that require at least two vehicles to be serviced, but are currently serviced by less than two.

For a set S, two checks have to be performed: 1) $k(S) > 1$ and 2) can the customers be serviced by a single vehicle? The first check is easy, but the second requires the solution of the TSPTW *feasibility* problem. Since this problem is NP-hard the separation algorithm can only be applied to small sets. This is done heuristically using a simple greedy algorithm based on Laporte, Nobert and Desrochers (1985).

The 2-path cuts outperformed the branch-and-price method without 2-path cuts. The proportion of the integrality gap closed by the 2-path cuts varies from 100% to 10% in a few cases. Overall 12 new unsolved Solomon instances were closed.

Cook and Rich (1999) extended the above 2-path cut approach to k-path cuts involving the solution of a VRPTW with $(k-1)$ customers as part of the separation algorithm. The authors performed experiments with k up to 6. For larger k, the percentage of the integrality gap that is closed is of course larger, but the separation algorithm requires substantially more time and therefore it is not evident that it is preferable to use k larger than 2.

Recently, Bard, Kontoravdis and Yu (2002) have proposed a branch-and-cut algorithm for the arc formulation of the VRPTW. This development parallels the initial uses of this technique for the VRP (Naddef and Rinaldi, 2002). Based on the results obtained by Mak (2001), a new arc formulation of the VRPTW is presented in Kallehauge and Boland (2004). In this formulation the time and capacity restrictions are modeled using infeasible path elimination constraints (IPECS). This new class of inequalities can be viewed as a strengthening of the IPECS described in Ascheuer, Fischetti and Grötschel (2000), Ascheuer, Fischetti and Grötschel (2001); Bard, Kontoravdis and Yu (2002) and can also be incorporated at the master problem level in the path formulation considered in this chapter.

Another line of research involves valid inequalities derived from the precedence relationships established by the time windows. That is, if a set of customers is served by the same vehicle, the associated time windows create a precedence structure among the corresponding nodes (Ascheuer, Fischetti and Grötschel, 2001). In Kallehauge and Boland (2004), two classes of valid inequalities for the precedence-constrained asymmetric traveling salesman polytope (Balas, Fischetti and Pulleyblank, 1995) are transferred to the VRPTW.

8. Generalizations of the VRPTW model

The methods considered in this chapter can be generalized and applied to a number of related problems as discussed by Desrosiers, Dumas, Solomon and Soumis (1995). Here we will concentrate on routing generalizations and show how a number of more complex routing problems can be modeled based on the framework introduced in the previous sections.

8.1 Non-identical vehicles

In the general case vehicles may differ with respect to travel time, travel costs, capacity and possibly other characteristics. We define a class of vehicles as a set of identical vehicles. There may be a cost associated with the vehicles of a particular class, and there may be bounds on their availability as well. These bounds are modeled in to the master problem as supplementary constraints.

The subproblem must be solved separately for each class of vehicles. The marginal costs of the arcs originating at the depot of the subproblem for a particular vehicle class must be modified by the simplex multiplier of the constraints on the availability of this class in the master problem. One can chose to solve one or more of the subproblems between each master iteration. The LP optimality criterion is that no subproblem generates columns with negative reduced costs. It is likely to be efficient to branch on the number of vehicles of a particular class if this number is fractional.

A special case occurs if vehicles do not differ with respect to traveling time, travel cost and time windows, but only have different capacities and possible availability and fixed costs. This problem is clearly solvable as described above, but it can also be transformed into the identical vehicle problem described earlier in this chapter. The advantage of this transformation is that only one subproblem must be solved at each iteration. To illustrate how the transformation works consider a problem with two classes of vehicles, with vehicle capacities q_1 and q_2 respectively, where $q_1 < q_2$. The fixed costs of using the vehicles are c_1 and c_2, respectively. Two extra nodes are inserted in parallel between the depot and the customers and any path must go through exactly one of these nodes. The two arcs from the depot to the new nodes are priced c_1 and c_2, respectively. If node 1 is chosen, the capacity is reduced by $q_2 - q_1$ since the resource window of node 1 starts at this quantity. Since the resource window of the depot is $[0, q_2]$, a path going through node 1 cannot service customers with accumulated demand of more than $q_2 - (q_2 - q_1) = q_1$. If there are bounds on the availability of the vehicles, these are inserted in the master problem and the simplex multipliers modify the cost of the two new arcs between the depot and the new nodes.

8.2　　Multiple depots

If the vehicles are based at different depots, one subproblem must be solved for each depot. Constraints on the availability of vehicles at a particular depot are kept in the master problem, and the associated simplex multiplier modifies the cost of arcs originating at the depot. This is equivalent to the general non-identical vehicle case discussed above.

One may assume that the vehicles are allowed to finish their routes at a depot different from the one the vehicles started, but that the number of vehicles starting and ending at any depot remains constant. In this particular case it is sufficient to solve one subproblem. One extra node per depot is created "before" the customers and one "after" the customers. For each depot there will be a constraint r in the master problem requiring the number of vehicles housed at that depot be kept constant. The right hand side will be zero, and the left hand side coefficient (r, p) will be 1 if route p starts at the depot associated with constraint r and ends at another depot, -1 if the route starts at another depot and ends at the depot associated with constraint r, and zero otherwise. The corresponding simplex multipliers modify the cost of arcs originating at the depot (with opposite sign). It is also easy to introduce different fixed costs associated with the vehicles housed at the depots.

8.3   Multiple or soft time windows

Customers may have several (disjoint) time intervals in which they can be serviced. A vehicle arriving between two time windows must wait until the beginning of the next time window. This doesn't truly complicate the problem since the usual dominance criterion in the sub-problem remains valid. A vehicle arriving at a particular node at time t_1 can do everything a vehicle arriving at time t_2 can, provided that $t_1 < t_2$.

If there exist a cost $c(s_i)$ dependent on the time s_i service at customer i begins, the time window is said to be soft. If the cost is non-decreasing with increasing time this is not problematic, since the dominance criteria remain valid. The most general case where $c(s_i)$ is a general function is not efficiently solvable. Ioachim, Gélinas, Desrosiers and Soumis (1998) present an algorithm for the linear case.

9.   Computational experiments

Almost from the first computational experiments, a set of problems became the test-bed for both heuristic and exact investigations of the VRPTW. Solomon (1987) proposed a set of 164 instances that have remained the leading test set ever since. For the researchers working on heuristic algorithms for the VRPTW a need for bigger problems made Homberger and Gehring (1999) propose a series of extended Solomon problems. These larger problems have as many as 1000 customers and several have been solved by exact methods.

9.1   The Solomon instances

The test sets reflect several structural factors in vehicle routing and scheduling such as geographical data, number of customers serviced by a single vehicle and the characteristics of the time windows (e.g., tightness, positioning and the fraction of time-constrained customers in the instances). Customers are distributed within a $[0, 100]^2$ square.

The instances are divided into 6 groups (test-sets) denoted R1, R2, C1, C2, RC1 and RC2. Each of the test sets contain between 8 and 12 instances. In R1 and R2 the geographical data is randomly generated by a random uniform distribution. In the test sets C1 and C2 the customers are placed in clusters, and finally in the RC1 and RC2 test-sets some customers are placed in clusters while others are placed randomly. In the test sets R1, C1 and RC1 the scheduling horizon is short permitting approximately 5 to 10 customers to be serviced on each route. The R2, C2 and RC2 problems have a long scheduling horizon allowing routes with more than 30 customers to be feasible. This makes the problems very hard to solve exactly and they have not been used until recently to test exact methods. The time windows for the test sets C1 and C2 are generated to permit good, maybe even optimal, cluster-by-cluster solutions. For each class of problems the geographical position of the customers is the same in all instances whereas the time windows are changed.

Each instance has 100 customers, but by considering only the first 25 or 50 customers, smaller instances can easily be generated. It should be noted that for the RC-sets this results in the customers being clustered since the clustered customers appear at the beginning of the file. Travel time between two customers is usually assumed to be equal to the travel distance plus the service time at the predecessor customer.

9.2   Computational results

This section reviews the results obtained by the best exact algorithms for the VRPTW. All are based on the column generation approach. The tables 3.1 through 3.6 present the solutions for the six different sets of the Solomon instances that have been solved to optimality. Column K indicates the number of vehicles used in the optimal solution while the column "Authors" give reference to the first publication(s) of the optimal solution for the problem: Kohl, Desrosiers, Madsen, Solomon and Soumis (1999) (KDMSS), Larsen (1999) (L), Kallehauge, Larsen and Madsen (2000) (KLM), Cook and Rich (1999) (CR), Irnich and Villeneuve (2005) (IV), Chabrier (2005) (C), and Danna and Le Pape (2005) (DLP). It should be noted that Desrochers, Desrosiers and Solomon (1992) prior to Kohl, Desrosiers, Madsen, Solomon and Soumis (1999) solved 50 of the 87 Solomon problems with narrow time windows, but with different travel times. Whereas all the above mentioned papers compute the travel times using one decimal point precision and truncation, time and cost is computed differently in Desrochers, Desrosiers and Solomon (1992). Furthermore, solutions to all C1 instances were reported for the first time by Kohl and Madsen (1997), who used a Lagrangian relaxation approach.

As discussed in Cordeau, Desaulniers, Desrosiers, Solomon, and Soumis (2002), the optimal algorithm of Kohl, Desrosiers, Madsen, Solomon and Soumis (1999) solved 69 of the 87 Solomon benchmark short horizon problems to optimality. Eleven additional problems were solved by

Table 3.1.   Optimal solutions for the R1 instances.

Problem	K	Dist.	Authors	Problem	K	Dist.	Authors
R101.25	8	617.1	KDMSS	R107.25	4	424.3	KDMSS
R101.50	12	1044	KDMSS	R107.50	7	711.1	KDMSS
R101.100	20	1637.7	KDMSS	R107.100	11	1064.6	CR+KLM
R102.25	7	547.1	KDMSS	R108.25	4	397.3	KDMSS
R102.50	11	909	KDMSS	R108.50	6	617.7	CR+KLM
R102.100	18	1466.6	KDMSS	R108.100			
R103.25	5	454.6	KDMSS	R109.25	5	441.3	KDMSS
R103.50	9	772.9	KDMSS	R109.50	8	786.8	KDMSS
R103.100	14	1208.7	CR+L	R109.100	13	1146.9	CR+KLM
R104.25	4	416.9	KDMSS	R110.25	5	444.1	KDMSS
R104.50	6	625.4	KDMSS	R110.50	7	697	KDMSS
R104.100	11	971.5	IV	R110.100	12	1068	CR+KLM
R105.25	6	530.5	KDMSS	R111.25	4	428.8	KDMSS
R105.50	9	899.3	KDMSS	R111.50	7	707.2	CR+KLM
R105.100	15	1355.3	KDMSS	R111.100	12	1048.7	CR+KLM
R106.25	5	465.4	KDMSS	R112.25	4	393	KDMSS
R106.50	8	793	KDMSS	R112.50	6	630.2	CR+KLM
R106.100	13	1234.6	CR+KLM	R112.100			

Larsen (1999); Cook and Rich (1999); Kallehauge, Larsen and Madsen (2000). Recently, Irnich and Villeneuve (2005) were successful in closing three additional instances. Four 100-customer instances are still open.

As also reported in Cordeau, Desaulniers, Desrosiers, Solomon, and Soumis (2002); Larsen (1999); Cook and Rich (1999); Kallehauge, Larsen and Madsen (2000) also provided exact solutions to 42 of the 81 Solomon long horizon problems. Since then, Irnich and Villeneuve (2005); Chabrier (2005); Danna and Le Pape (2005) have solved an additional 21 instances, leaving 18 problems still unsolved.

10.   Conclusions

In this chapter we have highlighted the noteworthy developments for optimal column generation approaches to the VRPTW. To date, such methods incorporating branching and cutting on solutions obtained through Dantzig-Wolfe decomposition are the best performing algorithms. Valid inequalities have proved an invaluable tool in strengthening the LP relaxation for this class of problems.

Table 3.2. Optimal solutions for the C1 instances

Problem	K	Dist.	Authors	Problem	K	Dist.	Authors
C101.25	3	191.3	KDMSS	C106.25	3	191.3	KDMSS
C101.50	5	362.4	KDMSS	C106.50	5	362.4	KDMSS
C101.100	10	827.3	KDMSS	C106.100	10	827.3	KDMSS
C102.25	3	190.3	KDMSS	C107.25	3	191.3	KDMSS
C102.50	5	361.4	KDMSS	C107.50	5	362.4	KDMSS
C102.100	10	827.3	KDMSS	C107.100	10	827.3	KDMSS
C103.25	3	190.3	KDMSS	C108.25	3	191.3	KDMSS
C103.50	5	361.4	KDMSS	C108.50	5	362.4	KDMSS
C103.100	10	826.3	KDMSS	C108.100	10	827.3	KDMSS
C104.25	3	186.9	KDMSS	C109.25	3	191.3	KDMSS
C104.50	5	358	KDMSS	C109.50	5	362.4	KDMSS
C104.100	10	822.9	KDMSS	C109.100	10	827.3	KDMSS
C105.25	3	191.3	KDMSS				
C105.50	5	362.4	KDMSS				
C105.100	10	827.3	KDMSS				

Table 3.3. Optimal solutions for the RC1 instances.

Problem	K	Dist.	Authors	Problem	K	Dist.	Authors
RC101.25	4	461.1	KDMSS	RC105.25	4	411.3	KDMSS
RC101.50	8	944	KDMSS	RC105.50	8	855.3	KDMSS
RC101.100	15	1619.8	KDMSS	RC105.100	15	1513.7	KDMSS
RC102.25	3	351.8	KDMSS	RC106.25	3	345.5	KDMSS
RC102.50	7	822.5	KDMSS	RC106.50	6	723.2	KDMSS
RC102.100	14	1457.4	CR+KLM	RC106.100			
RC103.25	3	332.8	KDMSS	RC107.25	3	298.3	KDMSS
RC103.50	6	710.9	KDMSS	RC107.50	6	642.7	KDMSS
RC103.100	11	1258	CR+KLM	RC107.100	12	1207.8	IV
RC104.25	3	306.6	KDMSS	RC108.25	3	294.5	KDMSS
RC104.50	5	545.8	KDMSS	RC108.50	6	598.1	KDMSS
RC104.100				RC108.100	11	1114.2	IV

Recent advances have stemmed from work on parallel implementations of the overall approach, acceleration strategies, primarily at the master problem level, and the subproblem. Solving the subproblem as a ESPPTWCC or a SPPTWCC-k-cyc has shown to be very beneficial.

Table 3.4. Optimal solutions for the R2 instances.

Problem	K	Dist.	Authors	Problem	K	Dist.	Authors
R201.25	4	463.3	CR+KLM	R207.25	3	361.6	KLM
R201.50	6	791.9	CR+KLM	R207.50			
R201.100	8	1143.2	KLM	R207.100			
R202.25	4	410.5	CR+KLM	R208.25	1	328.2	IV+C
R202.50	5	698.5	CR+KLM	R208.50			
R202.100				R208.100			
R203.25	3	391.4	CR+KLM	R209.25	2	370.7	KLM
R203.50	5	605.3	IV+C	R209.50	4	600.6	IV+C
R203.100				R209.100			
R204.25	2	355	IV+C	R210.25	3	404.6	CR+KLM
R204.50	2	506.4	IV	R210.50	4	645.6	IV+C
R204.100				R210.100			
R205.25	3	393	CR+KLM	R211.25	2	350.9	KLM
R205.50	4	690.1	IV+C	R211.50	3	535.5	IV+DLP
R205.100				R211.100			
R206.25	3	374.4	CR+KLM				
R206.50	4	632.4	IV+C				
R206.100							

Table 3.5. Optimal solutions for the C2 instances.

Problem	K	Dist.	Authors	Problem	K	Dist.	Authors
C201.25	2	214.7	CR+L	C205.25	2	214.7	CR+L
C201.50	3	360.2	CR+L	C205.50	3	359.8	CR+KLM
C201.100	3	589.1	CR+KLM	C205.100	3	586.4	CR+KLM
C202.25	2	214.7	CR+L	C206.25	2	214.7	CR+L
C202.50	3	360.2	CR+KLM	C206.50	3	359.8	CR+KLM
C202.100	3	589.1	CR+KLM	C206.100	3	586	CR+KLM
C203.25	2	214.7	CR+L	C207.25	2	214.5	CR+L
C203.50	3	359.8	CR+KLM	C207.50	3	359.6	CR+KLM
C203.100	3	588.7	KLM	C207.100	3	585.8	CR+KLM
C204.25	1	213.1	CR+KLM	C208.25	2	214.5	CR+L
C204.50	2	350.1	KLM	C208.50	2	350.5	CR+KLM
C204.100	3	588.1	IV	C208.100	3	585.8	KLM

Table 3.6. Optimal solutions for the RC2 instances.

Problem	K	Dist.	Authors	Problem	K	Dist.	Authors
RC201.25	3	360.2	CR+L	RC205.25	3	338	L+KLM
RC201.50	5	684.8	L+KLM	RC205.50	5	630.2	IV+C
RC201.100	9	1261.8	KLM	RC205.100	7	1154	IV+C
RC202.25	3	338	CR+KLM	RC206.25	3	324	KLM
RC202.50	5	613.6	IV+C	RC206.50	5	610	IV+C
RC202.100	8	1092.3	IV+C	RC206.100			
RC203.25	2	326.9	IV+C	RC207.25	3	298.3	KLM
RC203.50	4	490.122	IV+C	RC207.50	4	558.6	C
RC203.100				RC207.100			
RC204.25	3	299.7	C	RC208.25	2	269.1	C
RC204.50	3	444.2	DLP	RC208.50			
RC204.100				RC208.100			

Nevertheless, 25% of Solomon's problems are still unsolved. Additional research in each of these areas should lead to further advances. We expect that the further study of polyhedral structures, paralellism, acceleration strategies, and the subproblem will constitute the backbone of research in this area for the next several years. Master problem acceleration methods relying on local search heuristics is just beginning. Other strategies may consider the principle of *stabilization* for column generation discussed in du Merle, Villeneuve, Desrosiers and Hansen (1999) for the VRPTW. Speedup factors of 1 to 10 were achieved by using stabilized column generation on the airline crew pairing problem which closely related to the VRPTW.

Decomposition algorithms are also easily adaptable to other settings. This is because they comprise modules, such as dynamic programming, that can handle a variety of objectives. Lateness, for one, is becoming an increasingly important benchmark in today's supply chains that emphasize on time deliveries. Moreover, they can be run as optimization-based heuristics by means of early stopping criteria.

We hope that this chapter has shed sufficient light on current developments to lead to exciting further research.

Acknowledgments

The research of Marius M. Solomon was partially supported by the Patrick F. and Helen C. Walsh Research Professorship.

References

Ascheuer, N., Fischetti, M., and Grötschel, M. (2000). Polyhedral study of the asymmetric traveling salesman problem with time windows. *Networks* 36:69–79.

Ascheuer, N., Fischetti, M., and Grötschel, M. (2001). Solving the asymmetric travelling salesman problem with time windows by branch-and-cut. *Mathematical Programming* 90:475–506.

Balas, E., Fischetti, M., and Pulleyblank, W.R. (1995). The precedence-constrained asymmetric travelling salesman polytope. *Mathematical Programming* 68:241–265.

Bard, J., Kontoravdis, G., and Yu, G. (2002). A Branch-and-cut procedure for the vehicle routing problem with time windows. *Transportation Science* 36:250–269.

Chabrier, A. (2005). Vehicle routing problem with elementary shortest path based column generation. Forthcoming in: *Computers and Operations Research*.

Chabrier, A., Danna, E., and Claude Le Pape. (2002). Coopération entre génération de colonnes avec tournées sans cycle et recherche locale appliquée au routage de véhicules. In: *Huitièmes Journées Nationales sur la Résolution de Problèmes NP-Complets* (JNPC 2002).

Cook, W. and Rich, J.L. (1999). A parallel cutting-plane algorithm for the vehicle routing problem with time windows. *Technical report* TR99-04. Department of Computational and Applied Mathematics, Rice University.

Cordeau, J.-F., Desaulniers, G., Desrosiers, J., and Soumis, F. (2002). The VRP with time windows. In: *The vehicle routing problem* (P. Toth and D. Vigo, eds.) , pp. 157–193, SIAM Monographs on Discrete Mathematics and its Applications.

Crowder, H. and Padberg, M. (1980). Solving large-scale symmetric travelling salesman problems to optimality. *Management Science* 26:495–509.

Danna, E. and Le Pape, C. (2005). Accelerating branch-and-price with local search: A case study on the vehicle routing problem with time windows. This volume.

Desaulniers, G., Desrosiers, J., and Solomon, M.M. (2002). Accelerating Strategies in column generation methods for vehicle routing and crew scheduling problems. In: *Essays and Surveys in Metaheuristics* (C. Ribeiro and P. Hansen, eds.), pp. 309–324, Kluwer Academic Publishers.

Desrosiers, M., Desrosiers, J., and Solomon, M.M. (1992). A new optimization algorithm for the vehicle routing problem with time windows. *Operations Research* 40:342–354.

Desrochers, M. and Soumis, F. (1988). A generalized permanent labelling algorithm for the shortest path problem with time windows. *INFORMS* 26:191–212.

Desrosiers, J., Dumas, Y., Solomon, M. M., and Soumis, F. (1995). Time constrained routing and scheduling. In: *Handbooks in Operations Research and Management Sciences* (M. Ball, T. Magnanti, C. Monma and G. Nemhauser, eds.), vol. 8, Network Routing, pp. 35–139, North-Holland, Amsterdam.

Desrosiers, J., Sauvé, M., and Soumis, F. (1988). Lagrangean relaxation methods for solving the minimum fleet size multiple travelling-salesman problem with time windows. *Management Science* 34:1005–1022.

Dror, M. (1994). Note on the complexity of the shortest path models for column generation in VRPTW. *Operations Research* 42:977–978.

Dumitrescu, I., and Boland, N. (2003). Improved preprocessing, labeling and scaling algorithms for the weight-constrained shortest path problem. *Networks* 42:135–153.

Feillet, D., Dejax, P., Gendreau, M., and Gueguen, C. (2004). An exact algorithm for the elementary shortest path problem with resource constraints: Application to some vehicle routing problems. *Networks* 44:216–229.

Gélinas, S., Desrochers, M., Desrosiers, J., and Solomon, M.M. (1995). A new branching strategy for time constrained routing problems with application to backhauling. *Annals of Operations Research* 61:91–109.

Halse, K. (1992). Modeling and solving complex vehicle routing problems. *Ph.D Thesis*, Technical University of Denmark.

Homberger, J. and Gehring, H. (1999). Two evolutionary metaheuristics for the vehicle routing problem with time windows. *INFORMS* 37:297–318.

Ioachim, I., Gélinas, S., Desrosiers, J., and Soumis F. (1998). A dynamic programming algorithm for the shortest path problem with time windows and linear node costs. *Networks*, 31:193–204.

Irnich, S. and Villeneuve, D. (2005). The shortest path problem with k-cycle elimination ($k \geq 3$): Improving a branch-and-price algorithm for the VRPTW. Forthcoming in: *INFORMS Journal of Computing*.

Jörnsten, K., Madsen, O.B.G., and Sørensen, B. (1986). Exact solution of the vehicle routing and scheduling problem with time windows by variable splitting, *Research Report* 5/1986, Institute of Mathematical Statistics and Operations Research, Technical University of Denmark.

Kallehauge, B. (2000). Lagrangian duality and non-differentiable optimization—Applied on vehicle routing [in Danish]. *Master's Thesis*, Technical University of Denmark.

Kallehauge, B. and Boland, N. (2004). Infeasible path inequalities for the vehicle routing problem with time windows. *Technical Report*, Centre for Traffic and Transport, Technical University of Denmark.

Kallehauge, B., Larsen, J., and Madsen, O.B.G. (2000). Lagrangean duality and non-differentiable optimization applied on routing with time windows—Experimental results. *Technical Report* IMM-REP-2000-8, Department of Mathematical Modelling, Technical University of Denmark.

Kohl, N. (1995). Exact methods for time constrained routing and related scheduling problems. *Ph.D Thesis*, Technical University of Denmark.

Kohl, N., Desrosiers, J., Madsen, O.B.G., Solomon, M.M., and Soumis, F. (1999). 2-path cuts for the vehicle routing problem with time windows. *Transportation Science* 33:101–116.

Kohl, N. and Madsen, O.B.G. (1997). An optimization algorithm for the vehicle routing problem with time windows based on Lagrangean relaxation. *Operations Research* 45:395–403.

Kolen, A.W.J., Rinnooy Kan, A.H.G., and Trienekens, H.W.J.M. (1987). Vehicle routing with time windows. *Operations Research* 35:266–273.

Kontoravdis, G. and Bard, J.F. (1995). A GRASP for the vehicle routing problem with time windows. *ORSA Journal on Computing* 7:10–23.

Laporte, G., Nobert, Y., and Desrochers, M. (1985). Optimal routing under capacity and distance restrictions. *Operations Research* 33:1050–1073.

Larsen, J. (1999). Parallelization of the vehicle routing problem with time windows. *Ph.D Thesis*, Technical University of Denmark.

Larsen, J. (2004). Refinements of the column generation process for the vehicle routing problem with time windows. *Journal of Systems Science and Systems Engineering* 13(3):326–341.

Mak, V.H. (2001). On the asymmetric travelling salesman problem with replenishment arcs. *Ph.D Thesis*, Department of Mathematics and Statistics, The University of Melbourne.

du Merle, O., Villeneuve, D., Desrosiers, J., and Hansen, P. (1999). Stabilized column generation. *Discrete Mathematics* 199:229–237.

Naddef, D., and Rinaldi, G. (2002). Branch-and-cut algorithms for the capacitated VRP. In: *The Vehicle Routing Problem* (P. Toth and D. Vigo, eds.), SIAM Monographs on Discrete Mathematics and Applications, pp. 53–84, SIAM, Philadelphia.

Rousseau, L.-M., Gendreau, M., and Pesant, G. (2004). Solving VRPTWs with constraint programming based column generation. *Annals of Operations Research* 130:199–216.

Solomon, M.M. (1987). Algorithms for the vehicle routing and scheduling problems with time window constraints. *Operations Research* 35:254–265.

Chapter 4

BRANCH-AND-PRICE HEURISTICS: A CASE STUDY ON THE VEHICLE ROUTING PROBLEM WITH TIME WINDOWS

Emilie Danna
Claude Le Pape

Abstract   Branch-and-price is a powerful framework to solve hard combinatorial problems. It is an interesting alternative to general purpose mixed integer programming as column generation usually produces at the root node tight lower bounds (when minimizing) that are further improved when branching. Branching also helps to generate integer solutions, however branch-and-price can be quite weak at producing good integer solutions rapidly because the solution of the relaxed master problem is rarely integer-valued. In this paper, we propose a general cooperation scheme between branch-and-price and local search to help branch-and-price finding good integer solutions earlier. This cooperation scheme extends to branch-and-price the use of heuristics in branch-and-bound and it also generalizes three previously known accelerations of branch-and-price. We show on the vehicle routing problem with time windows (Solomon benchmark) that it consistently improves the ability of branch-and-price to generate good integer solutions early while retaining the ability of branch-and-price to produce good lower bounds.

1.   Introduction

Column generation is a powerful framework to solve hard optimization problems. It operates with a master problem that consists of a linear problem on the current set of columns, and a subproblem that iteratively generates improving columns. In case the master problem contains integrality constraints on some of its variables, column generation and branch-and-bound are combined: This is called branch-and-price, see Barnhart et al. (1998) for a general introduction. Branch-and-price provides the user both with a lower bound (when minimizing, as assumed throughout this paper) and integer solutions. Branch-and-price is known for providing tight lower bounds but it has sometimes difficulties to generate rapidly good solutions because the linear relaxation of the master problem rarely has an integer solution.

Local search (Aarts and Lenstra, 1997; Voß et al., 1999) is a completely different optimization technique with opposite properties. Local search algorithms use operators to define a neighborhood around a given solution or set of solutions. This subregion of the search space is then explored to iteratively generate better solutions and various strategies such as metaheuristics are used to move from one neighborhood to the next so as to escape local minima. Local search algorithms are notoriously effective at generating quickly excellent solutions. However, they do not provide the user with a lower bound on the objective. Hence the difference between the solution obtained and the optimal solution cannot be estimated and the user does not know if more time should be devoted to reach a better solution.

In this paper, we present a general cooperation scheme between branch-and-price and local search that improves the ability of branch-and-price to generate good integer solutions early while retaining the ability of branch-and-price to produce tight lower bounds.

In order to test this general hybrid scheme, we apply it to the vehicle routing problem with time windows (VRPTW). A number of industrial optimization problems are variations of the vehicle routing problem (VRP), which can be summarized as follows: Given a set of customers that each demand some amount of goods, a set of vehicles with given capacity that must start from and return to a depot, and known distances between all customers and the depot, and every pair of customers, the objective is to establish for each vehiclenn an ordered list of customers to visit so as to minimize the overall distance travelled and sometimes the number of vehicles needed. A classical additional constraint is to specify time windows that restrict the time of the day at which each customer can be served: This defines the vehicle routing problem with time windows. Cordeau et al. (2002) review different methods to solve it. Among exact methods, branch-and-price has recently been applied with success to this problem, see for example Desrochers et al. (1992); Kohl et al. (1999); Larsen (1999); Cook and Rich (1999); Kallehauge et al. (2001); Irnich (2001); Rousseau et al. (2002); Chabrier et al. (2002); Chabrier (2003); Irnich and Villeneuve (2003). Local search algorithms are also popular for solving the VRPTW, see for example Rochat and Taillard (1995); Homberger and Gehring (1999); Gambardella et al. (1999);

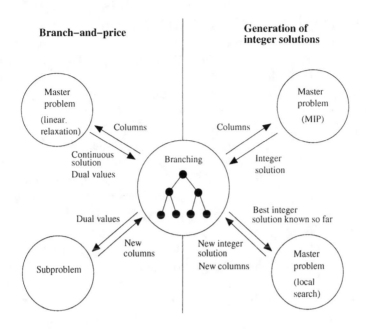

Figure 4.1. Cooperation scheme.

Cordeau et al. (2001); De Backer et al. (2000); Bräysy and Gendreau (2003a,b).

The remainder of the paper is organized as follows. Section 2 presents our general cooperation scheme between branch-and-price and local search. Section 3 details how our general scheme is applied to the vehicle routing problem with time windows. Section 4 gives computational results and discusses why our hybrid scheme works. Finally, Section 5 summarizes our conclusions.

2. General cooperation between branch-and-price and local search

2.1 Description of the algorithm and discussion

Figure 4.1 presents our cooperation scheme between column generation and local search. The left hand side of the figure shows the usual relaxed master problem and subproblem of branch-and-price. Note that the subproblem could be solved by any optimization technique. On the right hand side two components for obtaining integer solutions are specified. First, a mixed integer programming (MIP) solver is called regularly on the master problem with the current set of columns without relaxing the integrality constraints. If the MIP solver is called at the root node of the branch-and-price tree, the best integer solution found so far is used as the first solution of the MIP. If the MIP solver is called at a node further down the branch-and-price tree, the best integer solution found so far might not be valid for the branching decisions taken at that node, hence it cannot always be used as a first solution. The effort spent on solving the MIP is controlled with a time or node limit. When this limit is reached, the exploration of the branch-and-price tree is resumed. Secondly, local search is also called regularly to solve the master problem, its initial solution being the best integer solution found so far. Unlike the MIP solver, local search is not restricted to combining existing columns: Local search may not only provide better combinations of existing columns, but it may also introduce new columns. Hence the columns generated are more diverse which is likely to accelerate pricing, for example because it has thus greater chances to overcome degeneracy.

The strength of our hybrid scheme is diversification by means of different algorithms for solving the same problem. Branch-and-price obviously benefits from local search that is more effective at finding feasible solutions. But in turn, local search benefits from branch-and-price that provides it with diverse initial solutions. Indeed, the main difficulty of local search algorithms is to escape local minima. To overcome this difficulty, the strategy of various metaheuristics is to attempt to control a series of moves that increase the value of the objective function in order to reach a different and more promising region of the solution space. There exist even simpler diversification schemes that restart the same algorithm from the same initial solution but with different random seed initialization (see for example Alt et al., 1996, for a theoretical study of this strategy) or build a new initial solution as different as possible from the current local optimum in order to explore a hopefully different region of the solution space. In all cases, diversification is achieved at the cost of increasing the objective function. On the contrary, in our cooperation scheme, the mathematical programming component is a non-deteriorating diversification scheme for local search. Indeed, the upper bound for the master problem and the MIP cutoff are always updated with the value of the best feasible solution found so far. Hence, when the MIP solver finds a new integer solution or when the solution of the relaxed master problem is integer, it is by construction an improvement on the last local optimum found by local search: Diversification is achieved and the objective function is improved at the same time. This strategy has nonetheless a computational cost: solving a MIP is more expensive than classical diversification schemes.

Branch-and-price is an exact method, so it will in the end find the optimal integer solution. Therefore our cooperation scheme is not so useful when exploring the branch-and-price tree to optimality. However, this complete exploration may find good integer solutions only late in the computation. Our cooperation scheme helps to find good integer solutions at an earlier stage, which has numerous advantages. First, the user can stop optimizing as soon as satisfied with the quality of the integer solution found and use the truncated exploration of the branch-and-price tree as a powerful heuristic that also provides tight lower bounds. Next, good upper bounds are helpful to solve the subproblem more effectively, for example by allowing to eliminate arcs from the shortest path subproblem in the VRPTW case, see for example Hadjar et al. (2001); Irnich and Villeneuve (2003). A good upper bound may also reduce the number of iterations between master problem and subproblem at each node: Computing the so-called Lagrangean lower bound (LLB) while solving a tree node might allow to terminate the column generation process at that node before optimality, i.e. as soon as LLB is greater than the upper bound known so far, see for example Desrosiers and Lübbecke (2004). Finally, knowing a good upper bound early might help to explore only a relatively small number of nodes in the branch-and-price tree. Given a fixed branching strategy, a best-first exploration strategy guarantees that only the children nodes of a node with a lower bound smaller than the optimal objective value have to explored. In this sense, best-first search guarantees that a minimum number of nodes are explored. Knowing a good upper bound does not allow us to improve on this number. However, best-first search can fail to produce good integer solutions until the very end of the tree exploration, this is why other exploration strategies such as depth-first search are often preferred, although they lead to a higher number of explored nodes. Our cooperation scheme allows the user to choose a tree exploration strategy such as best-first search that explores a small number of nodes because our scheme doesn't rely only on branching to generate integer solutions.

2.2 Related work

The first algorithms related to the cooperation scheme just described are the so-called mixed integer programming heuristics, such as pivot-and-complement introduced by Balas and Martin (1980) or the diving heuristics described in Bixby et al. (2000). These heuristics are used in branch-and-bound (and branch-and-cut) to generate good integer solutions by taking heuristic decisions outside of the exploration of the tree when branching has difficulties in finding integer solutions. Our cooperation scheme between branch-and-price and local search can be seen as a generalization to branch-and-price of this use of heuristics in branch-and-bound in so far as it achieves the same goal (generating integer solutions early without interfering with the tree exploration strategy) and new columns are also introduced while generating integer solutions.

Our cooperation scheme also relates to the three following accelerating strategies for branch-and-price reviewed in Desaulniers et al. (2002): Using local search to generate initial primal and dual solutions, generating further integer solutions for the master problem by rounding to 1 or to the next integer the fractional variables of its continuous relaxation, and post-optimizing with local search the best known integer solution after a given time limit. Our cooperation scheme is a generalization of these accelerating strategies for the two following reasons. First, in our cooperation scheme, local search is called throughout the branch-and-price search and not only at the beginning or at the end of the optimization process. As explained in the previous section, this allows for fruitful interactions between the two components. Secondly, any local search method can be used in our cooperation scheme—Not just simple rounding techniques. Very effective domain-specific heuristics can be used, as we will show on the vehicle routing problem with time windows.

Finally, it should be mentioned that another existing strategy for combining local search and branch-and-price is to solve the subproblem with a local search algorithm. In our cooperation scheme, the local search algorithm generates new columns when solving the master problem. New columns can also be generated by directly solving the subproblem with a local search algorithm, as described for example in Savelsbergh and Sol (1998); Xu et al. (2003).

2.3   Parameters settings

A fair amount of tuning can be required so as to know when and for how long MIP and local search should be called. It obviously depends on the problem, but here are a few basic rules. Solving completely the MIP formulation of the master problem is time consuming so a time or node limit should be set and the MIP solver should preferably be called when we guess it has a good chance to find an improved integer solution, for example when the integrality gap between the best known integer solution and the value of the current continuous relaxation is high, or when the number of integer-infeasible variables in the relaxed master problem is small. Local search should be called for post-optimization at least each time a new integer solution is found by MIP or when the continuous relaxation of the master problem is integer. If local search turns out to find significantly more solutions than mathematical programming, it may be called more often, using as first solutions not only the solutions found by mathematical programming but also some of the solutions previously found by local search itself. A simple adaptive scheme can be used, decreasing or increasing the frequency and the computation time allotted to MIP and local search according to their respective success rates.

3.   Application to the vehicle routing problem with time windows

We now present the specific branch-and-price model and the heuristics we used for applying the general cooperation scheme just described to the vehicle routing problem with time windows.

3.1   Branch-and-price model and solution techniques

We used the following common model (Cordeau et al., 2002) where each column corresponds to a feasible route. Let $\{1, \ldots, n\}$ be the set of customers. For each feasible route r, let x_r be the variable defined by:

$$x_r = \begin{cases} 1 & \text{if route } r \text{ is used in the solution} \\ 0 & \text{otherwise} \end{cases}$$

and let c_r be the cost of route r. The VRPTW is then written as:

$$\min \sum_{r \in R} c_r x_r \quad \text{s.t.} \tag{4.1}$$

$$\sum_{r \in R} \delta_{ir} x_r = 1, \quad \forall i = 1, \ldots, n \tag{4.2}$$

$$x_r \in \{0, 1\}, \quad \forall r \in R \tag{4.3}$$

where R is the set of all feasible routes with respect to the capacity and time windows constraints, and $\delta_{ir} = 1$ if customer i is visited by route r, and 0 otherwise.

3.1.1   Decomposition into master problem and subproblem.

The first difficulty of this model is that the number of feasible routes grows exponentially with the number of customers. We hence use column generation to generate columns on the fly. The model is decomposed into a master problem and in a subproblem. The master problem is formulated as:

$$\min \sum_{r \in \widehat{R}} c_r x_r \quad \text{s.t.} \tag{4.4}$$

$$\sum_{r \in \widehat{R}} \delta_{ir} x_r = 1, \quad \forall i = 1, \dots, n \tag{4.5}$$

$$x_r \in \{0, 1\}, \quad \forall r \in \widehat{R} \tag{4.6}$$

where $\widehat{R}$ is the set of already generated columns. We solve in fact the continuous relaxation of the master problem, replacing (4.6) with the constraints

$$0 \le x_r \le 1, \quad \forall r \in \widehat{R}. \tag{4.7}$$

The subproblem is the following:

$$\min_{r \in R} c_r - \sum_{i=1}^{n} \pi_i \delta_{ir}$$

where $(\pi_i)_{i=1}^{n}$ is the dual price associated with (4.5). The subproblem is to be interpreted as a constrained shortest path problem on the original graph, where each arc (i, j) is valued by its original cost (distance) minus the dual value π_i associated with its starting extremity i.

3.1.2   Branch-and-price and branching strategy.   The second difficulty of this model is that, as stated in (4.6), x_r variables must take integer values in feasible solutions. Therefore the problem is solved with branch-and-price:

1 Start with an initial pool of columns, for example generated by a simple heuristic.

2 Solve the continuous relaxation of the master problem, replacing (4.6) with (4.7).

3 Solve the subproblem with the dual values updated at step 2, and attempt to generate several constrained shortest paths with negative reduced costs.

4 Iterate steps 2 and 3 until no more new routes with negative reduced cost can be generated.

5 If the solution of the continuous relaxation of the master problem is not integer, branch and iterate steps 2 and 3 at each node.

We use the following branching rule on arcs. Let x^* be the optimal solution of the relaxed master problem after the last subproblem iteration at the current node. If x^* is integer, no branching is necessary. If x^* is not integer, let $t_0 = \{i_0 = \text{depot}, i_1, i_2, \ldots, i_{p+1} = \text{depot}\}$ be the route such that x_{t_0} is the variable with most fractional value in x^*. $x_{t_0} < 1$, hence for each $k \in \{1, \ldots, p\}$, there exist other routes that cover i_k and take a non-zero value in x^*. For each route t such that $x_t^* > 0$ and that shares at least one node with t_0, there exists $q \in \{1, \ldots, p\}$ such that t covers i_q but does not take arc (i_q, i_{q+1}) or arc (i_{q-1}, i_q). Indeed, every route in $\widehat{T}$ is unique, hence t and t_0 can have a common subsequence of nodes but necessarily differ from each other by at least one arc (which initial or final extremity may be the depot). So, we enumerate the columns already generated and choose the first route t such that $x_t > 0$ and that shares at least one node with t_0. Then we choose (i_q, i_{q+1}) as branching arc where $q \in \{1, \ldots, p\}$ is the smallest index such that $(i_q, i_{q+1}) \notin t$ (or we choose arc $(i_0 = \text{depot}, i_1)$ if $(i_q, i_{q+1}) \in t \ \forall q \in \{1, \ldots, p\}$). The child nodes are then created as follows. In one branch, arc (i_q, i_{q+1}) is forbidden. In the other branch, i_q and i_{q+1} are allowed to be taken in a route only if they are linked by arc (i_q, i_{q+1}). In other words, in the second branch, every arc (i_q, r) with $r \neq i_{q+1}$ and every arc (s, i_{q+1}) with $s \neq i_q$ are forbidden. This branching rule is very practical because it is easy to incorporate in the master problem and in the subproblem.

3.1.3 Solving the subproblem.
The subproblem is solved with dynamic programming, with an adaptation of the label-based algorithm described in Desrochers (1988) so as to solve the *elementary* constrained shortest path problem. Details are given in a previous joint work with Alain Chabrier, see Chabrier et al. (2002); Chabrier (2003). The same idea was developed independently in Feillet et al. (2004). The motivation for generating only *elementary* constrained shortest paths in the subproblem is the following. If the distance used to compute the cost of routes conforms to the triangular inequality, then the optimal solution contains only elementary routes, whether cycles are allowed in the subproblem or not. Solving the non-elementary constrained shortest path is easier, so most column generation models in the literature allow cycles to be generated in the subproblem and sometimes add some mechanisms to partially eliminate non-elementary routes or improve the lower bound, see for example Houck et al. (1980); Kohl et al. (1999); Cook and Rich (1999); Irnich (2001); Irnich and Villeneuve (2003). These mechanisms are instrumental in solving instances with large time windows or with a large horizon because these instances are only loosely constrained by the

numerical data: It is entirely possible to build non-elementary routes and even to traverse cycles several times in the same route.

In short, the label-based algorithm used to solve the subproblem develops as follows. Partial paths starting from the depot and visiting a number of customers are built. As in a typical dynamic programming algorithm, dominated partial paths are gradually eliminated: If partial path p_1 and partial path p_2 both end at the same customer i, but p_1 arrives sooner at i, has smaller accumulated demand, and is less expensive than p_2, then partial path p_2 can be eliminated. Indeed, for every extension of p_2 to a complete path, p_1 could be extended in the same way and its extension would be less expensive than the extension of p_2. However, this dominance rule is no longer valid if we want to compute only *elementary* paths. Indeed, if the aforementioned extension of p_2 visits some customers that were already visited before i in p_1, then p_1 cannot be extended in the same way as p_2 because it would lead p_1 to visit these customers twice. We therefore change the dominance rule into the following: If partial path p_1 and partial path p_2 both end at the same customer i, but p_1 arrives sooner, has smaller accumulated demand, is less expensive than p_2, and if the set of customers visited by p_1 is a subset of the customers visited by p_2, then partial path p_2 can be eliminated. Refinements of this dominance rule are described in more details in Chabrier et al. (2002); Chabrier (2003). This implementation of elementary shortest path allowed us to solve to optimality 17 instances that were previously open (Chabrier et al., 2002; Chabrier, 2003), 9 of which have now been solved also by Irnich (2001); Irnich and Villeneuve (2003).

3.1.4   Acceleration strategies.

Various well-studied accelerations of the above branch-and-price model allowed us to improve computational times. In particular, (4.5) is replaced by a set covering inequality:

$$\sum_{r \in \widehat{R}} \delta_{ir} x_r \geq 1, \quad \forall i = 1, \ldots, n. \tag{4.8}$$

As a consequence, integer solutions may cover some customers more than once, especially at the beginning of the branch-and-price process. Therefore, each time the MIP solver finds a new integer solution or the relaxed master problem produces an integer solution, we use a greedy heuristic that iteratively removes each customer visited more than once from all routes except from the route from which its removal would yield the smallest cost saving. This allows us to improve the integer solution and the resulting columns are also added to the column pool.

3.2   Heuristics

3.2.1   Building an initial solution.   Heuristics are used for
two different purposes. A first heuristic is used to build an initial solution. This initial solution can be as simple as the trivial solution
"one customer per route". In our cooperation scheme and in the pure
local search scheme, we start with the solution generated with the *savings* heuristic (Clarke and Wright, 1964; Paessens, 1988) adapted to the
problem with time windows.

Heuristics are secondly used to improve on a given solution. We used
two local search algorithms, a relatively simple one and a more sophisticated one. Our computational results will demonstrate the effectiveness
of our cooperation scheme with these two different examples of local
search, which leads us to think that our cooperation scheme is likely to
be applied successfully in different settings. This will also allow us to
show that even a simple local search algorithm can improve the ability
of branch-and-price to generate good integer solutions early.

3.2.2   Large neighborhood search.   We first implemented a
Large Neighborhood Search (LNS) scheme based on constraint programming as described in Shaw (1998). Large Neighborhood Search proceeds
by iteratively fixing some variables of the problem to their value in the
current solution and solving a smaller subproblem on the rest of the variables. For the VRPTW, this amounts to removing a set of customers
from the current solution and inserting them back again to build a better
solution. First a small number of customers are released. If no better
solution is found during a given number of iterations, the subproblem is
enlarged, that is more customers are released simultaneously, see Shaw
(1998) for details.

LNS turned out to be too slow for neighborhoods consisting of more
than 20 customers. Therefore, in our pure LNS algorithm, we also use a
restart mechanism for further diversification. When the size of the LNS
neighborhood reaches 20, a quite different and possibly worse solution
is built by an insertion heuristic. The customers are inserted in the
"orthogonal" order of the current best solution: The first customer of
each route is inserted, then the second customer of each route, etc. Each
customer is inserted in its least expensive insertion point and additional
routes are opened as needed. In the end, some customers are randomly
moved from one route to another. The obtained solution is used as the
next starting point for a new complete run of LNS.

3.2.3   Guided tabu search.   In the second phase of our work,
we decided to use a highly effective implementation of local search for

vehicle routing problems: ILOG DISPATCHER. The neighborhood used in this case is the union of all possible 2-OPT (Croes, 1958; Lin, 1965), Or-OPT (Or, 1976), Relocate (insert a customer in another route), Exchange (swap two customers of two different routes), and Cross moves (exchange the ends of two routes). The exact implementation is detailed in De Backer et al. (2000) and in Ilog Dispatcher User's Manual (2002). As for metaheuristics, we use in this case guided tabu search which is a mix of guided local search (GLS) and tabu search. Its implementation is described in De Backer et al. (2000); Ilog Dispatcher User's Manual (2002). GLS introduced by Voudouris (1997) is a meta-heuristic that helps hill-climbing algorithms to escape local optima. It relies on optimizing an adaptively modified cost function based on the original cost function, but penalizing features that appear often in a solution. At each iteration, the penalized objective is first optimized using the hill-climbing algorithm. The penalized objective is then modified, increasing or decreasing the penalty of features according to their cost and to the number of iterations during which they have been penalized. This long-term memory mechanism enables to diversify the search.

However, in the cooperation scheme, the solutions found by the MIP solver or when the solution of the relaxed master problem is integer already ensure the long-term diversification of the local search component. A mechanism for short-term diversification is nonetheless needed and this is ensured by tabu search. Tabu search (Glover and Laguna, 1997) is a well-known effective metaheuristic. Basically, it escapes local optima by forbidding during a certain number of iterations (the tabu *tenure*) properties of moves recently performed or solutions recently visited, unless a certain *aspiration criterion* is validated—For example, the moves lead to a better solution than the best known solution so far. If the tabu tenure corresponds to a small number of iterations, this is a short-term memory mechanism.

De Backer et al. (2000) show that, as implemented in ILOG DISPATCHER for vehicle routing problems, Guided Tabu Search performs better than either simple Guided Local Search or simple Tabu Search.

4.   Computational results

4.1   Benchmark

All computational testing described in the next sections have been performed on the well-known Solomon VRPTW instances introduced in Solomon (1987), on which exact and heuristic methods for solving the VRPTW are often tested. We adopted the conventions used by most exact methods: The objective is to minimize overall distance independently

of the number of vehicles used; distances and traveling times are determined by the Euclidean distance rounded downward at the first decimal
place. The Solomon benchmark comprises two series of instances:

- series 1, the vehicle capacity is limited and the planning horizon is
  rather short;

- series 2 has vehicles with larger capacity and a longer planning
  horizon, which allows more customers to be served by the same
  route.

Hence, instances of series 1 are easier to solve because they are less
combinatorial: The total number of feasible routes for these instances is
smaller than for instances in series 2. Literature has so far concentrated
on series 1, with some notable exceptions: Larsen (1999); Cook and Rich
(1999); Kallehauge et al. (2001); Irnich (2001); Chabrier et al. (2002);
Chabrier (2003); Irnich and Villeneuve (2003)—See Cordeau et al. (2002)
for a general survey. Solomon instances are further divided in three
groups: For "R" instances, customers are geographically randomly distributed; for "C" instances, customers are geographically clustered; and
for "RC" instances, customers are alternatively random and clustered.
Each instance is a 100-customer problem, from which a smaller problem
is constructed taking into account only the first 50 customers.

Tables 4.1 and 4.2 give the solutions with which we will compare our
results in Sections 4.3 and 4.5. These reference solutions are the best
results known to us, taken from the literature or our own experiments
(see next section for a precise description of our methods), as indicated in
the *Origin* column. Solutions marked with * have been proved optimal.
When the optimum is not known, we take the best known lower bound
for series 1 (indicated in *italics*). But as no good lower bound is known
for the open instances of series 2, we take the best known upper bound
instead. The number of vehicles corresponding to each upper bound is
given in parentheses.

4.2   Methods

Recall that the approaches compared are:

1 Our cooperation scheme between branch-and-price and local search
  (BP+LNS, BP+DISPATCHER). The MIP solver is called every 4
  minutes with a 1-minute time limit. The local search algorithm is
  called for 10 seconds every 2 minutes and after a new and better
  integer solution has been found by branch-and-price. If the local
  search algorithm finds a new solution during a run, it is called
  immediately thereafter, again for 10 seconds. This very simple

Table 4.1. Reference solutions for series 1

Instance	50 customers		100 customers	
	Cost	Origin	Cost	Origin
C101	362.4 (5)*	Kohl et al. (1999)	827.3 (10)*	Kohl et al. (1999)
C102	361.4 (5)*	Kohl et al. (1999)	827.3 (10)*	Kohl et al. (1999)
C103	361.4 (5)*	Kohl et al. (1999)	826.3 (10)*	Kohl et al. (1999)
C104	358.0 (5)*	Kohl et al. (1999)	822.9 (10)*	Kohl et al. (1999)
C105	362.4 (5)*	Kohl et al. (1999)	827.3 (10)*	Kohl et al. (1999)
C106	362.4 (5)*	Kohl et al. (1999)	827.3 (10)*	Kohl et al. (1999)
C107	362.4 (5)*	Kohl et al. (1999)	827.3 (10)*	Kohl et al. (1999)
C108	362.4 (5)*	Kohl et al. (1999)	827.3 (10)*	Kohl et al. (1999)
C109	362.4 (5)*	Kohl et al. (1999)	827.3 (10)*	Kohl et al. (1999)
R101	1044.0 (12)*	Kohl et al. (1999)	1637.7 (20)*	Kohl et al. (1999)
R102	909.0 (11)*	Kohl et al. (1999)	1466.6 (18)*	Kohl et al. (1999)
R103	772.9 (9)*	Kohl et al. (1999)	1208.7 (14)*	Kohl et al. (1999)
R104	625.4 (6)*	Kohl et al. (1999)	962.3 (11)*	Irnich and Villeneuve (2003)
R105	899.3 (9)*	Kohl et al. (1999)	1355.3 (15)*	Kohl et al. (1999)
R106	793.0 (5)*	Kohl et al. (1999)	1234.6 (13)*	Cook and Rich (1999),Kallehauge et al. (2001)

Table 4.1 (continued).

Instance	50 customers		100 customers	
	Cost	Origin	Cost	Origin
R107	711.1 (7)*	Kohl et al. (1999)	1064.6 (11)*	Cook and Rich (1999), Kallehauge et al. (2001)
R108	617.7 (6)*	Kohl et al. (1999)	*919.9 (-)*	*Irnich and Villeneuve (2003)*
R109	786.8 (8)*	Kohl et al. (1999)	1146.9 (13)*	Cook and Rich (1999), Kallehauge et al. (2001)
R110	697.0 (7)*	Kohl et al. (1999)	1068 (12)*	Cook and Rich (1999), Kallehauge et al. (2001)
R111	707.2 (7)*	Cook and Rich (1999), Kallehauge et al. (2001)	1048.7 (12)*	Cook and Rich (1999),Kallehauge et al. (2001)
R112	630.2 (6)*	Cook and Rich (1999), Kallehauge et al. (2001)	*935.1 (-)*	*Cook and Rich (1999)*
RC101	944.0 (8)*	Kohl et al. (1999)	1619.8 (15)*	Kohl et al. (1999)
RC102	822.5 (7)*	Kohl et al. (1999)	1457.4 (14)*	Cook and Rich (1999), Kallehauge et al. (2001)
RC103	710.9 (6)*	Kohl et al. (1999)	1258 (11)*	Cook and Rich (1999), Kallehauge et al. (2001)
RC104	545.8 (5)*	Kohl et al. (1999)	1132.3 (10)*	Irnich and Villeneuve (2003)
RC105	855.3 (8)*	Kohl et al. (1999)	1513.7 (15)*	Kohl et al. (1999)
RC106	723.2 (6)*	Kohl et al. (1999)	*1356.1 (-)*	*Cook and Rich (1999)*
RC107	642.7 (6)*	Kohl et al. (1999)	1207.8 (12)*	Irnich and Villeneuve (2003)
RC108	598.1 (6)*	Kohl et al. (1999)	1114.2 (11)*	Irnich and Villeneuve (2003)

Table 4.2.   Reference solutions for series 2

Instance	50 customers		100 customers	
	Cost	Origin	Cost	Origin
C201	360.2 (3)*	Cook and Rich (1999), Larsen (1999)	589.1 (3)*	Cook and Rich (1999), Kallehauge et al. (2001)
C202	360.2 (3)*	Cook and Rich (1999), Kallehauge et al. (2001)	589.1 (3)*	Cook and Rich (1999), Kallehauge et al. (2001)
C203	359.8 (3)*	Cook and Rich (1999), Kallehauge et al. (2001)	588.7 (3)*	Kallehauge et al. (2001)
C204	350.1 (2)*	Kallehauge et al. (2001)	588.1 (3)*	Irnich and Villeneuve (2003)
C205	359.8 (3)*	Cook and Rich (1999), Kallehauge et al. (2001)	586.4 (3)*	Cook and Rich (1999), Kallehauge et al. (2001)
C206	359.8 (3)*	Cook and Rich (1999), Kallehauge et al. (2001)	586.0 (3)*	Cook and Rich (1999), Kallehauge et al. (2001)
C207	359.6 (3)*	Cook and Rich (1999), Kallehauge et al. (2001)	585.8 (3)*	Cook and Rich (1999), Kallehauge et al. (2001)
C208	350.5 (2)*	Cook and Rich (1999), Kallehauge et al. (2001)	585.8 (3)*	Kallehauge et al. (2001)
R201	791.9 (6)*	Cook and Rich (1999), Kallehauge et al. (2001)	1143.2 (8)*	Kallehauge et al. (2001)
R202	698.5 (5)*	Cook and Rich (1999), Kallehauge et al. (2001)	1029.6 (8)	BP+DISPATCHER
R203	605.3 (5)*	Irnich (2001), Chabrier et al. (2002), Chabrier (2003)	871.4 (6)	ILOG DISPATCHER
R204	506.4 (2)*	Irnich and Villeneuve (2003)	733.0 (5)	Røpke (2003)
R205	690.1 (5)*	Larsen (1999), Kallehauge et al. (2001)	951.9 (5)	Røpke (2003)
R206	632.4 (4)*	Irnich (2001), Chabrier et al. (2002), Chabrier (2003)	880.6 (4)	Røpke (2003)
R207	575.5 (3)	ILOG DISPATCHER, Røpke (2003)	794.0 (4)	Røpke (2003)

Table 4.2 (continued).

Instance	50 customers		100 customers	
	Cost	Origin	Cost	Origin
R208	487.7 (2)	ILOG DISPATCHER, Røpke (2003)	701.2 (3)	Røpke (2003)
R209	600.6 (4)*	Irnich (2001), Chabrier et al. (2002), Chabrier (2003)	855.7 (5)	Røpke (2003)
R210	645.6 (4)*	Irnich (2001), Chabrier et al. (2002), Chabrier (2003)	900.8 (6)	Røpke (2003)
R211	535.5 (3)*	Irnich and Villeneuve (2003), BP, BP+DISPATCHER	751.7 (3)	Røpke (2003)
RC201	684.8 (5)*	Larsen (1999), Kallehauge et al. (2001), Irnich (2001), Chabrier et al. (2002), Chabrier (2003)	1261.8 (9)*	Kallehauge et al. (2001)
RC202	613.6 (5)*	Irnich (2001), Chabrier et al. (2002), Chabrier (2003)	1092.3 (8)*	Chabrier et al. (2002), Chabrier (2003)
RC203	555.3 (4)*	Chabrier et al. (2002), Chabrier (2003)	923.7 (5)	Røpke (2003)
RC204	444.2 (3)*	BP+DISPATCHER	783.5 (4)	Røpke (2003)
RC205	630.2 (5)*	Irnich (2001), Chabrier et al. (2002), Chabrier (2003)	1154.0 (7)*	Chabrier et al. (2002), Chabrier (2003)
RC206	610.0 (5)*	Irnich (2001), Chabrier et al. (2002), Chabrier (2003)	1051.1 (6)	Chabrier et al. (2002)
RC207	558.6 (4)*	Chabrier et al. (2002), Chabrier (2003)	966.3 (5)	Chabrier et al. (2002)
RC208	476.7 (3)	Chabrier et al. (2002)	777.3 (3)	Røpke (2003)

adaptive scheme allows us to call the local search algorithm more often if it succeeds, without slowing down too much the completion of the optimality proof after the optimal solution has been reached.

2 Almost the same branch-and-price method as in the hybrid scheme, but used without local search. The minor differences with the branch-and-price and MIP scheme used in the cooperation are twofold:

- The MIP solver is called more often (every 3 minutes instead of every 4 minutes) with the same 1-minute time limit for each run, so as to compensate for the lack of other heuristics to generate integer solutions.

- In "BP 1", the initial pool of columns is the trivial solution built with one customer per route. In "BP 2", branch-and-price starts from the solution generated with the *savings* heuristic, as in the hybrid scheme.

3 The same local search method as in the hybrid scheme (LNS or ILOG DISPATCHER), but used alone.

The parameters were chosen experimentally. We found out that it was more effective to call the MIP solver often and with a small time limit than less often with a longer time limit because failures of the MIP solver to produce new integer solutions appeared to come from the inexistence of improving columns rather than from the inability of the solver to optimize successfully the MIP model. Note that the allocation of time limits and call frequencies to the different components of the hybrid scheme renders the execution of the overall algorithm non-deterministic and computer-dependent. Yet such an allocation does make sense when, as is often the case in practice, the main objective is to obtain the best possible result in limited CPU time.

All results were obtained with a one hour time limit for each instance, on a Pentium IV-1.5 GHz with 256 Mb of RAM, using ILOG CPLEX 8.1.0, ILOG SOLVER 5.3 and ILOG DISPATCHER 3.3.

4.3   Quality of integer solutions

We now present the main results for the methods we have just described. Table 4.3 shows the quality of solutions obtained by each algorithm on each series of the Solomon benchmark, for 50-customer and 100-customer instances. The quality of solutions obtained by each algorithm is measured as the mean relative deviation (in %) between the reference solution of Tables 4.1 and 4.2, and the upper bound obtained

Table 4.3.   Quality of solutions obtained.

Algorithm	Number of customers	Mean relative deviation (%)							Number of times optimality is reached (and proved)
		C1	R1	RC1	C2	R2	RC2	All	
BP 1	50	0.00	0.15	1.15	0.74	3.73	2.05	1.33	40 (37)
	100	0.00	2.40	6.17	6.50	8.63	5.26	4.77	17 (15)
BP 2	50	0.07	0.25	1.32	0.68	3.12	2.50	1.32	41 (37)
	100	0.29	2.37	6.05	7.80	7.76	5.60	4.86	17 (15)
Pure LNS	50	0.00	0.11	0.20	0.00	1.69	1.25	0.56	41 (-)
	100	0.00	2.96	3.93	2.61	5.85	7.18	3.74	12 (-)
Pure DISPATCHER	50	0.00	0.03	0.12	0.40	0.89	1.13	0.42	33 (-)
	100	0.09	1.06	2.36	0.00	0.62	1.66	0.94	13 (-)
BP+LNS	50	0.00	0.00	0.25	0.12	3.27	0.38	0.75	47 (34)
	100	0.00	1.70	3.62	3.04	6.08	3.93	3.07	19 (15)
BP+DISPATCHER	50	0.00	0.12	0.51	0.75	1.15	0.36	0.48	45 (38)
	100	0.00	1.47	4.46	0.26	4.13	2.56	2.17	20 (16)

by this algorithm. Recall that on series 2 the reference solutions are
possibly sub-optimal upper bounds, hence the numbers given for this
series are not necessarily upper bounds for the distance to the optimal
solution. Table 4.3 also gives the number of instances for which each
algorithm reaches and also proves (in parentheses) optimality. Recall
that only pure branch-and-price and our cooperation scheme are able to
produce optimality proofs. Note for comparison purposes that there are
56 instances in each 50-customer and 100-customer category. The optimal solution is known for 53 instances of the 50-customer category, and
38 instances of the 100-customer category. Figures 4.2 through 4.4 show
the evolution of solution quality over time for all 100-customer instances.

Our first conclusions from this experimental data are the following.
Combining local search and branch-and-price is consistently more effective than branch-and-price alone at obtaining good feasible solutions.
On all series, on 100-customers and 50-customer instances, both our
hybrids combining branch-and-price and local search (LNS or ILOG
DISPATCHER) are as effective as or, in most cases, more effective
than branch-and-price alone. The improvement obtained by combining
branch-and-price and local search is most often correlated with the performance of the local search algorithm used alone. As expected, our simple LNS algorithm performs worse than the more sophisticated guided
tabu search from ILOG DISPATCHER. In the same way, the cooperation
scheme combining branch-and-price and LNS is outperformed in most
series by the cooperation scheme combining branch-and-price and ILOG
DISPATCHER. However, even a simple local search algorithm such as
LNS can improve significantly the performance of branch-and-price.

On the contrary, for pure branch-and-price algorithms, starting from a
pool of columns built by a simple heuristic ("BP 2") does not consistently

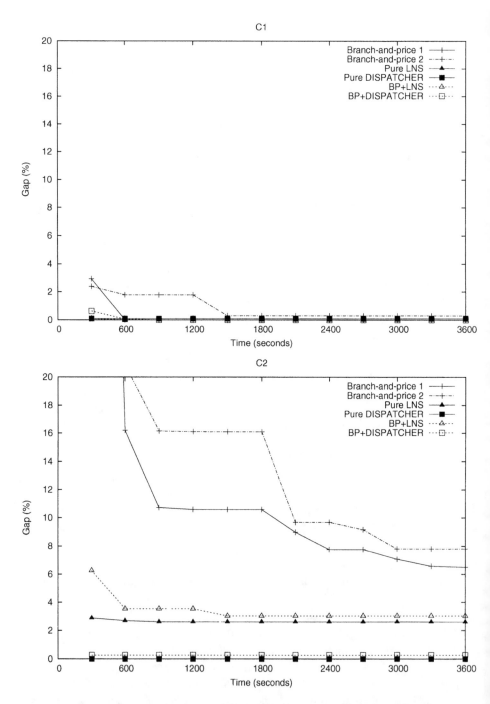

Figure 4.2.   Evolution of solution quality over time for C series (100-customer instances).

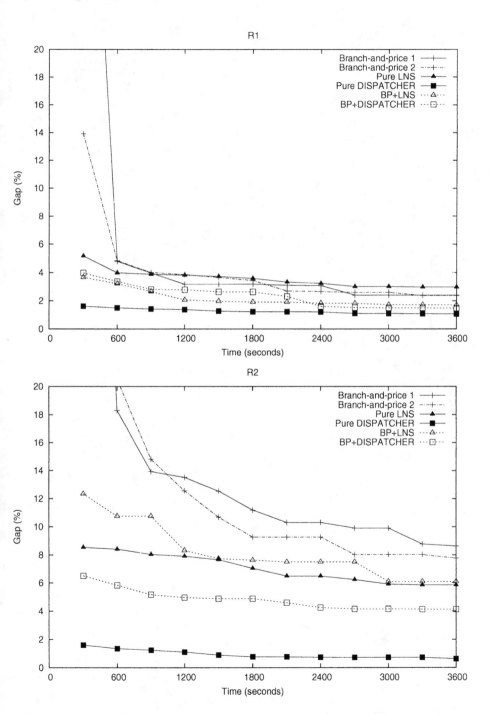

Figure 4.3. Evolution of solution quality over time for R series (100-customer instances).

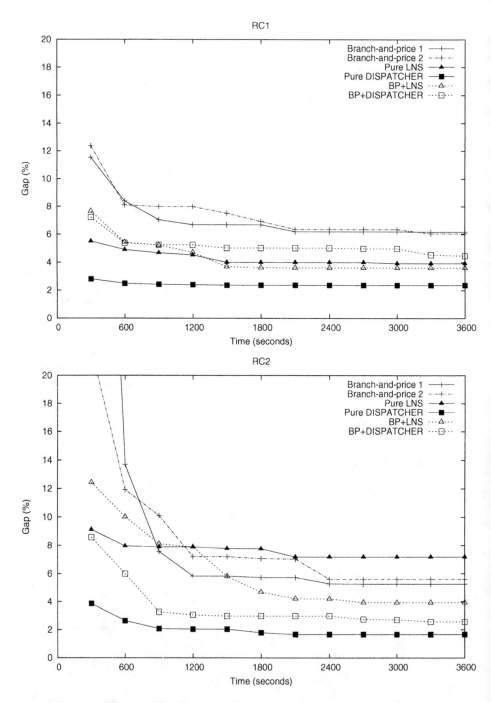

Figure 4.4. Evolution of solution quality over time for RC series (100-customer instances).

improve on the same algorithm starting from a trivial solution ("BP 1"). Our tentative explanation is that the simple *savings* heuristic is not powerful enough to foster a significant difference in performance.

Let us examine a few examples of the 100-customer instances for more specific remarks. On series R1, RC1 and most significantly RC2, our co-operation scheme between branch-and-price and LNS outperforms both pure LNS and the two variants of pure branch-and-price. This illustrates the useful interaction between the two components of the hybrid scheme, each optimizing the solutions found by the other component. On series RC2, branch-and-price outperforms pure LNS and our cooperation scheme performs nonetheless better than both. This shows the robustness of our cooperation scheme. Note that although BP+DISPATCHER significantly improves on pure branch-and-price BP 1 and BP 2, it does not give quite as good results as ILOG DISPATCHER used alone: ILOG DISPATCHER is especially effective at providing rapidly very good solutions for all series, hence it is difficult to outperform. But our cooperative scheme has nonetheless the advantage of additionally providing the user with a tight lower bound on the objective.

4.4 What component finds integer solutions in the cooperation scheme?

Let us now give more detailed results for each component of our hybrid scheme (branch-and-price and local search) so as to better understand why they succeed in finding good integer solutions early. Tables 4.4 and 4.5 show for each component of the cooperation scheme the number of times it succeeds in finding a new and better solution, divided by the number of times this component was called (column *%s* for success), or divided by the total number of integer solutions found by all components (column *%c* for contribution). Statistics are aggregated for 100-customer instances of Solomon series 1 and 2.

Table 4.4. Integer solutions found with the cooperation BP+LNS.

Component	Series 1		Series 2	
	%s	%c	%s	%c
Relaxed master problem integer	0.01	4.65	0.01	1.85
MIP	5.92	7.90	7.96	7.40
Multiple visits heuristic	62.96	7.90	80.00	7.40
Total LNS	27.22	79.53	29.03	83.33
... when optimizing a solution found by:				
Branch-and-price	85.00	23.72	64.70	20.37
LNS	21.12	55.81	24.63	62.96

Table 4.5.   Integer solutions found with the cooperation BP+DISPATCHER.

Component	Series 1		Series 2	
	%s	%c	%s	%c
Relaxed master problem integer	0.04	11.81	0.09	7.14
MIP	5.40	7.27	3.75	2.38
Multiple visits heuristic	35.71	6.81	45.83	4.36
Total ILOG DISPATCHER	24.84	74.09	35.05	86.11
...when optimizing a solution found by:				
Branch-and-price	63.73	26.36	50.54	18.25
ILOG DISPATCHER	18.58	47.72	32.38	67.85

The "Multiple visits heuristic" line refers to the greedy heuristic transforming a solution of the set covering formulation into a set partitioning solution, as described at the end of Section 3.1. It is mostly useful at the beginning of the optimization process: Afterwards, the upper bound is too tight to allow for customers to be visited more than once. Both our local search algorithms work on a model where each customer is visited exactly once. Therefore, the multiple visits heuristic cannot improve solutions found by local search. The line "Local search...when optimizing a solution found by branch-and-price" refers to the results of local search starting from a solution found when the relaxed master problem was integer, or from a solution found by the MIP solver, or from the output of the multiple visits heuristic optimizing a solution found by the two former methods.

On all series, local search finds the majority of solutions: The good results of our cooperation scheme are naturally obtained first thanks to the great ability of local search to find good feasible solutions. The success rate of local search is much higher when starting from a solution found by branch-and-price than when starting from a solution found by local search itself. This illustrates the diversification mechanism: When branch-and-price finds a solution, it is far from the last local search local optimum, hence it is more likely to be improved by local search.

4.5   Quality of lower bounds

In this section, we compare the ability of each method to provide lower bounds and evaluate whether our cooperative algorithms retain the ability of branch-and-price to generate good lower bounds. Recall that local search algorithms do not provide lower bounds nor optimality proofs.

Table 4.6 gives the mean relative deviation between the lower bound obtained by each studied algorithm and the reference solutions of Tables 4.1 and 4.2 on the 100-customers instances of Solomon series 1 and

Table 4.6.   Quality of lower bounds (100-customer instances).

Instances	Series 1	Series 2
BP 1	-0.76%	-0.34%
BP 2	-0.77%	-0.37%
BP+LNS	-0.77%	-0.34%
BP+DISPATCHER	-0.75%	-0.33%

series 2. Recall that, on series 1, reference solutions are either optimal solutions or possibly sub-optimal lower bounds, therefore Table 4.6 does not indicate for this series upper bounds on the distance between the lower bounds obtained and the optimum. Note also that on several instances (1 instance in series 1, 15 instances in series 2), none of the algorithms studied produces a lower bound on the objective: The one-hour time limit is too short to terminate pricing at root node. These instances are not taken into account for the computation of the mean relative deviation in Table 4.6. The overall conclusion of Table 4.6 is that both of our hybrids between branch-and-price and local search (LNS or ILOG DISPATCHER) retain the ability of branch-and-price to generate good lower bounds. Note also from Table 4.3 that our cooperation scheme between branch-and-price and local search produces approximately the same number of optimality proofs as pure branch-and-price.

4.6   Proof of optimality for previously unsolved instances

We finally present result for two previously open instances which we solved to optimality: R211 and RC204, both with 50 customers. Note that R211.50 was recently solved independently to optimality, as reported in Irnich and Villeneuve (2003). Table 4.7 gives for each instance the minimal distance and the corresponding number of vehicles. Table 4.8 gives the time in seconds needed to reach the optimal solution (T_{opt}) and subsequently prove optimality (T_{total}), and the number of nodes explored in the branch-and-price tree (nodes). We provide results for pure branch-and-price (BP 1 and BP 2) and for our cooperation scheme between branch-and-price and local search (BP+LNS and BP+DISPATCHER).

Note that BP+DISPATCHER is the only algorithm that solved instance RC204.50 to optimality. BP+LNS reached the optimal solution of RC204.50 but was not able to prove optimality within a week of CPU time. Pure branch-and-price (BP 1 and BP 2) also failed to solve

Table 4.7. Optimal values for two previously open instances.

Instance	R211.50	RC204.50
Cost	535.5	444.2
Number of vehicles	3	3

Table 4.8. Proof of optimality for two previously open instances.

Algorithm	R211.50			RC204.50		
	T_{opt}	T_{total}	nodes	T_{opt}	T_{total}	nodes
BP 1	115,100	196,868	257	-	-	-
BP 2	103,600	126,648	85	-	-	-
BP+LNS	214,100	300,184	281	152,100	-	-
BP+DISPATCHER	25,900	94,411	85	50,200	84,059	1

RC204.50 to optimality when given a week of CPU time. On this problem with a long horizon and large time windows, it appears to be extremely time consuming to compute interesting elementary constrained shortest paths. On the contrary, ILOG DISPATCHER succeeds in finding near-optimal routes within a reasonable time and branch-and-price can then find the optimal solution and prove optimality. This illustrates the fact that diversification for generating solutions but also individual columns is a key point of our cooperation scheme.

BP+DISPATCHER solves R211.50 to optimality faster than pure branch-and-price. The acceleration is especially visible when comparing the time needed to reach the optimal solution, but the total time including the optimality proof is also reduced. Note however that BP+LNS slows down the resolution and explores more nodes that BP 2. The phenomenon is witnessed for BP 1. Recall that our branching strategy is not fixed: The branching arc depends on the pool of columns already generated, this is why the number of nodes explored can vary from one branch-and-price variant to the next.

5.   Conclusion

In this paper we introduced a new general strategy for combining local search and branch-and-price. We showed with extensive computational experiments on the vehicle routing problem with time windows that our cooperation scheme consistently improves the ability of pure branch-and-price to find good integer solutions early, while retaining the ability of branch-and-price to generate good lower bounds. The quality improvement of integer solutions generated is most often correlated with the effectiveness of the local search algorithm used, but significant improvements can be obtained even with a simple local search algorithm. It remains to be seen if our cooperation scheme will be applied successfully to different and more complex problems. We believe nonetheless that our results on a quite difficult problem and with two local search algorithms of varied effectiveness are encouraging.

Our cooperation scheme generalizes three previously known accelerations for branch-and-price and can be applied to any branch-and-price model. It can also be seen as the generalization to branch-and-price of the use of heuristics in branch-and-bound. However, unlike most interesting heuristics for branch-and-bound, it is not domain-independent: A specific local search algorithm tailored to the problem at hand has to be written each time a different branch-and-price model is to be solved. We believe nonetheless that it is a step toward extending existing advanced strategies from branch-and-bound to branch-and-price.

Acknowledgements   The work presented in this paper was performed while the first author was a PhD student at ILOG.

We wish to thank the anonymous referee whose detailed comments helped us to significantly improve this chapter. We are also thankful to Jacques Desrosiers, Stefan Røpke, and Dominique Feillet for their helpful comments on a preliminary version of this chapter.

References

Aarts, E. and Lenstra, J. (1997). Local Search in Combinatorial Optimization. Wiley.

Alt, H., Guibas, L., Mehlhorn, K., Karp, R., and Wigderson, A. (1996). A method for obtaining randomized algorithms with small tail probabilities. *Algorithmica*, 16(4-5):543–547.

Balas, E. and Martin, C. (1980). Pivot and complement—A heuristic for $0 - 1$ programming. *Management Science*, 26(1):89–96.

Barnhart, C., Johnson, E., Nemhauser, G., Savelsbergh, M., and Vance, P. (1998). Branch-and-price: Column generation for solving huge integer programs. *Operations Research*, 46:316–329.

Bixby, R., Fenelon, M., Gu, Z., Rothberg, E., and Wunderling, R. (2000). MIP: Theory and pratice—Closing the gap. *System Modelling and Optimization: Methods, Theory, and Applications*, pp. 19–49, Kluwer Academic Publishers.

Bräysy, O. and Gendreau, M. (2003a). Vehicle routing with time windows, part I: Route construction and local search algorithms. *Technical*

Report, SINTEF Applied Mathematics, Department of Optimization, Oslo, Norway.

Bräysy, O. and Gendreau, M. (2003b). Vehicle routing with time windows, part II: Metaheuristics. *Technical Report*, SINTEF Applied Mathematics, Department of Optimization, Oslo, Norway.

Chabrier, A. (2003). Vehicle routing problem with elementary shortest path based column generation. Forthcoming in: *Computers and Operations Research*.

Chabrier, A., Danna, E., and Le Pape, C. (2002). Coopération entre génération de colonnes avec tournées sans cycle et recherche locale appliquée au routage de véhicules *Huitièmes Journées Nationales sur la résolution de Problèmes NP-Complets* (JNPC'2002), pp.83–97.

Clarke, G. and Wright, J. (1964). Scheduling of vehicles from a central depot to a number of delivery points. *Operations Research*, 12:568–581.

Cook, W. and Rich, J. (1999). A parallel cutting-plane algorithm for the vehicle routing problem with time windows. *Technical Report* TR99-04, Department of Computational and Applied Mathematics, Rice University.

Cordeau, J.-F., Desaulniers, G., Desrosiers, J., Solomon, *The Vehicle Routing Problem* (M., and Soumis, F. (2002). The VRP with time windows. In: Toth, P. and Vigo, D., eds.), pp. 157–193, SIAM Monographs on Discrete Mathematics and Applications.

Cordeau, J.-F., Laporte, G., and Mercier, A. (2001). A unified tabu search heuristic for vehicle routing problems with time windows. *Journal of the Operational Research Society*, 52:928–936.

Croes, G. (1958). A method for solving traveling-salesman problems. *Operations Research*, 6:791–812.

De Backer, B., Furnon, V., Shaw, P., Kilby, Ph., and Prosser, P. (2000). Solving vehicle routing problems using constraint programming and metaheuristics. *Journal of Heuristics*, 6:501–523.

Desaulniers, G., Desrosiers, J., and Solomon, M. (2002). Accelerating strategies for column generation methods in vehicle routing and crew scheduling problems. In: *Essays and Surveys in Metaheuristics* (C. Ribeiro and P. Hansen, eds.) , pp. 309–324, Kluwer Academic Publishers.

Desrochers, M. (1986). La fabrication d'horaires de travail pour les conducteurs d'autobus par une méthode de génération de colonnes. *Ph.D
Thesis*, Université de Montréal, Canada.

Desrochers, M., Desrosiers, J., and Solomon, M. (1992). A new optimization algorithm for the vehicle routing problem with time windows.
Operations Research, 40:342–354.

Desrosiers, J. and Lübbecke, M. (2004). A primer in column generation.
Les Cahiers du GERAD, G-2004-02, HEC, Montréal, Canada.

Feillet, D., Dejax, P., Gendreau, M., and Gueguen, C. (2004). An exact algorithm for the elementary shortest path problem with resource
constraints: Application to some vehicle routing problems. *Networks*,
44(3):216–229.

Gambardella, L., Taillard, E., and Agazzi, G. (1999). MACS-VRPTW:
A multiple ant colony system for vehicle routing problems with time
windows. In: *New Ideas in Optimization* (D. Corne, M. Dorigo, , and
F. Glover, eds.), pp. 63–76, McGraw-Hill.

Glover, F. and Laguna, M. (1997). Tabu Search, Kluwer Academic Publishers.

Hadjar, A., Marcotte, O., and Soumis, F. (2001). A branch-and-cut algorithm for the multiple depot vehicle scheduling problem. *Les Cahiers
du GERAD*, G-2001-25, HEC, Montréal, Canada.

Homberger, J. and Gehring, H. (1999). Two evolutionary metaheuristics
for the vehicle routing problem with time windows. *INFOR*, 37:297–
318.

Houck, D., Picard, J., Queyranne, M., and Vemuganti, R. (1980). The
travelling salesman problem as a constrained shortest path problem:
Theory and computational experience. *Opsearch* 17:93-109.

ILOG, S.A. (2002). ILOG DISPATCHER 3.3 User's Manual.

Irnich, S. (2001). The shortest path problem with k-cycle elimination
($k \geq 3$): Improving a branch and price algorithm for the VRPTW.
Technical Report, Lehr- und Forschungsgebiet Unternehmensforschung
Rheinish-Westfälishe Technische Hochshule, Aachen, Germany.

Irnich, S. and Villeneuve, D. (2003). The shortest path problem with
resource constraints and k-cycle elimination for $k \geq 3$. *Les Cahiers du
GERAD*, G-2003-55, HEC, Montréal, Canada.

Kallehauge, B., Larsen, J., and Madsen, O. (2001). Lagrangean duality and non-differentiable optimization applied on routing with time windows—Experimental results. *Technical Report IMM-TR-2001-9*, Department of Mathematical Modelling, Technical University of Denmark, Lyngby, Denmark.

Kohl, N., Desrosiers, J., Madsen, O.B.G., Solomon, M., and Soumis, F. (1999). 2-path cuts for the vehicle routing problem with time windows. *Transportation Science*, 1(13):101–116.

Larsen, J. (1999). Parallelization of the vehicle routing problem with time windows. *Ph.D Thesis*, Informatics and Mathematical Modelling, Technical University of Denmark, DTU.

Lin, S. (1965). Computer solutions of the traveling salesman problem. *Bell System Technical Journal*, 44:2245-2269.

Or, I. (1976). Traveling salesman-type problems and their relation to the logistics of regional blood banking. *Ph.D Thesis*, Department of Industrial Engineering and Management Sciences, Northwestern University.

Paessens, H. (1988). The savings algorithm for the vehicle routing problem. *European Journal of Operational Research*, 34:336–344.

Rochat, Y. and Taillard, E. (1995). Probabilistic diversification and intensification in local search for vehicle routing. *Journal of Heuristics*, 1:147–167.

Røpke, S. (2003). A General Heuristic for Vehicle Routing Problems. *International Workshop on Vehicle Routing and Multi-modal Transporation* (ROUTE'2003).

Rousseau, L.-M., Gendreau, M., and Pesant, G. (2002). Solving small VRPTWs with constraint programming based column generation. In: *Proceedings of the Fourth International Workshop on Integration of AI and OR Techniques in Constraint Programming for Combinatorial Optimization Problems* (CP-AI-OR'02) (N. Jussien and F. Laburthe, F.,eds.), pp. 333–344.

Savelsbergh, M. and Sol, M. (1998). Drive: Dynamic routing of independent vehicles. *Operations Research*, 46(4):474–490.

Shaw, P. (1998). Using constraint programming and local search methods to solve vehicle routing problems. In: *Proceedings of the Fourth International Conference on Principles and Practice of Constraint Pro-*

gramming (CP'98), pp. 417–431. Forthcoming in: INFORMS Journal of Computing.

Solomon, M. (1987). Algorithms for the vehicle routing and scheduling problem with time window constraints. *Operations Research*, 35:254–265.

Voß, S., Martello, S., Osman, I., and Roucairol, C. (1999). Meta-heuristics: Advances and Trends in Local Search Paradigms for Optimization. Kluwer Academic Publishers.

Voudouris, C. (1997). Guided local search for combinatorial optimization problems. *Ph.D Thesis*, Department of Computer Science, University of Essex, Colchester, UK.

Xu, H., Chen, Z., Rajagopal, S., and Arunapuram, S. (2003). Solving a practical pickup and delivery problem. *Transportation Science*, 37(3):347–364.

Chapter 5

CUTTING STOCK PROBLEMS

Hatem Ben Amor
José Valério de Carvalho

Abstract   Column generation has been proposed by Gilmore and Gomory to solve
cutting stock problem, independently of Dantzig–Wolfe decomposition.
We survey the basic models proposed for cutting stock and the corresponding solution approaches. Extended Dantzig–Wolfe decomposition
is surveyed and applied to these models in order to show the links to
Gilmore–Gomory model. Branching schemes discussion is based on the
subproblem formulation corresponding to each model. Integer solutions
are obtained by combining heuristics and branch-and-price schemes.
Linear relaxations are solved by column generation. Stabilization techniques such as dual-optimal inequalities and stabilized column generation algorithms that have been proposed to improve the efficiency of
this process are briefly discussed.

1.   Introduction

The cutting stock problem (CSP) was one of the problems identified by
Kantorovich in his paper entitled "Mathematical methods of organizing
and planning production", that first appeared in 1939, in Russian, and
was later published in Management Science (1960). The problem consist
of determining the best way of cutting a set of large objects into smaller
items. There are large potential economic savings resulting from the
optimization of this kind of problems. CSPs are encountered in a wide
variety of industrial applications, such as in the steel, wood, glass and
paper industries, and in service sector applications, such as cargo loading
and logistics.

In this paper, we focus on one-dimensional problems. Since Gilmore
and Gomory proposed the use of column generation (CG) to solve its
linear programming (LP) relaxation (Gilmore and Gomory, 1961, 1963),
several solution approaches for this problem were based on algorithms

using column generation complemented by heuristics. The nineties mark a turning point in this field: several algorithms combining column generation and branch and bound were proposed to solve the CSP. It was also recognized that those algorithms were also useful to solve instances with many items of many different sizes, yielding a low average demand, a problem that is usually denoted as the bin packing problem (BPP) or the Binary Cutting Stock Problem (BCSP). In this problem, items are assigned to bins in such a way that the capacity of the bins is not exceeded and the number of bins is minimized.

The one-dimensional CSP and BPP are essentially the same problem, even though, under Dyckhoff's system (Dyckhoff, 1990), they were classified as 1/V/I/R and 1/V/I/M, respectively. The reason that possibly motivated the use of a different classification for them was that different solution methods had been traditionally used to address them.

Literature reviews often include not only the CSP and the BPP, but also other problems closely related to them, as knapsack, vehicle loading and pallet loading problems, as well as many others. Examples are Sweeney and Paternoster (1992); Dowsland and Dowsland (1992). The book by Dyckhoff and Finke (1992) identifies more than 700 papers on Cutting and Packing, and classifies them. An annotated bibliography was proposed by Dyckhoff et al. (1997). Many papers refer to case studies where column generation is used to get solutions for real world applications in the aluminum industry (Stadtler, 1990; Helmberg, 1995), in the steel industry (Valério de Carvalho and Guimarães Rodrigues, 1995), in the paper industry (Goulimis, 1990), and in the forest industry (Sessions et al., 1988, 1989).

The aim of this paper is to provide a comprehensive overview on mathematical programming models for the CSP and the BPP, and to show their links to the Gilmore–Gomory model using extended Dantzig–Wolfe decomposition. It reviews recent solution methodologies for the exact solution of CSP and BPP, using branch-and-price, and discusses branching schemes based on the subproblem formulation corresponding to each model. Stabilization techniques, such as dual-optimal inequalities and stabilized column generation algorithms, that can be used to speed the convergence of column generation, are also presented.

This paper is organized as follows. In Section 2, we review the cutting stock models introduced by Kantorovich and by Gilmore and Gomory, a model based on arc-flows, and an acyclic network based multicommodity flow (MCF) model. We also comment on the quality of the bounds that result from their linear programming relaxations. Models with strong linear programming relaxations are of crucial importance in solving integer programming problems. When extended to integer programming problems, Dantzig–Wolfe decomposition is a tool that can be used to obtain stronger bounds. We discuss the application to Cutting Stock. In Section 3, we address the integer solution of cutting stock problems, using heuristics and column generation combined with branch and bound. We review the main issues, several branching schemes, and comment on computational results. In Section 4, we show that stabilization techniques can improve the behavior of column generation for solving the linear programming relaxation of the Gilmore–Gomory model. In Section 5, we present some extensions of the one-dimensional cutting stock problem and, in Section 6, some directions for future research.

2.   Mathematical programming models

The one-dimensional CSP consists of determining the smallest number of rolls of width W that have to be cut in order to satisfy the demand of m clients with orders of b_i rolls of width $w_i, i = 1, 2, \ldots, m$. W, w_i and b_i $(i = 1, \ldots, m)$ are assumed to be positive integers.

2.1   Kantorovich model

Kantorovich (1960) introduced the following mathematical programming model for the CSP to minimize the number of rolls used to cut all the items:

$$\min \sum_{k=1}^{K} x_0^k \qquad (5.1)$$

$$\text{st} \sum_{k=1}^{K} x_i^k \geq b_i, \quad i = 1, \ldots, m \qquad (5.2)$$

$$\sum_{i=1}^{m} w_i x_i^k \leq W x_0^k, \quad k = 1, \ldots, K \qquad (5.3)$$

$$x_0^k = 0 \text{ or } 1, \quad k = 1, \ldots, K \qquad (5.4)$$

$$x_i^k \geq 0 \text{ and integer }, \quad i = 1, \ldots, m, \ k = 1, \ldots, K \qquad (5.5)$$

where K is a known upper bound on the number of rolls needed, $x_0^k = 1$, if roll k is used, and 0, otherwise, and x_i^k is the number of times item i is cut in roll k. Constraints (5.2) enforce the satisfaction of the demand for the items, and constraints (5.3) guarantee that the items cut in a roll do not exceed its capacity. The latter group shall be called the knapsack constraints. Indeed, when $x_0^k = 1$, (5.3) is exactly a knapsack constraint with capacity W; and when $x_0^k = 0$, (5.3) is a knapsack constraint with capacity 0 (i.e. all variables x_i^k equal 0 and the kth bin is not used).

A lower bound for the integer optimum can be obtained from the solution of the linear programming relaxation, which results from substituting the two last constraints for $0 \leq x_0^k \leq 1$ and $x_i^k \geq 0$, respectively. This bound can be very weak. It is equal to the minimum amount of space that is necessary to accommodate all the items, $\lceil \sum_{i=1}^{m} b_i w_i / W \rceil$, and can be very poor for instances with large waste. In the limit, as W increases and all the items have a size $w_i = \lfloor W/2 + 1 \rfloor$, the integer optimal value is $\sum_{i=1}^{m} b_i$ whereas the lower bound approaches $(1/2) \sum_{i=1}^{m} b_i$ (Martello and Toth, 1990). This is a drawback of the model. However, computational experiments show that the instances with very small loss have linear relaxations that are the most difficult to solve exactly (Ben Amor, 1997). Furthermore, its solution space has symmetry. Different solutions to the model, with the same cutting patterns swapped in different rolls, will correspond to the same global cutting solution. Branching directly on individual x_0^k- variables or x_i^k- variables will not eliminate the current fractional solution. This means that any efficient branching has to make decisions independently from index k, unless some symmetry breaking rule is used.

2.2   Gilmore–Gomory model

Gilmore–Gomory proposed a model in which the possible cutting patterns are described by the vector $A^p = (a_1^p, \ldots, a_i^p, \ldots, a_m^p)^T$, where the element a_i^p represents the number of items of width w_i obtained in cutting pattern p. A cutting pattern p is valid if

$$\sum_{i=1}^{m} a_i^p w_i \leq W, \tag{5.6}$$

$$a_i^p \geq 0 \text{ and integer.} \tag{5.7}$$

Define P as the set of all feasible patterns and let λ^p be a decision variable that denotes the number of rolls cut according to cutting pattern p, for all $p \in P$. The CSP is modelled as follows:

$$\min \sum_{p \in P} \lambda^p \tag{5.8}$$

$$\text{st} \sum_{p \in P} a_i^p \lambda^p \geq b_i, \quad i = 1, 2, \ldots, m \tag{5.9}$$

$$\lambda^p \geq 0 \text{ and integer,} \quad \forall p \in P. \tag{5.10}$$

The number of columns in formulation (5.8)–(5.10) may be very large even for moderately sized problems. As it is usually impractical to enumerate all the columns, Gilmore and Gomory proposed column generation to solve its LP relaxation (Gilmore and Gomory, 1961). The problem is initialized with a set of cutting patterns (for instance, each one with multiple copies of the same item in quantities $\lfloor W/w_i \rfloor, \forall i$), and the dual information is used to price the columns out of the master problem. Let $\pi = (\pi_1, \ldots, \pi_m)$ be the dual variables associated to the constraints (5.9), and $\bar{\pi}$ the dual optimal solution at a given column generation iteration. The "most attractive" column is given by the solution of the following knapsack problem:

$$\max \sum_{i=1}^{m} \bar{\pi}_i a_i \tag{5.11}$$

$$\text{st} \sum_{i=1}^{m} a_i w_i \leq W \tag{5.12}$$

$$a_i \geq 0 \text{ and integer,} \tag{5.13}$$

which corresponds to finding the column $A^{\min}$ with the minimum reduced cost $\bar{c}_{\min} = 1 - \bar{\pi} A^{\min} = \min_{p \in P}(1 - \bar{\pi} A^p)$. If the reduced cost is negative, the column is added to the restricted master problem, which is re-optimized; otherwise, the current solution solves the linear relaxation of (5.8)–(5.10).

A lower bound can be easily calculated at any iteration, using duality. In matrix form, the dual of the CSP is $\max\{\pi b : \pi A^p \leq 1, \pi \geq 0\}$. As seen, $1 - \bar{\pi} A^{\min} \leq 1 - \bar{\pi} A^p, \forall p \in P$, which is equivalent to $(\bar{\pi}/\bar{\pi} A^{\min})$ $A^p \leq 1, \forall p \in P$. This means that $(\bar{\pi}/\bar{\pi} A^{\min})$ is a feasible solution to the dual of the CSP (with all valid columns enumerated). The value of this feasible dual solution, $(\bar{\pi}/\bar{\pi} A^{\min})b$, is a lower bound to the value of the primal problem z_{lp}, and is equal to the value of the optimal current solution, $\bar{\pi} b$, divided by the optimal value of the knapsack subproblem, $\bar{\pi} A^{\min}$ (see also Farley, 1990).

This bound can also be obtained using Lagrangean duality. Let c_p be the reduced cost of variable λ_p and K be an upper bound on $\sum_{p \in P} \lambda_p$ at the optimality of the linear relaxation of (5.8)–(5.10). The Lagrangean function of the linear relaxation of the problem is

$$b^T \bar{\pi} + \sum_{p \in P} \bar{c}_p \lambda_p.$$

Using $\lambda_p \geq 0$ and $\bar{c}_p \geq \bar{c}_{\min}$ ($\forall p \in P$), we have

$$b^T \bar{\pi} + \sum_{p \in P} \bar{c}_p \lambda_p \geq b^T \bar{\pi} + \bar{c}_{\min} \sum_{p \in P} \lambda_p.$$

Then given that $0 \leq \sum_{p \in P} \lambda_p \leq K$ and $\bar{c}_{\min} \leq 0$, we deduce the following relation

$$b^T \bar{\pi} + \sum_{p \in P} \bar{c}_p \lambda_p \geq b^T \bar{\pi} + \bar{c}_{\min} K.$$

From Lagrangean duality, we have

$$z_{lp} \geq \text{Min}_{\lambda \geq 0} \, b^T \bar{\pi} + \sum_{p \in P} \bar{c}_p \lambda_p.$$

Hence, we obtain

$$b^T \bar{\pi} + K \bar{c}_{\min} \leq z_{lp}.$$

The leftmost term of this relation is called the Lagrangean bound. Substituting the optimal value of the linear relaxation of CSP z_{lp} to K (actually, z_{lp} is *equal* to the optimal value of $\sum_{p \in P} \lambda_p$), we obtain the alternative expression of Farley's bound $\bar{z}/(1 - \bar{c}_{\min})$ where $\bar{z} = b^T \bar{\pi}$.

This bound can be used to cut-off the tails of column generation processes. However in practice, it does rarely cut more than one iteration (Ben Amor, 1997). This may be explained as follows. In order to have $\bar{z}/(1 - \bar{c}_{\min})$ close to $\bar{z}$, $\bar{c}_{\min}$ should be close enough to 0 which is more likely to happen for the very last CG iterations.

Farley's lower bound may be improved in the case of BCSP (Ben Amor, 1997). The same idea can be generalized in the following way. Let $U = \{i \colon w_i > W/2\}$ the set of items that need to be cut in separate bins, $P_i \subset P$ ($i \in U$) the set of patterns corresponding to such items, and $P_U = \cup_{i \in U} P_i$. The Lagrangean function of the linear relaxation may be written as

$$b^T \bar{\pi} + \sum_{i \in U} \sum_{p \in P_i} \bar{c}_p \lambda_p + \sum_{p \in P \backslash P_U} \bar{c}_p \lambda_p.$$

For each $i \in U$, $\bar{c}_{i \min}$ is the corresponding minimum reduced cost. Noting that for $i \in U$, $\sum_{p \in P_i} \lambda_p = b_i$, and $\sum_{p \in P \backslash P_U} \lambda_p = z_{lp} - \sum_{i \in U} b_i$, an analogous reasoning to the one above leads to the improved lower bound

$$\frac{\bar{z} + \sum_{i \in U} b_i (\bar{c}_{i \min} - \bar{c}_{\min})}{1 - \bar{c}_{\min}}.$$

Since $b_i \geq 0$ and $\bar{c}_{i \min} \geq \bar{c}_{\min}$ ($i \in U$), this bound is better than Farley's bound. A more general bound may be obtained by using $\bar{c}_{o \min}$, the minimum reduced cost among variables corresponding to patterns $p \in P \backslash P_U$. The bound is expressed as

$$\frac{\bar{z} + \sum_{i \in U} b_i (\bar{c}_{i \min} - \bar{c}_{o \min})}{1 - \bar{c}_{o \min}}.$$

The bound given by the optimal solution of the LP relaxation of Gilmore–Gomory's model is known to be very tight. Most of the one-dimensional cutting stock instances have gaps smaller than one, and we say that the instance has the integer round-up property, but there are instances with gaps equal to 1 (Marcotte, 1985, 1986), and as large as 7/6 (Rietz and Scheithauer, 2002). It has been conjectured that all instances have gaps smaller than 2, a property denoted as the modified integer round-up property (Scheithauer and Terno, 1995).

2.3   Decomposition of Kantorovitch formulation

Applying standard Dantzig–Wolfe decomposition principle to the linear relaxation of Kantorovitch formulation (5.1)–(5.5) leads to a lower bound that is weaker than the one implied by Gilmore and Gomory formulation (see Section 2.4). Alternatively, one can apply extended Dantzig–Wolfe decomposition using either a convexification or a discretization approach (see Vanderbeck, 2000b; Ben Amor, 2002). We apply both approaches to show that, although there exists a subtle difference between the two formulations that are obtained, they both lead to the same continuous bound which is the same as the one implied by Gilmore and Gomory formulation.

CSP has a block-diagonal structure and gives raise to $|K|$ subproblems. Define the set of integer feasible points of subproblem k ($k = 1, \ldots, K$),

$$S_k = \left\{ x^k = (x_0^k, \ldots, x_m^k)^T : \sum_{i=1}^{m} x_i^k w_i \leq W x_0^k, \quad x^k \geq 0, \right.$$
$$\left. x^k \text{ integer}, \; x_0^k \in \{0, 1\} \right\}$$

and let C_k be the convex hull of S_k.

First, all subproblems are identical. Hence, we can use a single set of integer feasible points of subproblems

$$S = \left\{ x = (x_0, \ldots, x_m)^T : \sum_{i=1}^{m} x_i w_i \leq W x_0, \quad x \geq 0, \; x \text{ integer}, \right.$$
$$\left. x_0 \in \{0, 1\} \right\}$$

and its convex hull C.

Note that S and C are bounded. If the binary component x_0 of x takes the value 0, all other components equal 0; this is the empty pattern. When it takes value 1, any feasible pattern (including the empty one) may be represented by the values of the other components. Because the number of used rolls is minimized, the empty pattern will never be part of the solution with component $x_0 = 1$. Let P_0 and P denote the set of all feasible patterns and the set of nonempty patterns, respectively. In a same manner we define Ω_0 and Ω the set of all feasible patterns and the set of nonempty patterns that are extreme points of C. The index 0 is associated with the empty pattern in P_0 and Ω_0.

We first present the convexification approach. In Kantorovitch formulation, constraints (5.3) can be replaced with $x^k \in C_k$ while keeping the same integer solutions set for the whole problem. Being $\{x^p = (x_{pi})_{i=1,\dots,m}\}_{p\in\Omega_0}$ the set of extreme points of C, any point x^k in C_k can be expressed as a convex combination of these points:

$$x^k = \sum_{p\in\Omega_0} \lambda_p^k x^p, \qquad \sum_{p\in\Omega_0} \lambda_p^k = 1, \qquad \lambda_p^k \geq 0, \ \forall p \in \Omega_0.$$

It is worth to point out at this level that the new constraint $x^k \in C_k$ takes into account integrality requirements whereas the kth constraint of (5.3) does not.

Substituting in (5.1)–(5.5), we obtain the formulation (Ben Amor, 1997; Vance, 1998)

$$\min \sum_{p\in\Omega} \sum_{k=1}^{K} \lambda_p^k \tag{5.14}$$

$$\text{st} \sum_{p\in\Omega} \sum_{k=1}^{K} x_{pi}\lambda_p^k \geq b_i, \quad i = 1,\dots,m \tag{5.15}$$

$$\sum_{p\in\Omega_0} \lambda_p^k = 1, \quad k = 1,\dots,K \tag{5.16}$$

$$\lambda_p^k \geq 0, \quad \forall p \in \Omega_0, \ k = 1,\dots,K \tag{5.17}$$

$$x^k = \sum_{p\in\Omega} \lambda_p^k x^p, \quad x^k \text{ integer}, \ k = 1\dots,K. \tag{5.18}$$

A slightly different formulation can be obtained by applying the discretization approach. Let $\{x^p = (x_{pi})_{i=1,\dots,m}\}_{p\in P_0}$ be the set of all points of S (we use the same notation since the set of extreme points of C is included in S). Any point $x^k \in S_k$ is written as a binary convex combination of all elements of S:

$$x^k = \sum_{p\in P_0} \lambda_p^k x^p, \qquad \sum_{p\in P_0} \lambda_p^k = 1, \qquad \lambda_p^k \in \{0,1\}, \forall p \in P_0.$$

Substituting in (5.1)–(5.5), we obtain the formulation

$$\min \sum_{p \in P} \sum_{k=1}^{K} \lambda_p^k \tag{5.19}$$

$$\text{st} \sum_{p \in P} \sum_{k=1}^{K} x_{pi} \lambda_p^k \geq b_i, \quad i = 1, \ldots, m \tag{5.20}$$

$$\sum_{p \in P_0} \lambda_p^k = 1, \quad k = 1, \ldots, K \tag{5.21}$$

$$\lambda_p^k \geq 0, \quad \forall p \in P_0, \quad k = 1 \ldots, K \tag{5.22}$$

$$\lambda_p^k \in \{0, 1\}, \quad \forall p \in P_0, \ k = 1 \ldots, K \tag{5.23}$$

where integrality may equivalently be required for variables x^k.

In order to eliminate the symmetry due to identical subproblems, an aggregation procedure is used to obtain index k-free formulations. We present it in a general manner that may be applied to any other IP problem. Define the integer variables λ_p for $p \in \Omega$ (or $\in P$) as

$$\lambda_p = \sum_{k=1}^{K} \lambda_p^k. \tag{5.24}$$

Such variable counts in fact the number of rolls cut owing to a nonempty pattern p, and the number of unused rolls when $p = 0$. Replacing in either formulation, index k disappears from the objective function and covering constraints. Computing a λ_p^k-solution from a λ_p-solution is always possible. We have in fact a transportation problem with $|\{p \colon \lambda_p > 0\}| + 1$ supply ($= \lambda_p$ when $\lambda_p > 0$ and $K - \sum_{q \neq 0} \lambda_q$ when $p = 0$) nodes and K demand ($= 1$) nodes. This problem is feasible *if and only if* $\sum_{p \in \Omega_0} \lambda_p = K$ (or $\sum_{p \in P_0} \lambda_p = K$) which follows trivially from (5.16) (or (5.21)) and (5.24).

As a consequence, in the discretization approach, integrality may be required for variables λ_p ($p \in P$) leading to the formulation of Gilmore and Gomory (5.8)–(5.10). In the convexification approach, the formulation obtained:

$$\min \sum_{p \in \Omega} \lambda_p \tag{5.25}$$

$$\text{st} \sum_{p \in \Omega} x_{pi} \lambda_p \geq b_i, \quad i = 1, \ldots, m \tag{5.26}$$

$$\sum_{p \in \Omega_0} \lambda_p = K \tag{5.27}$$

$$\lambda_p \geq 0, \quad \forall p \in \Omega_0 \tag{5.28}$$

$$\lambda_p = \sum_{k=1}^{K} \lambda_p^k, \quad p \in \Omega_0 \tag{5.29}$$

$$\lambda_p^k \geq 0, \quad \forall p \in \Omega_0, \ k = 1, \dots, K \tag{5.30}$$

$$x^k = \sum_{p \in \Omega} \lambda_p^k x^p, \quad x \text{ integer}, \ k = 1, \dots, K. \tag{5.31}$$

is basically different from the one of Gilmore and Gomory. It considers a subset of the set of columns used in (5.8)–(5.10) and integrality is still required for variables x^k.

Linear relaxations obtained by dropping integrality constraints are solved by column generation. If $\bar{\pi}$ is the dual optimal solution of the restricted master problem at some column generation iteration, the reduced cost of any column x^p is $1 - \sum_{i=1}^{m} \bar{\pi}_i x_{pi}$. Since this objective function is linear, minimizing it over S will produce an extreme point of C (or eventually a boundary point of C). Computing the most attractive column aims at finding the minimum reduced cost column, that is an integer feasible knapsack solution, is equivalent to solving a knapsack problem which is the same subproblem as (5.11)–(5.13) of Gilmore and Gomory formulation.

It is important to note that only feasible points that are extreme points (or on the boundary) of $C = \text{conv}(S)$ can be generated by solving the knapsack problem. Integer feasible points of S that are interior to C are not needed for linear relaxation optimality and both formulation lead to the same LP bound.

The extreme points are enough to produce an integer solution of CSP if branching decisions are taken on x^k-variables (convexification approach) while keeping the same subproblem, whereas they are not sufficient if decisions are taken directly on λ-variables (discretization approach) without altering the subproblem structure. This is illustrated by the following example (Ben Amor, 1997). Let the CSP having roll length $W = 6$, two items of lengthes $w_1 = 2$ and $w_2 = 3$ with corresponding demands $b_1 = 4$ and $b_2 = 3$. The feasible patterns that may be generated are $(3, 0)$ and $(0, 2)$. The empty pattern $(0, 0)$ cannot be generated by the subproblem and may be trivially taken into account. The optimal solution consists of using patterns $(3, 0)$, $(0, 2)$ and $(1, 1)$ exactly once. But the last pattern is an interior point to the set of feasible patterns and will never be generated without modifying the subproblem. However, this pattern can be expressed as a convex combination of extreme pattern as follows: $(1, 1) = (1/6)(0, 0) + (1/3)(3, 0) + (1/2)(0, 2)$. Even though this can be generalized, symmetry still remains a critical issue

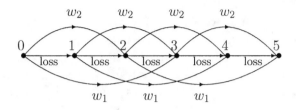

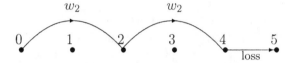

Figure 5.1.   Graph and a cutting pattern.

when dealing with branching schemes based on x^k-variables. The ways of obtaining an integer solution to CSP are addressed later in the paper.

2.4   Arc-flow model

Valério de Carvalho (1999) proposed an arc-flow model for the integer solution of BPP and CSP. A valid packing solution to a single bin (or a cutting pattern) can be modelled as the problem of finding a path in an acyclic directed graph, $G = (V, A)$, with $V = \{0, 1, 2, \ldots, W\}$ and $A = \{(i, j): 0 \leq i < j \leq W$ and $j - i = w_d$ for every $d \leq m\}$, meaning that there exists a directed arc between two vertices if there is an item of the corresponding size. Consider additional arcs between $(k, k+1), k = 0, \ldots, W - 1$, corresponding to unoccupied portions of the bin. There is a packing in a single bin *if and only if* there is a path between vertices 0 and W. The lengths of the arcs in the path define the item sizes to be packed.

EXAMPLE 5.1 Figure 5.1 shows the graph associated with an instance with bins of capacity $W = 5$ and items of sizes 3 and 2. In the same figure, a path is shown that corresponds to 2 items of size 2 and 1 unit of loss.

Shapiro (1968) used this kind of formulation to model the knapsack problem as the problem of determining the longest path in a directed graph. Likewise, it can be used to model BPP and CSP. If a solution to a single bin corresponds to the flow of one unit between vertices 0 and W, a path carrying a larger flow will correspond to using the same packing solution in multiple bins.

By the flow decomposition properties (see Ahuja et al., 1993), non-negative flows can be represented by paths and cycles. The graph G is acyclic, and any flow can be decomposed in directed paths connecting vertex 0 to vertex W. A solution with integer values of flow in every arc, can be transformed into an integer solution to the BPP or the CSP.

The problem is formulated as the problem of determining the minimum flow between vertex 0 and vertex W with additional constraints enforcing that the sum of the flows in the arcs of each item size must be greater than or equal to the value required. Decision variables x_{ij}, associated with the arcs defined above, correspond to the number of items of size $j - i$ placed in any bin at the distance of i units from the beginning of the bin. The number of variables is $O(mW)$. The model is as follows:

$$\min z \tag{5.32}$$

$$\text{st} \sum_{(i,j)\in A} x_{ij} - \sum_{(j,k)\in A} x_{jk} = \begin{cases} -z, & \text{if } j = 0 \\ 0, & \text{if } j = 1,\ldots,W-1 \\ z, & \text{if } j = W \end{cases} \tag{5.33}$$

$$\sum_{(k,k+w_d)\in A} x_{k,k+w_d} \geq b_d, \quad d = 1,2,\ldots,m \tag{5.34}$$

$$x_{ij} \geq 0, \quad \forall (i,j) \in A \tag{5.35}$$

$$x_{ij} \text{ integer}, \quad \forall (i,j) \in A. \tag{5.36}$$

If we apply Dantzig–Wolfe decomposition to (5.32)–(5.35) keeping (5.32) and (5.34) in the master problem, the subproblem defined by (5.33) and (5.35) is a flow problem with a solution space that correspond to the valid flows between vertex 0 to vertex W.

Actually, the variable z can be seen as a feedback arc from vertex W to vertex 0 (could also be denoted as x_{W0}) and the solutions to the subproblem as circulation flows, which include a path between vertices 0 and W and the arc x_{W0}. There is a one-to-one correspondence between circulations and paths. If we see the subproblem solutions as circulations, the subproblem constraints define a homogeneous system, and is unbounded. Therefore, the corresponding polyhedron has a single extreme point, the null solution, and a finite set of extreme rays, which are the directed circulations, each corresponding to a valid pattern. The subproblem will only generate extreme rays, and the substitution of the patterns in (5.32) and (5.34) results in the Gilmore–Gomory model (5.8)–(5.10), which is a nonnegative linear combination of the patterns, with no convex combination constraint. Since the subproblem has the integrality property, both original and decomposed formulations have the same LP bound. Finally, note that either convexification or discretization approaches results lead to Gilmore and Gomory formulation.

2.5   An acyclic capacitated network based MCF model

Ben Amor (1997) considers feasible patterns as paths in an acyclic capacitated network (see Figure 5.2 for an illustration). First since the order in which items are cut on a roll has no effect on the cost function, items are ordered in non-increasing order of length, i.e. $w_i \leq w_{i+1}$, $i = 1, \ldots, m - 1$. Each roll has a corresponding subnetwork. A pair of fictive origin and destination depots represent the start $(o(k))$ and end $(d(k))$ nodes of a path (pattern) in subnetwork (roll) k $(k = 1, \ldots, K)$. To item (client) i corresponds a set of $n_i = \lfloor W/w_i \rfloor$ nodes $i_1, \ldots, i_{n_i}$ and a supplementary node i_0. Two types of arcs are used within this set of nodes: (i_v, i_{v+1}) $(v = 1, \ldots, n_i - 1)$ and (i_v, i_0) $(v = 1, \ldots, n_i)$. Any path (pattern) visiting item i, begins with node i_1 and leaves at node i_0. An inter-task arc joining item i and item j $(j > i)$ exists if $w_i + w_j \leq W$. Such an arc starts at node i_0 and ends at nodes j_1. The first, respectively the last, arc of a path takes the form $(o(k), i_1)$ $(i = 1, \ldots, m)$, respectively $(i_0, d(k))$ $(i = 1, \ldots, m)$. An additional arc $(o(k), d(k))$ corresponds to the empty pattern. To represent a feasible pattern, a path R has to respect the knapsack constraint $\sum_{i \in R} w_i \leq W$. A resource is used to compute the load of a roll at each node of the network. Arcs of types $(o(k), i_1)$ $(i = 1, \ldots, m)$, (j_0, i_1), and (i_v, i_{v+1})

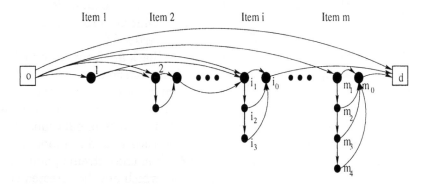

Figure 5.2.   An acyclic network for CSP.

$(v = 1, \ldots, n_i - 1)$ have resource consumption w_i. Other arcs have 0 resource consumption. An upper limit of W on the resource amount used up to any node, except the origin $o(k)$, is imposed. Let N the set of task nodes, i.e. $N = \cup_{i=1}^{m}\{i_v, v = 0, \ldots, n_i\}$, and A_k the set of arcs corresponding to subnetwork k. To each arc $(i, j) \in A_k$ we associate a binary variable x_{ij}^k. CSP is then formulated as a multicommodity flow problem where the number of nonempty paths (patterns) is minimized.

$$\min \sum_{k=1}^{K} \sum_{j=1}^{m} x_{o(k)j_1}^k \tag{5.37}$$

$$\text{st} \sum_{k=1}^{K} \left(\sum_{v=1}^{n_i} \sum_{(j,i_v)\in A_k} x_{ji_v}^k \right) \geq b_i, \quad i = 1, \ldots, m \tag{5.38}$$

$$\sum_{(o(k),j)\in A^k} x_{o(k)j}^k = 1, \quad k = 1, \ldots, K \tag{5.39}$$

$$\sum_{(i,j)\in A^k} x_{ij}^k - \sum_{(j,i)\in A^k} x_{ji}^k = 0, \quad \forall i \in N, \ k = 1, \ldots, K \tag{5.40}$$

$$\sum_{\left(i,d(k)\right)\in A^k} x_{id(k)}^k = 1, \quad k = 1, \ldots, K \tag{5.41}$$

$$\sum_{i=1}^{m} w_i \left(\sum_{v=1}^{n_i} \sum_{(j,i_v)\in A^k} x_{ji_v}^k \right) \leq W \left(\sum_{j=1}^{m} x_{o(k)j_1}^k \right), \quad k = 1, \ldots, K \tag{5.42}$$

$$x_{ij}^k \in \{0, 1\}, \quad k = 1, \ldots, K, \ (i, j) \in A^k. \tag{5.43}$$

The objective function (5.37) aims at minimizing the number of nonempty paths (patterns). Each constraint (5.38) requires that the number of items i cut on all rolls satisfy the demand b_i and constraints (5.39)–(5.41) enforce flow conservation for each commodity k while requiring that 1 unit of flow be shipped from $o(k)$ to $d(k)$. Finally, (linear) constraints (5.42) ensure that capacity limit for any used roll is not overpassed and constraints (5.43) require that all variables be binary.

Applying either the convexification or the discretization approach of extended Dantzig–Wolfe decomposition to this formulation, followed by commodity aggregation as in Section 2.3, leads to Gilmore and Gomory model, with a capacitated shortest path as subproblem. Integrality may equivalently be required either for λ-variables or x^k-variables. It is more natural and easier to develop a branching scheme based on x^k-variables. However, there is a major issue here, symmetry between rolls, that must be taken in account. Branching schemes developed for several models are discussed later in the paper.

Formulation (5.37)–5.43 may be seen as a disaggregated form of Kantorovitch formulation (5.1)–(5.4). In the latter case, when branching constraints based on original variables are added to the master problem (and the corresponding dual variables added to arc costs in the subproblem), only patterns that are extreme points of the knapsack feasible domain may be generated (see Section 2.3). However in the former case, all patterns are extreme points of the subproblem feasible domain (5.39)–(5.43). And hence, adding branching constraints based on the original variables to the master problems is sufficient to generate any feasible pattern that may be needed in an optimal integer solution.

The binary case, i.e. all demands b_i equal 1 and any item may not be cut more than once on a pattern, present an interesting particularity. Each item is represented by a single node and must be visited exactly once by a single path. This allows the aggregation of all commodities and index k is no longer needed in the formulation. As a consequence, the corresponding branching becomes simpler as will be seen later.

3.   Integer solutions

Up to the nineties, it was recognized that it was not easy to combine column generation with tools to obtain integer solutions to the CSP. We quote the final comments in Gilmore (1979):

> "A linear programming formulation of a cutting stock problem results in a matrix with many columns. A linear programming formulation of an integer programming problem results in a matrix with many rows. Clearly a linear programming formulation of an integer cutting stock problem results in a matrix that has many columns and many rows. It is not surprising that it is difficult to find exact solutions to the integer cutting stock problem."

Therefore, up to the nineties, much research has been devoted to the development of heuristics, most of them are based on column generation, when integer solutions for the CSP were searched.

3.1   Heuristics

Gilmore and Gomory (1961) suggest a rounding procedure to obtain integer solutions for the CSP. After solving the linear programming relaxation of their column generation model, rounding up the fractional variables guarantees a heuristic solution of value $z_H : z_H \leq z_{\mathrm{LP}} + m$, where z_{LP} is the optimum of the LP relaxation. As, in the CSP, the values of the demands are generally high, the integer solutions thus obtained are of good quality.

Rather than simply rounding up, more elaborate heuristics based on column generation were also devised to improve effectiveness. Wäscher and Gau (1996) present an extensive computational study with a comparison of several heuristics. The most effective are based on procedures that involve variable fixing (using different strategies) and the solution of residual problems: after finding the optimal solution of the Gilmore–Gomory model, some of the fractional optimal values are fixed (rounded up or down), thus satisfying an integer part of the demand. The remaining unsatisfied demand yields a CSP instance with integer demands, called the residual problem. When only few items are left unsatisfied, the residual problem is much smaller than the original CSP, and can be tackled with an exact bin packing algorithm; otherwise, heuristics are used. The solution for the original problem is the combination of the patterns that were fixed with the solution of the residual problem.

In extensive computational experiments with instances with average demands of 10 and 50, the optimal solutions were found in almost all cases. However, when the average demand is very low, as in the BPP, where it can be close to one or even equal to one, the optimal variables are often a fraction of unity, and it may not be so easy for heuristics to find good solutions for the residual problems. In this case, it is possible to resort to other combinatorial enumeration techniques (see, for instance, Martello and Toth, 1990 or Scholl et al., 1997).

The First Fit Decreasing (FFD) and the Best Fit Decreasing (BFD) heuristics are often used to build solutions for the BPP, and for the CSP. They can be used to obtain starting solutions for column generation, to solve residual problems, and also to get upper bounds in the nodes of the search tree, when looking for an integer solution for the problem.

In these heuristics, the items are sequentially assigned to bins according to the following rules. In the FFD heuristic, the largest unplaced item is assigned to the bin with smallest index already used that has a sufficient remaining capacity; if there is none, a new bin is started. In the BFD heuristic, the largest unplaced item is assigned to the bin with smallest remaining capacity, but still sufficient to accommodate the item; if there is none, a new bin is started. Both these heuristics have an absolute performance ratio of $3/2$, i.e., $z_H \leq 3/2 \, z^*$, where z^* is the value of the optimum (Simchi–Levi, 1994).

To obtain starting solutions for column generation, some authors resort to pseudo-polynomial heuristics, based on the solution of a series of knapsack problems. These are greedy procedures that iteratively select the best cutting pattern using the items in a list. Initially, the list has all the items; at each iteration, the items in the knapsack solution are removed, and the process is repeated until the list is exhausted. Their computation time is not significant in the framework used, and they generally provide better starting solutions, with, at least, some very good cutting patterns, even though the last patterns may be very poor. Two examples will be given. In both cases, the heuristics do favor the choice of knapsack solutions with larger items in the first iterations, leaving the smaller items, which should be easier to combine, to the last iterations.

Vanderbeck (1999) solves a knapsack problem that finds, among the solutions with maximum capacity usage, the one that is lexicographically smaller when considering a solution vector where the items are ordered by non-increasing sizes. Valério de Carvalho (2003) uses a knapsack problem with weights w_j equal to the item sizes, and profits $p_j = \left(1 - (j-1)/n\right)w_j$, $j = 1, \ldots, n$. The values of the profits were chosen to be different from the item sizes to avoid having a subset-sum problem, which is, in practice, difficult to solve (Martello and Toth, 1990).

3.2 Branch-and-price

In the nineties, several attempts to combine column generation with branch-and-bound succeeded in obtaining the optimum integer solution of larger instances of some integer programming and combinatorial optimization problems. This technique has been denoted as branch-and-price. Stronger LP models and good quality lower bounds are of vital importance when using LP based approaches to solve integer problems. Branch-and-price also proved to be a useful framework for the solution of quite large instances of both CSP and BPP, that other combinatorial enumeration branch-and-bound algorithms failed to solve to optimality.

In branch-and-price, it is desirable to ensure compatibility between the restricted master problem and the subproblem. First, the branching rule should not induce intractable changes in the structure of the subproblem. Desirably, it should remain the same optimization problem both during the solution of the LP relaxation and the branch-and-price phase. The second issue is symmetry. Branching strategies should be devised that partition the solution space in mutual exclusive sets, which are explored in different nodes of the branch-and-bound search tree. Symmetry is detrimental, because the same solution may exist (in a different form) in several different nodes of the branch-and-price tree. Other issues are also important to obtain more robust algorithms. Balanced partition rules should be selected: the branching constraints should partition the solution set evenly among subtrees. It is also desirable to select the branching constraints so that stronger decisions are taken at lower levels of the branch-and-bound tree. Most applications of branch-and-price are for problems with binary variables (for a review, see Barnhart et al., 1998 and Desaulniers et al., 1998). Finally, note that the CSP has general integer variables, not restricted to be binary.

The strategy of imposing branching constraints directly on the variables of the reformulated model poses the following difficulty: a column that is restricted by a branching constraint in the master problem may turn out to be the most attractive column generated by the subproblem. To deal with this problem, some authors keep track of the columns already present in the master problem that must not be regenerated. The subproblem has to be solved by an enumeration scheme that rejects the forbidden columns (Degraeve and Schrage, 1999; Degraeve and Peeters, 2003; Belov, 2003). To cope with this difficulties, Desrosiers et al. (1995) argue that, for most integer programming problems, the best strategy to combine column generation with branch-and-bound is to use branching constraints based on the variables of the original model.

3.3   Branching schemes

Branching strategies are related to the original formulation that is decomposed to produce the IP master problem, or at least to the kind of subproblem used to generate columns. For CSP, there are three possible original formulations: Kantorovitch formulation (5.1)–(5.4), the arc-flow formulation (5.32)–(5.36), and the MCF formulation (5.37)–(5.43).

Kantorovitch formulation

Even if it appears to be more natural, efficiency of branching on variables x^k is compromised due to the symmetry between rolls. For instance, fixing any x_0^k to 1 or 0, meaning that roll k is used or not, has no effect on the problem because all rolls are identical. Once the number of rolls (or a lower and/or upper bound) at optimality is known, one may fix x_0^k variables corresponding to these rolls to 1 and others to 0, so that many variables are eliminated from the problem. Moreover, bounding any x_i^k ($k = 1, \ldots, K, i = 1, \ldots, m$) by any integer value does not eliminate any fractional solution because such a solution may be retrieved by using another roll k for item i.

The solution of CSP is composed of patterns and each pattern is composed of a set of items to be cut together on a same roll. Hence at items level, the information that is useful to build a solution should give either a set of items to be/not to be cut on a same set of rolls, or a maximum or/and minimum numbers of copies of an item to be cut on a set of rolls.

Vance's rule.   Vance (1998) first proposed a straightforward branching scheme based on integrality requirements on variables x_p^k in model (5.14)–(5.18) where no rule is used for symmetry breaking between rolls.

This strategy suffers however from the symmetry between rolls, as said before, and was beaten by a quite complex branching strategy based on Gilmore and Gomory model variables for medium size problems. This latter strategy is based on the total flow of columns containing a certain minimum number of copies of items from a set S. It suffers two major drawbacks: there is no guarantee on the size of S and the subproblem is no longer a knapsack since many additional variables are needed. Consequently, it should become untractable even for small depth nodes of the branching tree. An improving strategy is based o the use of maximal patterns, but it amounts to branching directly on λ_p variables. The structure of the subproblem can be preserved for one branch thanks to a specialized algorithm and is highly affected for the other branch (even the maximal pattern property is lost). This last strategy has been successfully used to solve medium size problems.

Vanderbeck's rule.   Vanderbeck (1999) compared several branching schemes and suggested the use of one that is based on the binary representation of the cutting pattern columns of the Gilmore–Gomory model. The binary representation of a column A^p is a 0-1 vector $A^{p'}$ with size $m' = \sum_{d=1}^{m} m_d$, with $m_d = \lceil \log_2(l_d^{\max} + 1) \rceil$, where $l_d^{\max} = \min(b_d, \lfloor W/w_d \rfloor)$ is the upper bound on the number of copies of item d in a cutting pattern, which is limited by the demand for item d and the size of the roll. We will denote the elements of $A^{p'}$ as a_k', $k = 1, \ldots, m'$, dropping the index p, for the sake of clarity. The binary representation is such that the number of items produced for order d is $a_d = \sum_{j=0}^{m_d-1} 2^j a_{p_d+j}'$, $d = 1, \ldots, m$, where $p_d = 1 + \sum_{j=1}^{d-1} m_j$.

Given a fractional solution of Gilmore–Gomory model, it is always possible to find subsets of rows O and $P \subset \{1, \ldots, m'\}$, and a subset of columns

$$\widehat{P} = \{p \in P \colon a_k' = 0, \forall k \in O \text{ and } a_k' = 1, \forall k \in P\}$$

such that $\alpha = \sum_{p \in \widehat{P}} \lambda^p$ is fractional. The branching constraints are: $\sum_{p \in \widehat{P}} \lambda^p \geq \lceil \alpha \rceil$ and $\sum_{p \in \widehat{P}} \lambda^p \leq \lfloor \alpha \rfloor$. If it is possible to identify sets, such that $|O| + |P| = 1$, the branching rule leads to very easy modifications in the subproblem, both in the left and in the right branches. This branching scheme has been used along with a combination of heuristics (the best of BFD, FFD, and a pseudopolynomial) and Martello and Toth (1990) lower bounds. Experiments with CSP instances, where the values of the demands are large, show that the procedure is quite robust and powerful. However, when demands are very small, as in the BPP, it may be necessary to select sets with $|O| + |P| \geq 2$, leading to a subproblem that is no longer a knapsack problem, but an extended knapsack problem with new extra binary variables, needed to identify the attractive columns correctly. Computational experiments show results comparable to the ones obtained with the arc-flow model for the instances under study.

Arc-flow model

The arc-flow model provides a branching scheme for a branch-and-price algorithm for the CSP that preserves the structure of the subproblem, which is as a longest path problem in an acyclic digraph with modified costs, that can be solved using dynamic programming.Valério de Carvalho (1999) implemented a column generation algorithm based on a master problem with variables of the arc-flow model. It starts with a subset of arcs corresponding to an initial solution, and new arcs are added to the original formulation if they belong to an attractive path and are not already used. In the arc-flow model, there are flow conservation constraints for the nodes. If the new arcs are incident into nodes not previously considered, new constraints are explicitly added to the formulation. When an arc-flow variable x_{ij} takes a fractional value α, branching constraints of the following type are imposed: $x_{ij} \leq \lfloor \alpha \rfloor$ and $x_{ij} \geq \lceil \alpha \rceil$.

The model has symmetry, because different paths may correspond to the same cutting pattern. Reduction criteria are used to eliminate some arcs, reducing symmetry, but still keeping all the valid paths. After this operation, in instances with a small average number of items per bin, which happen to be rather difficult instances, the symmetry is low, and its undesirable effects are not so harmful. Computational results show the optimal solutions of all the bin packing instances of the OR-Library (see Beasley, 1990). These instances have demands of few items of each size. For example, the larger instances, of the $t501$ class, have about 200 different item sizes and a total of 501 items, yielding an average demand of about 2.5.

A different strategy is to use a master problem with the columns of the reformulated model of Gilmore–Gomory and branching constraints based on the arc-flow variables, which are explicitly added to the formulation (Ben Amor, 1997; Valério de Carvalho, 1999). This strategy, used in Alves and Valério de Carvalho (2003), eliminates symmetry and preserves the structure of the subproblem, as follows. Any column of Gilmore–Gomory's model involves a unique set of arc-flow variables, if we consider that items are placed by decreasing value of width. Branching constraints imposed on a given arc-flow will constrain the value of a definite set of columns of the reformulated model. Penalties and prizes

resulting from branching constraints of the type greater-than-or-equal-to
or less-than-or-equal-to, respectively, only affect the reduced cost of the
corresponding arc in the subproblem.

Let G^w and H^w be the sets of branching constraints corresponding
to the left and the right branches, respectively, at a given node w of
the branch-and-bound tree, and let $G^w_{(i,j)} \subseteq G^w$ and $H^w_{(i,j)} \subseteq H^w$ be
the sets of branching constraints imposed on the specific arc (i, j). Let
π_d, $d = 1, \ldots, m$, be the dual variables associated with the demand
constraints, and μ and ν the vectors of dual variables associated with the
branching constraints. The reduced cost of variable x_{ij} at node w is $\bar{c}_{ij} =$
$\pi_d - \sum_{l \in G^w_{(i,j)}} \mu_l + \sum_{l \in H^w_{(i,j)}} \nu_l$, where d is the item that corresponds to
arc (i, j). Only the costs change, and the subproblem structure remains
unchanged. It may be solved using dynamic programming as a knapsack
problem that only selects cutting patterns with items placed by non-
increasing width.

This branching scheme can be easily extended to branching constraints
based on sets of arcs incident on a given node. In this case, the dual
variable of a branching constraint acts on all arcs in the set. Nevertheless, the computational burden is heavier for instances with larger values
of roll widths.

MCF model

Integrality may be required equivalently either for variables x^k_{ij}, λ^k_p,
or λ_p. In the first two cases symmetry is very harmful to branching
scheme efficiency. Ben Amor (1997) proposed a branching scheme based
on aggregating arc flow variables

$$x_{ij} = \sum_{k=1}^{K} x^k_{ij}$$

that count the number of columns (rolls) using arc (i, j). These variables
must be integer for any integer solution to CSP. Hence, for any solution
such that $x_{ij} = \alpha$ is fractional for some arc (i, j) one creates two nodes by
adding either the constraint $x_{ij} \leq \lfloor \alpha \rfloor$ or the constraint $x_{ij} \geq \lceil \alpha \rceil$ to the
original formulation. The dual variables associated to these constraints
will affect the arc costs in the subproblem at each column generation
iteration and the subproblem structure remains unchanged. Since all
columns (patterns) are extreme points of the subproblem, any column
that is needed to attain integer optimality can be generated in this way.

However, because a node may be visited by more than one path in
an optimal integer solution, one can obtain a fractional solution (λ_p
fractional) while all x_{ij} are integer. In this case, there exists at least

one item i for which an associated node is visited more than once in the solution. A new item with demand equal to 1 is created while the demand of i is decreased by 1. The master problem covering constraints are modified following these changes. The worst case happens when all items are completely disaggregated and the problem becomes a BCSP where all items have demand equal to 1. In this case the branching scheme using aggregated flow variables is convergent since

$$x_{ij} \in \{0, 1\}, \quad \forall(i, j) \Leftrightarrow \lambda_p \in \{0, 1\}, \forall p.$$

This branching scheme preserves the structure of the subproblem as a constrained shortest path and decisions are directly enforced in the subproblem while the master problem size remains unchanged. Constraints of the type $x_{ij} = 0$ simply amount to removing the arc from the subproblem. On the other hand, the constraints of the type $x_{ij} = 1$ amount to removing all arcs out of node i and all arcs into node j except arc (i, j). It was successfully used to solve the vehicle routing problem with time windows using branch-and-price (Desrochers et al., 1992) and proved to be efficient for BCSP. The main difficulty in BCSP is the solution of the linear relaxation by column generation. But this may be done as efficiently as for the corresponding CSP by aggregating identical size items.

Branching on aggregated flow variables has many interesting properties. First, it allows the use of several score functions to choose the branching variable. Scores take into account the weights of items. Lower bounds and preprocessing procedure of Martello and Toth (1990) can be extended to the problems obtained at each branch-and-bound node. The experiments carried out show that no disaggregation has been necessary and the branching scheme turns out to be efficient for usually used test problems. Even for the binary case, branching on aggregated flow variables turns out to be very efficient. Also fixing several variables at once proved to be efficient in a depth first strategy and no backtracking has shown to be necessary. This is due to the fact that nearly all solved problems have the integer round-up property, i.e. the optimal integer value is obtained by rounding up the linear relaxation optimal value.

General comments

An efficient branching scheme for column generation should have a smaller tree than the one resulting from branching directly on λ-variables. The rules of Vanderbeck and Vance have the drawback of significantly modifying the subproblem structure. Moreover, the number of possible branching nodes is very high. Branching scheme based on MCF model has the advantage of preserving a constrained shortest path as subproblem. Possible disaggregation may lead to subproblem of larger size deeper in the branching tree which breaks the rule that deeper in the branching tree, subproblems and desirably master problems should become easier to solve. Compared to this method, the branching strategy based on the arc-flow model presents the advantage that no disaggregation is needed. Moreover, both strategies allow using more sophisticated branching rules based on x_{ij} variables. Besides all these intrinsic differences, all branching schemes lead to efficient solution of classical test problems. This is due to the fact that these problems have zero gap, the use of depth first strategy, and the use of several heuristics.

4.   Stabilization

Column generation processes are known to have a slow convergence and degeneracy difficulties. There are often large oscillations in the values of the dual variables from one iteration to the next. Primal degeneracy also arises: in many iterations, adding new columns to the restricted master problem does not help to improve the objective value. Recent computational experiments show that these problems can be mitigated using *dual-optimal inequalities* and *stabilization methods*.

From the dual standpoint, column-generation processes can be viewed as dual cutting-plane algorithms (Kelly, 1961), in which the restricted set of variables used to initialize the restricted master problem provides a first relaxation of the dual space. Clearly, better heuristics for the starting solution provide tighter relaxations. Then, at each iteration, dual feasibility cuts are added to the model to eliminate the previous undesired dual solution. The dual-space relaxation is successively tightened, until a feasible dual solution to the entire problem is found.

Consider the LP relaxation of the CSP, $\min\{cx \colon Ax = b, x \geq 0\}$, where the columns of A correspond to valid cutting patterns, whose dual is $\max\{\pi b \colon \pi A \leq c\}$. The following inequalities are a family of *dual-optimal inequalities* (Ben Amor et al., 2003) meaning that they are valid inequalities for the optimal dual space of the CSP. Any optimal dual solution will obey:

$$-\pi_i + \sum_{s \in S} \pi_s \leq 0, \quad \forall i, S, \tag{5.44}$$

for any given width w_i, and a corresponding set S of items, indexed by s, such that $\sum_{s \in S} w_s \leq w_i$. Intuitively, from the primal point of view, these columns mean that an item of a given size w_i can be split, and used to fulfill the demand of smaller orders, provided the sum of their widths is smaller than or equal to the initial size.

If we add at initialization time a set of dual-optimal inequalities to the dual problem, $\pi D \leq d$, we get the following primal-dual pair:

$$
\begin{array}{ll}
\min & cx + dy \\
\text{st} & \\
& Ax + Dy = b \\
& x, y \geq 0
\end{array}
\qquad \qquad
\begin{array}{ll}
\max & \pi b \\
\text{st} & \\
& \pi A \leq c \\
& \pi D \leq d.
\end{array}
$$

The motivation for the use of dual-optimal inequalities is the following: the dual space is restricted during all the column generation process to get a better convergence. From the primal point of view, new columns are added to the master problem. By combining these columns with already generated columns, the master problem implicitly considers patterns corresponding to columns that have not been generated yet by the subproblem. As a consequence, the primal objective decreases faster without the need to generate those columns. The added columns allow not only for covering items of set S in (5.44) with item i but also for implicitly taking into account columns resulting from the use of larger sets S. For example, combining two columns corresponding to pairs (S_1, i_1) and (S_2, i_2) with $|S_1| = |S_2| = 2$ and $i_2 \in S_1$ results in a column corresponding to the pair (S_3, i_1) where $S_3 = (S_1 \setminus \{i_2\}) \cup S_2$ has its cardinality equal to 3. Being taken into account implicitly, the corresponding column need not to be added to the master problem (see Ben Amor, 2002).

Even though the primal space is relaxed, it is always possible to retrieve a valid solution for the original CSP with the same cost. Obviously, it will be an optimal solution for the original problem. Valério de Carvalho (2003) provides an algorithm to drive the y variables to 0, when they take a positive value in the optimal solution, to achieve that purpose. Basically, the algorithm amounts to picking selected valid cutting patterns in the optimal solution, and to performing the splitting operation defined by the dual-optimal inequalities.

It is interesting to see that, in the space of the arc-flow variables, the dual-optimal inequalities correspond to cycles where exactly one arc (the one corresponding to item i in (5.44)) is traversed in the direction opposite to its orientation. Combining a path and a cycle produces a new valid path. Actually, the algorithm to retrieve a solution to the original problem is not needed, if a perturbation technique is used, that amounts to giving a cost of $\varepsilon > 0$ to the cycles (Ben Amor et al., 2003).

Valério de Carvalho (2003) used only dual-optimal inequalities obtained from sets S of small cardinality ($|S| \leq 2$). From the sets of cardinality 1, there are the inequalities $-\pi_i + \pi_{i+1} \leq 0$, $i = 1, \ldots, m-1$. From sets of cardinality 2, only one inequality of the type $-\pi_i + \pi_j + \pi_k \leq 0$, was used for each value of i. It is selected using the smallest index j (corresponds to largest width w_j), if such value exists, for which there is a k such that $w_i \geq w_j + w_k$. Therefore, the total number of dual-optimal inequalities is kept small (less than $2m$). Computational experiments show a sensible reduction in the number of columns generated and degenerate iterations. The savings are more impressive in larger, more difficult instances, when there is an explosion in the number of possible columns. For some instances, the speed-up factor is approximately 4.5, and the percentage of degenerate iterations falls from approximately 39.8% to about 8.5%. Ben Amor (2002) conducted a similar study on the classical test problems and another set of more difficult test problems. Results show an impressive reduction in the master problem cpu time and especially the number of column generation iterations.

Even better results can be obtained if one uses *deep dual-optimal inequalities* that cut portions of the dual optimal space, but preserve, at least, one dual optimal solution (Ben Amor et al., 2003). If a dual-optimal solution for the problem, $\tilde{\pi}^*$, is known in advance, the following stabilized primal and dual problems can be used:

$$
\begin{array}{ll}
\min & cx - (\tilde{\pi}^* - \Delta)y_1 + (\tilde{\pi}^* + \Delta)y_2 \\
\text{st} & Ax - y_1 + y_2 = b \\
& x \geq 0, \; y_1 \geq 0, \; y_2 \geq 0
\end{array}
\qquad
\begin{array}{ll}
\max & \pi b \\
\text{st} & \pi A \leq c \\
& \tilde{\pi}^* - \Delta \leq \pi \leq \tilde{\pi}^* + \Delta.
\end{array}
$$

where $\Delta > \mathbf{0} \in \mathbb{R}^m$.

The stabilized dual problem is constructed in such a way that the dual solution is restricted to a non-empty box strictly containing the known optimal dual solution. The stabilization method amounts to penalizing dual variables when they lie outside the predefined box, and enforces the selection of a valid optimal primal solution of the original problem (see Ben Amor, 2002). Computational experiments were run with instances, denoted as triplets (Beasley, 1990), because the optimal solution has exactly three items per bin, which fulfill exactly its capacity. If an instance of the CSP has no loss at optimality, the solution $\pi_i^* = w_i/W$, $i \in I$, is an optimal dual solution, because assigning these values to the dual constraints $\sum_{i \in I} a_{ip}\pi_i \leq 1$ simply replicates the knapsack constraint used to build the feasible patterns, and the corresponding dual objective function reaches the optimal value $\sum_{i=1}^m b_i w_i/W$ (Ben Amor, 1997). The computational results are impressive. The speed-up factor is approximately 10.

Computational results also show that convergence for different equivalent primal models is similar provided that their dual optimal spaces are equally restricted. Ben Amor et al. (2003) compares two models for the CSP problem: in the first, the aggregated CSP, items of the same size were aggregated in the same constraint, as is usually done, while, in the second, the binary disaggregated CSP, items are considered in separate constraints, but dual inequalities impose equal dual values for items of the same size. The number of column generation iterations, the master problems cpu times, and the number of columns generated in both cases is remarkably similar, even though the models have very different sizes of mater problems and subproblems.

Finally a proximal stabilized column generation algorithm proved to be very efficient in solving CSP linear relaxation (Ben Amor, 2002). The key issue is that the dual vector π^* defined above is a good initial solution for difficult problems (those ones with very small loss), and even it is not very close to a dual optimal solution for other problems, it may still have the nice property that the distribution of its components is close to the one of an optimal solution.

5.   Extensions

One extension of the CSP is the multiple size lengths cutting stock problem (MLCSP) and its counterpart, the variable sized bin packing problems (VSBPP). They are variants of the standard problem in which large objects of different capacities are allowed. The arc flow model can also be extended to formulate these problems. Using a Dantzig–Wolfe decomposition, one obtains the machine balance problem formulation of Gilmore–Gomory (Valério de Carvalho, 2002). A branch-and-price algorithm for a version with limited availability of the bins, based on a master problem with columns of the reformulated model and branching constraints based on the arc-flow variables, proved to be adequate for both the MLCSP and the VSBPP (Alves and Valério de Carvalho, 2003).

Again, the advantage of aggregating items of the same size into a single group, as well as of the aggregation of bins, reducing symmetry, enables solving in a few seconds all the VSBPP instances proposed in (Monaci, 2002), in which a combinatorial enumeration algorithms failed to solve 65 out of the 300 instances within a time limit of 900 seconds. The larger instances have a maximum of 5 different types of bins and up to 500 items. Other classes of instances with about 25 different item sizes and 14 different bin capacities were also solved in, on average, one second, approximately. The algorithm has been applied to MLCSP instances, proposed in (Belov, 2003), which uses a forbidden columns dynamic programming enumeration scheme, providing comparable results; it was able to solve 44 out of 50 instances with 4 different types of bins with capacities ranging from 5000 to 10000.

Another extension of the one-dimensional CSP is a version in which the number of setups is minimized. It is a problem of great practical importance, because there are usually significant setup cost associated to changing from one cutting pattern to another. This problem is much more complex than the classical one-dimensional CSP, because additional binary variables are needed to indicate when a given cutting pattern is selected, and the resulting model has much larger duality gaps. Therefore, most approaches are based on heuristics. Vanderbeck (2000a) developed a branch-and-price-and-cut algorithm, and applied it to instances with up to 32 item sizes. Computational results show that optimal solutions were obtained in 12 out of 16 instances, while in the remaining, solutions were found within one unit of optimality.

6.   Future research

The stabilization of the solution of the LP relaxation of the CSP produces impressive results. Instances with a much larger number of different item sizes can now be tackled, and their LP solutions and the corresponding bounds found in reasonable time. The application of similar ideas to the branch-and-price phase requires further investigation. The experience with the CSP also shows that it may be worthwhile to investigate and characterize the structure of the dual optimal space of other integer-programming and combinatorial-optimization problems.

Models with original variables provide additional insight, that can be used to derive more balanced and powerful branching rules, as the ones that result from hyperplane branching, and primal cuts expressed in terms of the original variables. Many integer programming and combinatorial optimization problems can be represented as pure network models, or network models with side constraints. Branch-and-price with branching constraints based on the variables of the original model seem to be a promising approach for this type of problems.

Acknowledgments

We thank two anonymous referees for their constructive comments, which led to a clearer presentation of the material. We are indebted to Jacques Desrosiers for the fruitful discussion that lead to some of the results we co-authored, and are cited in this paper. The second author was partially supported by FCT–Fundação para a Ciência e a Tecnologia (Projecto POSI/ 1999/ SRI/ 35568) and by Centro de Investigação Algoritmi da Universidade do Minho (Grupo de Engenharia Industrial e de Sistemas).

References

Ahuja, R., Magnanti, T., and Orlin, J. (1993). Network Flows: Theory, Algorithms and Applications. Prentice Hall, Englewood Cliffs, New Jersey.

Alves, C. and Valério de Carvalho, J.M. (2003). A stabilized branch-and-price algorithm for integer variable sized bin-packing problems. Technical Report, Universidade do Minho, Portugal; http://www.dps.uminho.pt/cad-dps.

Barnhart, C., Johnson, E.L., Nemhauser, G., Savelsbergh, M., and Vance, P. (1998). Branch-and-Price: Column generation for solving huge integer programs. *Operations Research*, 46:316–329.

Beasley, J.E. (1990). Or-library: Distributing test problems by electronic mail. *Journal of the Operational Research Society*, 41:1060–1072.

Belov, G. (2003). Problems, models and algorithms in one- and two-eimensional cutting. *Ph.D Thesis*, Fakultät Mathematik und Naturwissenschaften der Technischen, Universität Dresden, Dresden, Germany.

Ben Amor, H. (1997). Résolution du problème de découpe par génération de colonnes. *Master's Thesis*, École Polytechnique de Montréal, Canada.

Ben Amor, H. (2002). Stabilisation de l'algorithme de génération de colonnes. *Ph.D Thesis*, École Polytechnique de Montréal, Canada.

Ben Amor, H., Desrosiers, J., and Valério de Carvalho, J.M. (2003). Dual-optimal inequalities for stabilized column generation. *Les Cahiers du GERAD* G-2003-20, HEC, Montréal, Canada.

Degraeve, Z. and Peeters, M. (2003). Optimal integer solutions to industrial cutting-stock problems: Part 2, benchmark results. *INFORMS Journal on Computing*, 15:58–81.

Degraeve, Z. and Schrage, L. (1999). Optimal integer solutions to industrial cutting stock problems. *INFORMS Journal on Computing*, 11:406–419.

Desaulniers, G., Desrosiers, J., Ioachim, I., Solomon, M.M., Soumis, F., and Villeneuve, D. (1998). A unified framework for deterministic time constrained vehicle routing and crew scheduling problems. In: *Fleet Management and Logistics* (Crainic, T.G. and Laporte, G., eds.), pp. 57–93. Kluwer, Norwell, MA.

Desrochers, M., Desrosiers, J., and Solomon, M. (1992). A new optimization algorithm for the vehicle routing problem with time windows. *Operations Research*, 40:342–354.

Desrosiers, J., Dumas, Y., Solomon, M., and Soumis, F. (1995). Time constrained routing and scheduling. In: *Handbooks in Operations Research & Management Science 8, Network Routing*, pp. 35–139. Elsevier Science B. V.

Dowsland, K.A. and Dowsland, W.B. (1992). Packing problems. *European Journal of Operational Research*, 56:2–14.

Dyckhoff, H. (1990). A typology of cutting and packing problems. *European Journal of Operational Research*, 44:145–159.

Dyckhoff, H. and Finke, U. (1992). Cutting and Packing in Production and Distribution: a typology and bibliography. Physica-Verlag, Heidelberg.

Dyckhoff, H., Scheithauer, G., and Terno, J. (1997). Cutting and packing. In: *Annotated Bibliographies in Combinatorial Optimization*, pp. 393–413. John Wiley and Sons, Chichester.

Farley, A. (1990). A note on bounding a class of linear programming problems, including cutting stock problems. *Operations Research*, 38:992–993.

Gilmore, P. (1979). Cutting stock, linear programming, knapsacking, dynamic programming and integer programming, some interconnections. In: *Annals of Discrete Mathematics 4*, pp. 217–236. North Holland, Amsterdam.

Gilmore, P. and Gomory, R. (1961). A linear programming approach to the cutting stock problem. *Operations Research*, 9:849–859.

Gilmore, P. and Gomory, R. (1963). A linear programming approach to the cutting stock problem–Part 2. *Operations Research*, 11:863–888.

Goulimis, C. (1990). Optimal solutions to the cutting stock problem. *European Journal of Operational Research*, 44:197–208.

Helmberg, C. (1995). Cutting aluminum coils with high lengths variabilities. *Annals of Operations Research*, 57:175–189.

Kantorovich, L. (1960). Mathematical methods of organising and planning production (translated from a report in Russian, dated 1939). *Management Science*, 6:366–422.

Kelley Jr., J.E. (1961). The cutting-plane method for solving convex programs. *J. Soc. Ind. Appl. Math.*, 8(4):703–712.

Marcotte, O. (1985). The cutting stock problem and integer rounding. *Mathematical Programming*, 33:82–92.

Marcotte, O. (1986). An instance of the cutting stock problem for which the rounding property does not hold. *Operations Research Letters*, 4:239–243.

Martello, S. and Toth, P. (1990). Knapsack Problems: Algorithms and Computer Implementations. Wiley, New York.

Monaci, M. (2002). Algorithms for packing and scheduling problems. *Ph.D Thesis*, Università Degli Studi di Bologna, Bologna, Italy.

Rietz, J. and Scheithauer, G. (2002). Tighter bounds for the gap and non-IRUP constructions in the one-dimensional cutting stock problem. *Optimization*, 51(6):927–963.

Scheithauer, G. and Terno, J. (1995). The modified integer round-up property of the one-dimensional cutting stock problem. *European Journal of Operational Research*, 84:562–571.

Scholl, P., Klein, R., and Juergens, C. (1997). Bison: a fast hybrid procedure for exactly solving the one-dimensional bin-packing problem. *Computers and Operations Research*, 24:627–645.

Sessions, J., Layton, R., and Guanda, L. (1988). Improving tree bucking decisions: A network approach. *The Compiler*, 6:5–9.

Sessions, J., Olsen, E., and Garland, J. (1989). Tree bucking for optimal stand value with log allocation constraints. *Forest World*, 35:271–276.

Shapiro, J. (1968). Dynamic programming algorithms for the integer programming problem. I: The integer programming problem viewed as a knapsack type problem. *Operations Research*, 16:103–121.

Simchi-Levi, D. (1994). New worst-case results for the bin-packing problem. *Naval Research Logistics*, 41:579–585.

Stadtler, H. (1990). One–dimensional cutting stock problem in the aluminum industry and its solution. *European Journal of Operational Research*, 44:209–223.

Sweeney, P. and Paternoster, E. (1992). Cutting and packing problems: a categorized, application-orientated research bibliography. *Journal of Operational Research Society*, 43:691–706.

Valério de Carvalho, J.M. (1999). Exact solution of bin-packing problems using column generation and branch-and-bound. *Annals of Operational Research*, 86:629–659.

Valério de Carvalho, J.M. (2002). LP models for bin-packing and cutting stock problems. *European Journal of Operational Research*, 141(2):253–273.

Valério de Carvalho, J.M. (2003). Using extra dual cuts to accelerate column generation. Forthcoming in: *INFORMS Journal on Computing*.

Valério de Carvalho, J.M. and Guimarães Rodrigues, A.J. (1995). An LP based approach to a two–phase cutting stock problem. *European Journal of Operational Research*, 84:580–589.

Vance, P. (1998). Branch-and-Price algorithms for the one-dimensional cutting stock problem. *Computational Optimization and Applications*, 9:211–228.

Vanderbeck, F. (1999). Computational study of a column generation algorithm for binpacking and cutting stock problems. *Mathematical Programming*, Serie A, 86:565–594.

Vanderbeck, F. (2000a). Exact algorithm for minimizing the number of setups in the one-dimensional cutting stock problem. *Operations Research*, 48:915–926.

Vanderbeck, F. (2000b). On Dantzig–Wolfe decomposition in integer programming and ways to perform branching in a branch-and-price algorithm. *Operations Research*, 48:111–128.

Wäscher, G. and Gau, T. (1996). Heuristics for the integer one-dimensional cutting stock problem: a computational study. *OR Spektrum*, 18:131–144.

Chapter 6

LARGE-SCALE MODELS IN THE AIRLINE INDUSTRY

Diego Klabjan

Abstract   Operations research models are widely used in the airline industry. By using sophisticated optimization models and algorithms many airlines are able to improve profitability. In this paper we review these models and the underlying solution methodologies. We focus on models involving strategic business processes as well as operational processes. The former models include schedule design and fleeting, aircraft routing, and crew scheduling, while the latter models cope with irregular operations.

1.   Introduction

In the United States the Airline Deregulation Act of 1978 gave the airlines much more commercial freedom to compete. Since then, to leverage demand with capacity or sit inventory, the airlines have pioneered revenue management. Among other breakthroughs, to offer a variety of itineraries, major airlines have developed the so-called *hub-and-spoke* networks. On the other hand, to improve profitability they use sophisticated tools for reducing cost. In recent years, the raise of low-fare, no-frill airlines such as Southwest in the U.S. and Ryanair in Europe put additional pressure on the remaining carriers. To keep low fares, the airlines must maintain low cost per airline-sit-mile. This is commonly achievable through contract renegotiations and by using enhanced modeling and optimization techniques.

Since the 1950s the airlines are using operations research models in solving their complex planning and operational problems. These models have become increasingly complex. On the one hand, the airlines have become larger (through mergers or expending service) resulting into large-scale models. On the other hand, the continuing pressure to increase profitability resulted into more "accurate" models and better solution methodologies. For example, an excess crew cost of several percent was acceptable a decade ago but it is not today. In many cases, by using state-of-the-art crew scheduling decision support systems the excess cost has been pushed below one percent. Large-scale models have become computationally tractable due to algorithm, hardware, and software advances.

On the algorithmic front the most notable advance has been the introduction of *column generation*. In column generation, a model is given implicitly and is dynamically updated in order to improve the incumbent solution. Such an approach enables handling of the entire large-scale problem and at the same time it reduces the computational burden.

In this paper we review large-scale linear mixed integer models that are frequently encountered and used in the airline industry. We also outline in Section 2 the underlying methodologies for solving these models. We start in Section 3, by explaining business processes in airline planning and operations. In Section 4 we present models that concern the passenger service. Models for schedule planning and fleeting are given in Section 4.1, then we review aircraft scheduling in Section 4.2, and at the end we discuss crew scheduling in Section 4.3. For every problem we discuss planning and operational models. Recent trends are presented in Section 5.

2.   Solution methodologies for large-scale models

Here we briefly overview three most common techniques for solving large-scale linear mixed integer models: branch-and-price, Lagrangian relaxation, and Benders decomposition. We start with branch-and-price.

Large-scale linear programs are often solved by *delayed column generation*. In this algorithm, at every iteration, only a subset of columns is considered. The problem with only a subset of columns is called the restricted master problem. In every iteration of the algorithm, first the restricted master problem is solved and let π be the optimal dual vector, which for ease of discuss we assume it exists. Next the so called *subproblem* is solved. In subproblem solving we identify a set S of columns with the lowest reduced cost with respect to π. If we cannot find a column with negative reduced cost, then we stop since π is an optimal dual solution to the original problem and together with the optimal primal solution to the restricted master problem we have an optimal primal/dual pair. Otherwise, we append columns in S to the restricted master problem and the entire procedure is iterated. When the restricted master problem includes too many columns after several iterations, columns with large reduced cost are removed from the restricted master problem.

Frequently the most computationally intensive step in delayed column generation is subproblem solving since it needs to scan many columns and typically it is a complex task to generate a single one. When columns correspond to constrained paths in a network, an efficient algorithm known as *constrained shortest path* is often employed, (Desrosiers et al., 1995; Desaulniers et al., 1998). In this case, the task is to find the cheapest cost $s - t$ path (reduced cost in delayed column generation framework) among all paths with certain properties. We explain the algorithm by an example. Assume we want to find a shortest path with respect to the reduced cost in a network subject to the duration and the number of arcs in paths being below a given number. By duration we mean that every arc has an associated transit time and the duration of a path is the sum of transit times along the path. We introduce label vectors, which in this case have 3 coordinates. The first one corresponds to the reduced cost, the second one to the duration, and the last one to the number of arcs. With every node we associate a set of label vectors. For example, a label vector $(-45, 134, 4)$ at node i corresponds to an $s-i$ path with reduced cost -45, duration 134, and 4 arcs. The constrained shortest path algorithm uses the same framework as standard shortest path algorithms. Suppose the algorithm selects a node i for scanning. The constrained shortest path algorithm next scans all neighbors j of i and all label vectors $k = (k_1, k_2, k_3)$ at i. Each label vector k is updated by traversing the arc (i, j) and the updated label vector is appended to node j. In our example, the label update means that the new label vector has $k_3 + 1$ as the third component, k_2 plus the transit time of arc (i, j) as the second component, and k_1 plus the 'reduced cost' of arc (i, j) as the first component. The key observation is that under some realistic assumptions label vectors that are dominated can be discarded. If we have two label vectors $\overline{k}, \widetilde{k}$ at node j and $\overline{k} \leq \widetilde{k}$ component-wise, then the $s - j$ path corresponding to $\widetilde{k}$ is not going to be part of the shortest path. The efficiency of the algorithm depends heavily on the frequency at which dominance occurs. Note that if there is no dominance, the algorithm simply enumerates all paths. It turns out that dominance occurs often in practice and therefore the algorithm is computationally efficient. Two alternative algorithms for subproblem solving in presence of constrained shortest paths are sketched in Section 4.3.

Branch-and-price is a branch-and-bound algorithm, where LP relaxations at every node are solved by delayed column generation. Since subproblems are often combinatorial in nature, the standard variable dichotomy is not appropriate. When columns correspond to constrained paths in a network, the following branching strategy is frequently used. Let r, s be two adjacent nodes, which are selected based on the incumbent LP solution. Then in one branch only paths where node s immediately follows node r are considered. This is easily reflected in the network by removing all arcs from r, except the (r, s) arc. On the other branch we forbid all paths where s does follow r, which is captured by removing the (r, s) arc. This branching rule produces more balanced tree. Since LP relaxations tend to be computationally intensive, only few branch-and-bound nodes are evaluated. For this reason, a common strategy is to use depth-first search and abort after the first integer solution is obtained. An excellent survey on branch-and-price is given by Barnhart et al. (1998a).

Another common technique is *Lagrangian relaxation*, (Geoffrion, 1974; Fisher, 1981, 1985; Martin, 1999). Suppose we can partition constraints into "easy" and "difficult" constraints. The concept behind is that if the difficult constraints are removed, the resulting problem is easily solvable. In Lagrangian relaxation, every difficult constraint gets a linear penalty and it is moved to the objective function. The resulting problem is called the Lagrangian relaxation and it is a function of the penalties. Let us assume that we have a maximization problem. For any given values of penalties, the Lagrangian relaxation is computationally easy. It is easy to see that it always provides an upper bound on the optimal solution. The goal now is to find the best upper bound, i.e. to minimize the Lagrangian relaxation over all possible penalties. This is the Lagrangian dual problem, which is a nonlinear optimization problem. In practice it is solved by variants of the subgradient algorithm. One drawback of this approach is that there is no guarantee to find feasible solutions. They have to be constructed heuristically during the execution of the subgradient algorithm. The algorithm is very appealing since it is easy to implement and it handles complex (difficult) side constraints.

The *Benders decomposition*, (Benders, 1962; Minoux, 1986), is well suited for mixed integer programs with linking integer variables. It requires that for any fixed value of integer variables, the resulting problem is an LP, where the constraint matrix is often block diagonal. The algorithm at every iteration solves a mixed integer program (restricted master problem) with a single continuous variable that provides a bound on the optimal solution. Next the linear program resulting from the original problem by fixing integer variables to the values from the restricted master problem is solved. The optimal dual vector to this LP provides a Benders cut, which is added to the restricted master problem and the procedure is repeated. The same framework can be used for convex problems.

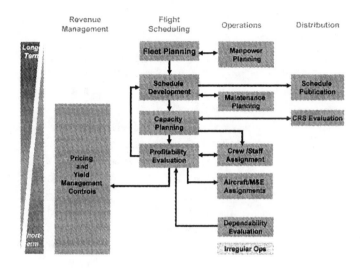

Figure 6.1.   Business processes (M&E=maintenance and engineering).

Another technique, called constraint programming, is paving its way to the large-scale linear programming. In this chapter we do not discuss constraint programming and relevant literature.

3.   Airline planning and day of operations

In this section we review typical business processes used by combination airlines, Figure 6.1. While every airline has its own processes and its own organization names, most of the airlines follow the depicted processes and terminology. The time frames can very significantly.

Long term fleet and manpower planning consists of making strategic decisions with respect to the number of aircraft and the fleet decomposition, and cockpit crew manpower planning. In fleet planning considerations such as the airline's mission (e.g., Southwest has a single fleet that allows relatively simple and efficient operations), aircraft utilization, route structure, cargo/passenger mix, etc., are taken into account.

The schedule development phase typically starts 12 months before the day of operations and it lasts up to 9 months. In the first phase the airline establishes the *service plan*, which is the set of services to operate in a given market. The service plan is either daily for domestic operations and weekly for international, long-haul service. The marketing group considers several factors such as traffic forecasts, status of competing carriers, internal resources, and marketing initiatives. Marketing initiatives are approved by upper management and involve decisions such as entering a new market. The designed service plan typically does not divert substantially from the current schedule. Following the service plan, the scheduling group generates a detailed *flight schedule*, i.e. a flight departure and arrival time. The flight schedule has to obey a set of operating constraints, e.g. maintenance planning, and given generic resources such as the number of aircraft. The schedule is then published. Next is capacity planning or fleeting. In fleeting an equipment type is assigned to each flight subject to available resources such as the number of aircraft. The goal of the *fleet assignment model* is to maximize profit. The schedule together with the sit capacity is then input to the computer reservation system (CRS). The produced fleeting solution is then evaluated with a profitability evaluation model and potential improvements are fed back to the schedule development group for possible minor adjustments. The schedule development phase and fleeting are discussed in Section 4.1.

Once the equipment types are assigned, *aircraft routing* and *crew scheduling* follow. In aircraft routing, called also maintenance routing, which is discussed in Section 4.2, a specific tail number is assigned to each flight subject to maintenance constraints. The objectives are usually incentives such as throughs and robustness. The goal of crew scheduling, see Section 4.3, is to assign crew members to individual flights in order to minimize the crew cost and maximize various objectives related to contractual obligations, quality of life, and crew satisfaction. Crew schedules have to satisfy complex regulatory and contractual rules. Potentially crew planners detect unfavorable connections and give feedback to schedule and fleet planners. The crew scheduling process typically starts three months before the day of operation and it is constantly updated until a few weeks before the day of operations.

Only minor changes to fleeting, aircraft routes and crew schedules are made during the last few weeks before the day of operations. To better match demand with capacity, some airlines perform dynamic fleet and aircraft swaps, known also as demand driven dispatch or D^3 for short, (Berge and Hopperstad, 1993; Talluri, 1996; Clark, 2000). If preferential bidding is used, approximately one month before the day of operations, crews bid for their monthly crew assignments and only minor changes such as two way trip swaps are performed in the last few weeks.

Throughout the strategic planning processes pricing and yield or revenue management are actively involved. In revenue management, the airline controls the sit inventory by adjusting fare prices, setting overbooking limits, and making decisions at any given time about selling particular fare classes on a given passenger itinerary. Since models and solution methodologies in revenue management and pricing are substantially different than the remaining models resulting from the aforementioned processes, they are not discussed here (see e.g. van Ryzin and Talluri, 2002, for a survey on revenue management). We also do not discuss the cargo side of planning and operations.

The actual day of operations, called also *execution scheduling*, consists of making final minor adjustments to the flight schedule (e.g., adjust arrival times based on the daily wind forecast), executing the pre-planned schedule (e.g., file the flight plan) and rescheduling for irregular operations or *disruption management*. The latter is carried out by operations controllers, which are typically located in the airline operations control center (see e.g. Clarke et al., 2002). Most frequent sources of irregular operations are weather, unscheduled maintenance, congestion, crew unavailability, security problems, etc. Disruption management is composed of three processes. When an irregular operation occurs, first the aircraft are rerouted, which is called *aircraft recovery*. In this stage in addition to rerouting the aircraft, decisions on delaying and canceling flights are made. Next is the *crew recovery process*, where crews are assigned new crew itineraries. The controllers can use original, standby, and reserve crews. At the end is the *passenger reaccommodation process*, where passengers are rerouted to alternative itineraries. Clearly the new schedule must conform to all regulatory and contractual rules. While the airlines often impose more stringent rules in planning, in operations they typically use precise rules. Contractual rules for operations are usually different from those in planning.

4.   Models for passenger service

In this section we focus on the passenger side of planning and operations.

4.1   Schedule planning and fleeting

For most of the airlines schedule planning is a manual process mostly driven by marketing requirements. On the other hand, decision support tools for fleeting are common. There are only few manuscripts on schedule planning but there is vast literature on fleeting. Since research papers that address schedule planning, cover fleeting as well, we start with fleeting.

The basic *fleet assignment model* (FAM), called also the leg-based fleet assignment model, is to find an optimal assignment of equipment types to flights. The input consists of a list of flights, which are given by the destination/origin station (airport) and departure/arrival time, a set of equipment types and the corresponding number of aircraft for each equipment type. Since each equipment type has its own sit capacity, on a given flight the equipment type decision can produce low load factor (lost revenue of using too large sit inventory) or a potential spill of passengers to competitors if the realized demand is higher than the sit capacity of the assigned equipment type. The typical objective function consists of the variable and fixed cost of operating a flight by a given equipment type and an estimate of potential revenue.

Next we formally describe the FAM, see e.g. Abara (1989); Hane et al. (1995). First we define the *flight time-space network*. The network has a node (u, i) for each time when an arrival or departure of leg i occurs at station u. If an event corresponds to a departure, then let t_i be the departure time of flight i. If it corresponds to an arrival, then t_i is the arrival time of flight i plus the minimum plane turn time (the so-called ready time). We assume that the activity times t_i are ordered in time, i.e. $t_1 \leq t_2 \leq t_3 \cdots \leq t_l$, where l is the number of activities at the station. There is a *flight arc* $\{(u, i), (v, j)\}$ for each leg that departs at station u at time t_i and arrives at station v at time t_j. In addition there are *ground arcs* $\{(u, i), (u, i+1)\}$ for each u and i, where we assume that we have a wraparound arc between the last and first node of the day. Each station has exactly one wraparound arc.

The model has two types of variables, the fleet assignment variables x and the ground arc variables y. For each leg i and for each fleet k there is a binary variable x_{ik}, which is 1 if and only if leg i is assigned to fleet k. For each ground arc g and for each fleet k we define a nonnegative variable y_{gk} that counts the number of planes in fleet k on the ground in the time interval corresponding to g. Let et be a fixed time typically corresponding to a time with low activity, e.g. 3 am. The FAM model reads

$$\min \sum_{\substack{i \in A \\ k \in K}} c_{ik} x_{ik}$$

$$\sum_{k \in K} x_{ik} = 1 \quad i \in A \tag{6.1}$$

$$\sum_{i \in O(v)} x_{ik} - \sum_{i \in I(v)} x_{ik} + y_{o(v)k} - y_{i(v)k} = 0 \quad v \in V, \ k \in K \tag{6.2}$$

$$\sum_{g \in W} y_{gk} + \sum_{i \in M} x_{ik} \leq b_k \quad k \in K \tag{6.3}$$

$$y \geq 0, \quad x \text{ binary},$$

where

$I(v)$: set of flight arcs to node v $\qquad A$: set of all flight arcs

$O(v)$: set of flight arcs from node v $\quad V$: set of nodes

M : set of flights in the air at *et* $\qquad K$: set of all fleets

W : set of ground arcs containing *et* $o(v)$: ground arc from node v

b_k : number of aircraft in fleet k $\qquad i(v)$: ground arc to node v

c_{ik} : cost of assigning fleet k to leg i.

Constraints (6.1) require that each leg is assigned to exactly one fleet, (6.2) express the flow conservation of aircraft, and (6.3) assure that we do not use more aircraft than there are in a fleet.

This basic FAM model is relatively easily solvable even for large flight networks by commercial integer programming solvers, (Hane et al., 1995). The model can be enhanced by incorporating some aircraft maintenance and crew requirements, (Clarke et al., 1996; Rushmeier and Kontogiorgis, 1997), explicitly modeling aircraft routes, (Barnhart et al., 1998b), and incorporating departure time decisions, (Rexing V, 2000; Desaulniers et al., 1997b; Bélanger et al., 2003). The biggest drawback of this model is the revenue component of the objective function. In a multi-leg passenger itinerary a capacity decision on a flight effects the number of passengers spilled from the itinerary and therefore the revenue contribution of other flights. Kniker (1998) explores several alternatives to compute the cost component c but none of them captures *network effects* accurately. Therefore the model has to be augmented to capture multi-leg passenger itineraries.

For ease of discussion we assume that every passenger itinerary has a single fare and we assume that passengers are not recaptured, i.e. if the booking demand exceeds the sit inventory on a given flight, the non-booked passengers are not captured on airline's alternative itineraries. Under these assumptions we next present the *passenger mix model*, which decides how many booked passengers to have in any itinerary given a fixed leg sit inventory. Let P be the set of all itineraries. The fare of itinerary $p \in P$ is denoted by f_p and let C_i be the available sit inventory of leg i. Let w_p, $p \in P$ be the decision variable that counts the number of booked passengers on itinerary p. The model for the optimal number of booked passenger reads

$$\max \sum_{p \in P} f_p w_p$$

$$\sum_{i \in p} w_p \leq C_i \quad \text{for every leg } i \qquad (6.4)$$

$$w_p \leq D_p \quad p \in P \qquad (6.5)$$

$$w \text{ integer.}$$

Here $i \in p$ represents that leg i is part of itinerary p. For every $p \in P$, the unconstrained demand is denoted by D_p and it can be obtained either by a direct O − D (origin-destination) forecasting method or by segregating leg based demand forecasts. Constraints (6.4) impose the sit capacity limits and (6.5) meet the forecasted demand. Enhancements and generalizations of this model are given in Kniker (1998).

A fleet assignment model that captures O − D itineraries, called the *origin-destination fleet assignment model*, is obtained by combining the leg based FAM and the passenger mix model. The only required modification is to replace the right-hand side of (6.4) by $\sum_{k \in K} \widetilde{C}_k x_{ik}$, where $\widetilde{C}_k$ is the sit capacity of equipment type k. While the leg based FAM is relatively easily solvable, this is not the case for the O − D fleeting model. The number of variables and therefore constraints in this model can be as high as 200,000. (Note that in the presence of multiple fare classes per itinerary, (6.5) are no longer the simple upper bound constraints.)

Barnhart et al. (2002b); Kniker (1998) solve the model by branch-and-price. The pricing step is not computationally intensive and it is done by a simple scan routine, i.e. there is no need for solving the constrained shortest path problem. They report computational times of several hours just to find the first integer solution. Indeed, even solving the LP relaxation of the model takes 2 hours and half for a realistic model consisting of 70,000 itineraries and approximately 2,000 legs. The authors enhance the solution methodology by employing sophisticated preprocessing techniques and valid inequalities. In order to improve tractability, Barnhart et al. (2002a) develop an alternative model. Instead of having decision variables that assign single legs to a fleet, the new model requires decision variables that assign a subset of legs to a fleet. Thus the assignment flight leg variables are grouped together. Clearly considering all possible subsets of legs is not tractable, however, the authors show that by carefully selecting subsets, the resulting model is tractable. Another alternative formulation to O − D fleeting is presented in Jacobs et al. (1999), where the underlying model is solved by Benders decomposition.

Next we discuss models that incorporate schedule design decisions. Until recently, algorithms for schedule design and fleeting were mostly iterative in nature, (Berge, 1994; Marsten et al., 1996; Etschmaier and Mathaisel, 1985). Given a schedule, first demands are estimated by a schedule evaluation model. Next the FAM is solved by using the computed demands and the resulting solution is evaluated. In order to modify the schedule, flights for addition or deletion are identified. The profit resulting from addition or deletion of these flights is then estimated and based on the resulting profit a subset of these flights is selected for addition and deletion. These flights are then added to the schedule and the procedure is repeated.

Recently models that consider fleeting or aircraft routes, and schedule design decisions simultaneously emerged. Most of them are still iterative as they dynamically generate passenger itineraries and evaluate schedules, however, given a subset of itineraries, the decision of which flights to use and the fleeting decision are made simultaneously. Lettovský et al. (1999) give a model that can construct a schedule from scratch. As part of the input are service frequencies, origin-point of presence, and demand information. The model then generates a schedule that maximizes revenue subject to basic operational constraints. Their objective function is nonlinear since they use a logit-based market-share model. They solve the model by Benders decomposition (no details are given in their publication). Lohatepanont and Barnhart (2002) present a linear model that given a set of mandatory and optional flights, selects a subset of optional flights that maximizes total revenue. They use the O − D fleeting model and augment it with optional flights. The nonlinear relation in flight demands is taken into account by solving several models iteratively and adjusting demands based on the incumbent solution. In each iteration the model is solved by branch-and-price. Yan and Tseng (2002); Yan and Wang (2001) present a similar model but they solve it by Lagrangian relaxation. They relax all constraints except flow conservation constraints of passenger and aircraft. Erdmann et al. (2001) present an approach for scheduling flights of charter carriers. They go a step further since they explicitly model aircraft routes. They solve the model by branch-and-price.

Antes (1997) presents common business processes used by the airlines in schedule planning. Berdy (2002) gives an excellent review on nuts-and-bolts of route generation. These two manuscripts do not present mathematical models.

4.2   Aircraft routing

4.2.1   The planning stage.

In tactical planning after each flight has an assigned equipment type, the aircraft routing problem follows. In this stage each individual aircraft or *tail number* is assigned to each flight in a given time period. Note that the fleeting solution decomposes the flight schedule and therefore there is an aircraft routing problem for every fleet (e.g. Boeing 737–300 and 737–400 fleets yield two separate routing problems).

In addition to the assignment requirement that each flight must be assigned a unique tail number, the routes should not use more than the available number of aircraft and they must meet maintenance requirements. In the U.S., the FAA requires four types of checks. The A-checks or line maintenance are routine checks (visual inspection of major systems), which have to be performed approximately every 65 block hours and a certain number of take-offs. Durations of A-checks are typically from 3 to 10 hours and they are usually performed during the night. B-checks are typically done once in several months and they require detailed visual inspection. For C- and D-checks an aircraft is taken out of service for a month and they are done once every one to four years. Since these two check are spaced at large intervals, they do not pose scheduling difficulties. For this reason aircraft routing solutions consider only A- and B-checks. Maintenance checks can only be performed at specific maintenance stations, which are typically separate for each fleet. In order to decrease unscheduled maintenance events, many airlines impose more stringent maintenance requirements, e.g. A-checks every 40 block hours and even frequent more stringent checks. In addition to these regulatory maintenance rules, some airlines impose equal utilization of aircraft, called also the big cycle constraint.

It is extremely difficult to assign a single cost attribute to an aircraft routing solution. Some airlines consider the routing problem as a pure feasibility problem. Often a value of a routing solution is a weighted sum of several attributes such as the contribution from throughs (the benefit of offering certain non-stop connections) and robustness measures to possibly decrease occurrences of unexpected events.

In the planning stage, typically several weeks or months in advance, first generic aircraft routes are constructed during a rolling time horizon. This is the *aircraft rotation problem*. These generic routes satisfy short maintenance requirements such as A-checks but do not consider, for example, B-checks and aircraft positions at the beginning of the time horizon. Only a few weeks or even days before the day of operations, the actual tail numbers are assigned to each flights, i.e. the *aircraft assign-*

ment problem is solved. The assignment follows generic routes as much as possible but it takes into account longer maintenance requirements and the actual aircraft position at the beginning of the horizon.

Both problems are modeled either as a multicommodity flow problem or a partitioning/packing problem. Next we present a partitioning formulation from Barnhart et al. (1998b) for the rotation problem, which assumes a rolling time horizon and only checks that have to be done periodically and the period is shorter than the time horizon.

Suppose we are given a flight schedule (of a single equipment type) in a time horizon. A *string* is an ordered sequence of flights that originates and terminates at a maintenance station. The arrival station of a flight in a string is equal to the departure station of the next flight in the string and the connection times are longer than or equal to the minimum plane turn time. In addition, a string is maintenance feasible, e.g. the sum of the block times of the flights in the string is less than the one imposed by A-checks and the number of flights in a string is less than the maximum number of takeoffs between two A-checks. An *augmented string* is a string with the maintenance time interval attached to the end of it. The maintenance is assumed to start as soon as possible and it lasts for the duration of the required check. For example, if an aircraft arrives at 4pm local time and the maintenance cannot start before 8pm, it is assumed that the maintenance indeed starts at 8pm and lasts for the required period. Let S be the set of all augmented strings. A decision variable x_s is 1 if augmented string $s \in S$ is in an aircraft route and 0 otherwise. To combine augmented strings together, we need ground arc variables at maintenance stations MS, which is similar to the FAM. As in Section 4.1, we define ground arcs $y_{i(v)}, y_{o(v)}$ for $v \in V$. V corresponds to activities at stations in MS and ready time is defined based on the termination time of augmented strings. The model reads

$$\min \sum_{s \in S} c_s x_s$$

$$\sum_{i \in s} x_s = 1 \quad \text{for every flight } i \tag{6.6}$$

$$\sum_{\substack{j \in O(v) \\ j \in s}} x_s - \sum_{\substack{\in I(v) \\ j \in s}} x_s + y_{o(v)} - y_{i(v)} = 0 \quad v \in V \tag{6.7}$$

$$\sum_{g \in W} y_g + \sum_{s \in S} r_s x_s \leq b \tag{6.8}$$

$$y \geq 0, \quad x \text{ binary}.$$

Here c_s is the cost of augmented string s, b is the number of aircraft in the fleet, and r_s counts how many times augmented string s crosses time et, where et is defined as in the FAM. Constraints (6.6) require that each flight be assigned to a string, flow balance at maintenance stations is guaranteed by (6.7), and (6.8) is the plane count constraint. Note that due to the flow balance constraints, strings can always be concatenated into an aircraft rotation. The big cycle constraint can be modeled in the similar way as the subtour elimination constraints in the traveling salesman problem (see Barnhart et al., 1998b, for details). Additional constraints such as capacities at the maintenance stations can easily be embedded. Barnhart et al. (1998b) solve this model by branch-and-price. The subproblem is solved by a constrained shortest path algorithm. For every maintenance requirement there is a label, i.e. we must maintain a label for block hours and number of takeoffs, and we must use labels for any nonlinear cost component. If the big cycle constraint is imposed, then row generation is required as well since this implies an exponential number of additional constraints.

Cordeau et al. (2001); Mercier et al. (2003) model the aircraft assignment problem as a multicommodity network flow with nonlinear resource constraints. The resource constraints model maintenance requirements. The model is solved by a combination of Benders decomposition and branch-and-price. The pricing problem is solved by a constrained shortest path algorithm. Sriram and Haghani (2003) use the multicommodity formulation as well. They model maintenance requirements as linear constraints and therefore their formulation is very complex. The solution methodology is a heuristic based on local search.

Clarke et al. (1997) consider the aircraft rotation problem. They modeled it as an Eulerian tour with side constraints. The side constraints capture maintenance requirements. Since the Eulerian tour problem is equivalent to the traveling salesman problem on the line graph, they actually solve the traveling salesman problem. This transformation enables them to capture the big cycle constraint as the subtour elimination constraints. The model is solved by Lagrangian relaxation, where the maintenance and subtour constraints are relaxed. The underlying master problem then becomes a simple assignment problem.

Feo and Bard (1989); Daskin and Panayoyopoulos (1989) model the assignment problem as the set partitioning problem. In such a formulation each aircraft route corresponds to a column in the formulation. The former work solves the underlying model heuristically. They first generate a set of routes for each aircraft independently. Next they solve the resulting partitioning problem by a greedy heuristic to obtain the solution. Daskin and Panayoyopoulos (1989) rewrite the formulation as a set packing model. One family of constraints require that each flight is in a route and the other one that each route is selected at most once. They solve the model by Lagrangian relaxation, where the latter constraints are relaxed.

Paoletti et al. (2000) give details on aircraft rotation and assignment at Alitalia. The rotation problem is solved as an assignment problem, where maintenance requirements are not considered. They maximize the throughs value and the aircraft turn times. The assignment problem is solved a day before the day of operations and is considerably more complex. It tries to follow the solution from the assignment problem as much as possible. Their model is string based but it has several additional operational constraints. They employ a constraint programming approach.

A completely different framework is given by Gopalan and Talluri (1998); Talluri (1996). They approach the rotation problem from a combinatorial point of view. They model the problem as the Eulerian tour problem. The former work considers 3-day maintenance checks and they show that if only these checks are required, the problem is polynomially solvable. The 4-day checks are addressed in the latter manuscript. In this case the problem becomes NP-hard and they propose several heuristics. In both cases the maintenance requirement means that an Eulerian tour must visit certain nodes (maintenance stations) every 3 or 4 arcs since their arcs correspond to lines of flying (day's activity of an aircraft).

4.2.2   Day of operations.   In this section we cover the execution part of aircraft routing. In a day of operations, due to unexpected events such as inclement weather or unscheduled maintenance, new aircraft routes have to be found.

As is the case in the planning stage, two types of models are found: The multicommodity ones and set partitioning models. The solution methodologies are either local search techniques or integer programming heuristics.

Early work on aircraft recovery is presented in Jarrah et al. (1993). They model the recovery problem on a time-space network. They consider cancellations and delays separately, i.e. for each one of them they have a different model. The underlying network is a pure minimum cost network optimization model and thus it does not include any side constraints. Yan and Young (1996); Yan and Lin (1997) consider delays, cancellations, and aircraft ferrying in a single multicommodity flow model with side constraints. They solve the model by Lagrangian relaxation. A quadratic programming formulation is presented by Cao and Kanafani (1997a,b). The underlying model is a multicommodity flow model with side constraints, however side constraints are moved to the objective function with quadratic penalty terms. Thengvall et al. (2000, 2003) present a multicommodity network flow model with side constraints. In the former, they solve the model with a commercial integer programming software while in the second they apply the bundle algorithm after relaxing the flight covering constraints. The former work introduces a new objective of deviation from the original schedule and they consider only minor disruptions. In most of the instances the LP relaxation gives an integer solution and if this were not the case, they use rounding to obtain a feasible solution. The delays are modeled by introducing several copies of a single flight, each one with a different departure time. Bard et al. (2001) present a similar model but they focus more on airport closures or reduced slot capacity (e.g., when the ground delay program is in effect).

In large disruptions it takes longer to return back to normal operations and therefore maintenance constraints become an issue. Partitioning formulations, where variables correspond to complete routes, are used in this case. Løve et al. (2002) present a local search heuristic approach for solving the problem. They minimize total delay, number of cancellations, and the number of aircraft swaps. Argüello et al. (1997a); Argüello et al. (1997b) present the underlying set partitioning formulation, which is then solved by the greedy randomized adaptive search procedure. Rosenberger et al. (2001) give a similar set partitioning formulation. Their decision variables correspond to flight cancellations and they have binary variables that assign an aircraft route to a specific tail number. The basic constraints are to assign each aircraft to a route (ferrying, diversions, and over-flying are allowed) and that each flight must be either covered by a route or cancelled. They model slot availability as well. The solution methodology consists of first selecting a subset of routes and then finding a solution over these routes by means of a commercial integer programming solver.

4.3 Crew scheduling

4.3.1 The planning stage.
In tactical planning, after the aircraft routes are obtained, crew scheduling is next. Crew scheduling itself is decomposed into two processes.

In the crew pairing phase crew pairings or itineraries are obtained. A *pairing* is a sequence of flights, where the destination station of a flight in the sequence corresponds to the origin station of the next flight. In addition, the origin station of the first flight and the destination station of the last flight must correspond to the same crew base. In the crew pairing stage, a pairing is not assigned to a particular crew member. The crew pairing problem is to find a least cost subset of pairings that partition all flights.

After pairings are obtained for a given time period (typically a month), individual crew members are assigned to these pairings. *Rostering* is a common process outside of North America. Given crew preferences for individual pairings and patterns, an assignment of pairings to crew members is sought in rostering. The objective consists of meeting as many preferences as possible and to minimize potential costs. The *bidline* process is commonly used by North American carriers. This process consists of first generating bidlines (generic monthly assignments) and then crew members based on seniority bid for bidlines. A third alternative is *preferential bidding*, where individual rosters are constructed sequentially (seniority implies the order) based on individual preferences.

Crew pairing   A *duty* is a subsequence of a pairing that comprises a working day of a crew. Connection times within a duty, called sit connections, are short (35 minutes to a few hours) whereas connection times between duties, called layovers or rests, are much longer (10 hours and more). A pairing must satisfy many regulatory rules. To name just a few of them, there is a minimum sit and rest time, the elapsed time and the flying time of a duty is upper bounded, and there is the complicated 8-in-24 rule imposed by the FAA. In addition to these rules, union rules complicate pairing structure even further (maximum number of days in a pairing, more complex duty elapsed times). On top of all this, the cost of a pairing is usually nonlinear. Often the cost of a pairing is the maximum of tree quantities: A fraction of the pairing elapsed time, sum of duty costs in the pairing, and the number of duties times the minimum guaranteed pay. Linear terms capture hotel and meal expenses. The cost of a duty is the maximum of three terms as well: A fraction of the flying time, a fraction of the elapsed time, and the minimum guaranteed pay. Some airlines offer a fixed salary to crews and therefore their objective is to minimize the number of crews.

Most often the problem is modeled as the set partition problem with side constraints. Let P be the set of all pairings and for a $p \in P$ let c_p be the cost of pairing p. The model reads

$$\min \sum_{p \in p} c_p x_p$$

$$\sum_{i \in p} x_p = 1 \quad \text{for every leg } i$$
$$x \text{ binary,}$$

where x_p is 1 if pairing p is selected and 0 otherwise. In practice side constraints are added, which most often model equal use of resources. For example, if at crew base cb there are only a given number of crews, then

$$l_{cb} \leq \sum_{p \in S_{cb}} x_p \leq u_{cb}$$

is added, where S_{cb} is the set of all pairings starting at crew base cb and l_{cb}, u_{cb} are the lower, upper bound on the number of available crews at cb, respectively. Other typical side constraints are to balance pairings across crew bases with respect to cost, the number of days of pairings, or the number of duties.

This problem is computationally challenging for the following two reasons. Each pairing has complex feasibility rules and cost structure. In addition, the number of pairings even for a medium size problem is enormous. Fleets with 200 flights can have billions of pairings. For this reason, whenever there is a repetition of flights in the time horizon, the crew pairing optimization is performed in three steps. In the first step the so called *daily problem* is solved. This is the crew problem solved over a single day time horizon and it is assumed that every flight is operated every day. Once a daily solution is repeated over the real time horizon, some pairings become infeasible (called broken pairings). The operational legs of these pairings are then considered in the *weekly exceptions problem*, (Barnhart et al., 1996). The final solution then consists of daily pairings without the broken pairings and the pairings from the weekly exceptions problem. The weekly exceptions problem is a special case of the so called *weekly problem*, where pairings from the end of the horizon wrap around to the beginning of the horizon. The main distinction between a daily problem and a weekly problem is that in the former problem a pairing cannot cover the same leg more than once while this is allowed in the latter problem. When transitioning from one (monthly) work schedule to another, the *dated problem* needs to be solved to account for pairings that span both months. In the dated problem, flights on specific dates are given and they have to be partitioned by pairings.

A standard approach is to view pairings as constrained paths in either the *flight network*, (Minoux, 1984; Desaulniers et al., 1999), or the *duty period network*, (Lavoie et al., 1988; Anbil et al., 1994; Vance et al., 1997; Desaulniers et al., 1997a). The flight network has a node associated with each departure and arrival. There is a flight arc connecting each departure node of a flight with the arrival node of the same flight. In addition, there are connection arcs between any two arrival and departure nodes with the arrival station of the first flight being equal to the departure station of the second flight and the connection time is within legal limits, i.e. the time is either between the minimum and maximum sit connection time or between the minimum and maximum rest time. In addition, the network has two artificial nodes s and t. Node s is connected to every departure node of a flight that can start a pairing. Similarly, every arrival node of a flight that can end a pairing is connected to node t. Every pairing is an $s - t$ path in the flight network. Due to various pairing feasibility rules that cannot be embedded in the flight network, every $s - t$ path is not necessarily a pairing. The duty period network is constructed in a similar way except that flight arcs are replaced by duty periods and connection arcs correspond to legal rest connections. It is assumed that duties are enumerated beforehand. The duty period network captures more feasibility rules since all duty legality rules are embedded in the network, however, it requires much more storage.

The literature on crew pairing optimization is abundant with Barnhart et al. (2003) providing more details and surveying the literature. Here we focus only on branch-and-price related aspects and we survey only branch-and-price related literature. In branch-and-price type algorithms subproblem solving is done on either of the two networks. There are three approaches to find a low reduced cost pairing (subproblem solving). The first one, pioneered in the context of urban transit by Desrochers and Soumis (1989), is by constrained shortest path, (Anbil et al., 1994; Vance et al., 1997; Desaulniers et al., 1997a). In this approach a label is maintained for every feasibility rule that is not embedded in the network, e.g. 8-in-24 rule, elapsed time rules, etc. In addition, if the cost of a pairing is nonlinear, then each component in the maximum needs to have a separate label. If the duty period network is used, fewer labels are needed as many rules are already satisfied in the duty construction phase. For U.S. domestic carriers, the number of labels on the flight network can be as high as 20. A second approach is used in the commercial crew pairing solver from Carmen Systems, (Galia and Hjorring, 2003). Their approach is based on finding the k th shortest path. They find a shortest path on the current network. If the path is not feasible, they modify the network so that the obtained path is no longer a path in the network. Once a feasible path is found, it corresponds to a k th shortest path in the original network for a k. The third approach is to perform a depth-first search enumeration of pairings on a network, (Marsten, 1994; Anbil et al., 1998; Klabjan et al., 2001b; Makri and Klabjan, 2004). Since there are too many pairings, the search has to be truncated by, for example, not considering all the duties and all connection arcs. Another enhancement is by prunning the search earlier due to some lower bounds on the reduced cost, (Anbil et al., 1998; Makri and Klabjan, 2004).

Crew pairing branch-and-price algorithms employed tailored branching rules, which are based on the branching rule designed for set partitioning, (Ryan and Foster, 1981). The most widely used rule is to branch on *follow-ons*. In this branching rule, two flights r, s are selected and branching follows the scheme presented in Section 2. Follow-on branching is used in Anbil et al. (1998); Vance et al. (1997); Anbil et al. (1994); Desaulniers et al. (1997a). An alternative branching rule, called *timeline branching* is proposed in Klabjan et al. (2001b). In timeline branching two flights r, s are selected and a connection time t. In one branch the rule requires that only pairings with the connection time between r and s less than t are considered and the other branch considers only pairings with the connection time larger than or equal to t.

Reserve crew planning and training scheduling is discussed in Sohoni and Johnson (2002a,b); Sohoni et al. (2003). Klabjan et al. (2001a) present a model and solution methodology to solve the weekly crew pairing problem that is not based on the traditional daily/weekly exceptions paradigm. All of the related material presented so far relates to scheduling of cockpit crews. The flight attendant problem, (Day and Ryan, 1997; Kwok and Wu, 1996), is similar except that several flight attendants are required to cover a flight. These problems tend to be larger since the flight attendants are cross qualified but, on the other hand, the feasibility rules are computationally easier.

Bidline process, rostering, and preferential bidding   Once a set of pairings is obtained that covers all flights in a month, these pairings and additional tasks such as reserve crew duties and flight training, are next assigned to individual crews. The problem decomposes further, not only based on the equipment type, but also based by the crew member rank (such as Captain, First Officer).

Feasibility rules in rostering are even more complicated than the pairing feasibility rules. The rules are imposed either by a regulatory agency such as the FAA, the airline itself, and there are contractual rules. Some of the basic requirements are: Limits on the rest time between two tasks, limits on a working period (working week) between 4 to 8 days, limits on the number of monthly and yearly block hours. Then there are restrictions with respect to task coverage, e.g. one captain and one first officer for a given task, two captains and one first officer for simulator training, etc. Rules involving several rosters are common as well, e.g. some crew members prefer to fly together (married couples) and language restrictions.

In rostering several objectives are possible. From the airline perspective, minimizing open time or unassigned activities is important. Open time consists of tasks that are not assigned to regular crews but they are covered either by reserve crews or overtime is used. Clearly these two options are costly to the airline. If the number of block hours of a crew member is larger than a certain limit, the airline has to additionally pay the crew member for the overtime. Therefore the airline's interest is in minimizing the overtime pay. The third objective of the airline is to optimize assignments to training on simulators. These type of training is mandatory and very expensive. The airline also tries to produce rosters that are equitable across crew members, e.g. the crew members should have equal flying time and the number of off days. On the other hand, crew members have their own goals and preferences. Each member has its own preferences such as starting duties early in the morning, favoring certain pairings, etc. A quality roster must meet as many preferences as possible. Additional details on rostering rules and examples are given in Kohl and Karisch (2004).

The rostering problem can be modeled in the following way, (Gamache and Soumis, 1998; Gamache et al., 1999; Kohl and Karisch, 2004). Let the decision variable x_s^k be one if roster s is selected for crew member k. The model reads

$$\min c_s^k x_s^k \tag{6.9}$$

$$\sum_{\substack{k \in K \\ i \in s}} x_s^k \geq n_i \quad \text{for every task } i \tag{6.10}$$

$$\sum_s x_s^k = 1 \quad \text{for every crew member } k \tag{6.11}$$

$$x \text{ binary,} \tag{6.12}$$

where c_s^k is the cost of assigning roster s to crew member k, and n_i is the number of crew members that are required for task i. Equations (6.10) guarantee task coverage and (6.11) assign a roster to every crew member. Rules that involve several rosters have to be explicitly modeled by adding cuts at the subproblem level, hence better controlling the generated columns.

Similarly to the crew pairing approach, there exists an underlying network such that a roster is a path in this network but not necessarily the other way around, (Gamache et al., 1998). To exploit individual preferences, it is actually convenient to construct a network for every individual crew member. This leads to branch-and-price approaches. Gamache et al. (1998); Gamache and Soumis (1998) are the first ones to describe a branch-and-price algorithm. An important observation from their work is that it is beneficial to construct crew rosters for individual crew members that are disjoint with respect to tasks. Subproblem solving is performed by constrained shortest path. Kohl and Karisch (2004); Kharraziha et al. (2003) use k th shortest path in subproblem solving and they present a general modeling language to capture feasibility rules and objectives. The same modeling language is used also in their crew pairing optimizer, (Galia and Hjorring, 2003). To warm start the algorithm, they construct rosters heuristically. Day and Ryan, 1997 describe cabin crew rostering at Air New Zealand in their short-haul operations. The problem is solved by first assigning off days and it is followed by assigning pairings and other tasks. This two phase approach simplifies the problem but it can lead to suboptimal solutions. In each phase they employ branch-and-price.

Many pure heuristic approaches to rostering have been developed by various airlines. They can be found in various proceedings of The Airline Group of the International Federation of Operational Research Societies (AGIFORS) meetings. A detailed description of a simulated annealing heuristic approach is given in Lučić and Teodorović, 1999.

Gamache et al. (1998) give an approach to preferential bidding. Their methodology consists of producing individual rosters sequentially one by one in a given order (e.g. seniority). Suppose rosters for the first $k-1$ crew members have already been obtained. The roster for the k th crew members is obtained by solving (6.9)–(6.12) with the following changes. The objective function considers only rosters of crew member k. Equations (6.10) are included only for those tasks that are not covered by the first $k-1$ crew members. There is (6.11) constraint for every non assigned crew member. Clearly only rosters feasible to unassigned crew members are considered. The model produces an optimal roster for crew member k and at the same time it guarantees a feasible solution for the remaining unassigned members. If there are m crew members, then m models are solved. To improve the execution time, each model is relaxed to allowing fractional solutions to rosters of crew members $k+1, k+2, \ldots, m$. They further improve the algorithm by adding cuts. Achour et al., 2003 enhance this work by combining branching decisions and cuts.

The literature on the bidline process is limited. Jarrah and Diamond (1997) present a heuristic approach to the bidline process. A crew planner sets parameters, e.g. the length of a working period, number of off days. If the parameters are restrictive enough, there are not many rosters to consider and the resulting set partitioning model is solved by explicitly enumerating all rosters. If there are too many rosters to consider, they employ a local search heuristic. A simulated annealing heuristic to the bidline process is given in Campbell et al. (1997).

4.3.2   Day of operations.   In disruption management, the crew recovery problem follows aircraft recovery. The input to the problem are the new aircraft routes together with the new departure times and flight cancellations. In crew recovery new crew assignments have to be obtained. Depending on the airline, non disrupted crew members can be involved in the reassignment or not. But clearly the number of such crew member should be minimized. Another objective is to return back to the original crew schedule as soon as possible. Then there is the objective of minimizing the cost, which can consist of the direct salary based cost, uncovered flight cost, crew deadheading, etc. In crew recovery standby and reserve crews can be used but the latter are costly.

Teodorović and Stojković (1990) develop a sequential approach based on a dynamic programming algorithm, using the first-in-first-out principle to minimize the crews' ground time. Wei and Yu (1997) present a heuristic-based framework for crew recovery. Song et al. (1998) present a multicommodity integer network flow model and a heuristic search algorithm to solve it. Stojković et al. (1998) present a column generation approach similar to the one used for crew pairing problems. Lettovskýet al. (2000); Lettovský (1997) base their column generation approach on the rostering model. They give details on how to quickly generate promising pairings. Stojković and Soumis (2001) incorporate flight scheduling decisions into the crew recovery nonlinear multicommodity flow model. Together with new crew assignments, their model produces new departure times. The model is solved by branch-and-price, where the subproblem is solved by constrained shortest path. Stojković and Soumis (2003) expand this model by allowing crews to split, i.e., if a first officer and a captain in planning are assigned to cover a given flight, the recovered schedule might keep the first officer at the same flight but it is paired with a different captain.

5.   Recent advances

In recent years models and optimization based methodologies that integrate the three planning areas started to emerge. Integration of aircraft routing and crew pairing is discussed in Cohn and Barnhart (2002); Klabjan et al. (2002); Cordeau et al. (2001); Mercier et al. (2003). Solving the combined fleeting and aircraft routing model, (Barnhart et al., 1998b), has already been discussed in this chapter. Barnhart et al. (1998c) take the first step towards a model for integrating fleeting and crew pairing. All of this integration efforts are in an early stage and most of the methodologies are not yet suited for large-scale problems. Another obstacle in adopting these models by the airlines is that they require changes in business processes. Legacy carriers are notorious for their unwillingness to change their internal processes. On the other hand, smaller, mostly low-cost carriers are more flexible and open to business process reengineering, (Garvin, 2000). This fact goes hand in hand with the current inability of solving large-scale integrated models. Clearly the airlines have to follow and embrace the advances in modeling and algorithms, and the researchers have to improve decision support systems to be more tractable.

The other emerging trend is in robustness. It is well documented that customer complaints, delays, and flight cancellations were on a rise every year from 1996 until 2000. They reached the top in summer 2000, where it even caught attention by the Congress. In early 2001 tactical models that embed robustness emerged. These models do not necessarily produce a cost/profit optimal solution but a suboptimal solution that fares better in operations under uncertainty. On the crew pairing side, approaches by Schaefer et al. (2000); Ehrgott and Ryan (2001); Yen and Birge (2000); Chebalov and Klabjan (2002) provide robust solutions. Robust fleeting solutions are discussed in Rosenberger et al. (2004); Kang and Clarke (2003); Listes and Dekker (2003). Ageeva (2000) presents an approach to robust airline routing. A robust approach to passengers rerouting in disruption management is given by Karow (2003). While many sources of frequent disruptions (congestion being the dominant one) have abated since the events of September 11, new ones are popping up (increased security measures). Nevertheless, delays have been drastically reduced due to a substantially lower demand and therefore the airlines have lost poise for robust solutions. However, the airline industry is recovering and not far in the future the demand will be at the pre September 11 level. So even though robust solutions have lost appeal in the industry, the researchers are seeing this direction as the next big step in improving profitability and customer satisfaction.

References

Abara, J. (1989). Applying integer linear programming to the fleet assignment problem. *Interfaces*, 19:20–38.

Achour, H., Gamache, M., and Soumis, F. (2003). Branch and cut at the subproblem level in a column generation approach: Application to the airline industry. *Les Cahiers du GERAD* G-2003-34, HEC, Montréal, Canada.

Ageeva, Y. (2000). Aproaches to Incorporating Robustness Into Airline Scheduling. Master's Dthesis, Massachusetts Institute of Technology.

Anbil, R., Barnhart, C., Johnson, E., and Hatay, L. (1994). A column generation technique for the long-haul crew assignment problem. In: *Optimization in Industry* II, pp. 7–24. John Wiley & Sons.

Anbil, R., Forrest, J., and Pulleyblank, W. (1998). Column generation and the airline crew pairing problem. In: *Proceedings of the International Congress of Mathematicians Berlin*, Extra Volume ICM 1998 of Documenta Mathematica. Journal der Deutschen Mathematiker-Vereinigung, pp. 677–686, Universität Bielefeld, Fakultät für Mathematik, Bielefeld.

Antes, J. (1997). Structuring the process of airline scheduling. In: *Operations Research Proceedings* (P. Kischka, ed.), 1997.

Argüello, M., Bard, J., and Yu, G. (1997a). A GRASP for aircraft routing in response to groundings and delays. *Journal of Combinatorial Optimization*, 5:211–228.

Argüello, M., Bard, J., and Yu, G. (1997b). Models and methods for managing airline irregular operations aircraft routing. In: *Operations Research in the Airline Industry* (G. Yu, ed.), pp. 1–45, Kluwer Academic Publishers.

Bard, J., Yu, G., and Argüello, M. (2001). Optimizing aircraft routings in response to groundings and delays. *IIE Transactions*, 33:931–947.

Barnhart, C., Johnson, E., Nemhauser, G., and Vance, P. (1996). Exceptions crew scheduling. *Technical Report*, The Logistics Institute, Georgia Institute of Technology.

Barnhart, C., Johnson, E., Nemhauser, G., Savelsbergh, M., and Vance, P. (1998a). Branch-and-price: Column generation for solving huge integer programs. *Operations Research*, 46:316–329.

Barnhart, C., Boland, N., Clarke, L., Johnson, E., Nehmauser, G., and Shenoi, R. (1998b). Flight string model for aircraft fleeting and routing. *Transportation Science*, 32:208–220.

Barnhart, C., Lu, F., and Shenoi, R. (1998c). Integrated airline schedule planning. In: *Operations Research in the Airline Industry* (G. Yu, ed.), pp. 384–403, Kluwer Academic Publishers.

Barnhart, C., Farahat, A., and Lohatepanont, M. (2002a). Airline fleet assignment: An enhanced revenue model. *Technical Report*, Massachusetts Institute of Technology.

Barnhart, C., Kniker, T., and Lohatepanont, M. (2002b). Itinerary-based airline fleet assignment. *Transportation Science*, 36:199–217.

Barnhart, C., Cohn, A., Johnson, E., Klabjan, D., Nemhauser, G., and Vance, P. (2003). Airline crew scheduling. In: *Handbook of Transportation Science* (R. W. Hall, ed.), Kluwer Scientific Publishers.

Bélanger, N., Desaulniers, G., Soumis, F., and Desrosiers, J. (2003). Periodic airline fleet assignment with time windows, spacing constraints, and time dependant revenues. Forthcoming in: *European Journal of Operational Research*.

Benders, J. (1962). Partitioning procedures for solving mixed-variables programming problems. *Numerische Mathematik*, 4:238–252.

Berdy, P. (2002). Developing effective route networks. In: *Handbook of Airline Economics* (D. Jenkins, ed.), McGraw-Hill.

Berge, M. (1994). Timetable optimization—Formulation, solution approaches, and computational issues. In: *AGIFORS proceedings*.

Berge, M. and Hopperstad, C. (1993). Demand driven dispatch: A method for dynamic aircraft capacity assignment, models and algorithms. *Operations Research*, 41:153–168.

Campbell, K., Durfee, R., and Hines, G. (1997). FedEx generates bid lines using simulated annealing. *Interfaces*, 27:1–16.

Cao, J. and Kanafani, A. (1997a). Real-time decision support for integration of airline flight cancellations and delays part I: Mathematical formulations. *Transportation Planning and Technology*, 20:183–199.

Cao, J. and Kanafani, A. (1997b). Real-time decision support for integration of airline flight cancellations and delays part II: Algorithms and computational experiments. *Transportation Planning and Technology*, 20:201–217.

Carmen Systems. http://www.carmen.se.

Chebalov, S. and Klabjan, D. (2002). Robust airline crew scheduling: Move-up crews. In: *Proceedings of the 2002 NSF Design, Service, and Manufacturing Grantees Research Conference*.

Clark, P. (2000). Dynamic fleet management. In: *Handbook of Airline Operations* (G. Butler and M. Keller, eds.), pp. 273–285, McGraw-Hill.

Clarke, L., Hane, C., Johnson, E., and Nemhauser, G. (1996). Maintenance and crew considerations in fleet assignment. *Transportation Science*, 30:249–260.

Clarke, L., Johnson, E., Nemhauser, G., and Zhu, Z. (1997). The aircraft rotation problem. In: *Annals of OR: Mathematics of Industrial Systems* II (R. E. Burkard, T. Ibaraki, and M. Queyranne, eds.), pp. 33–46, Baltzer Science Publishers.

Clarke, M. D., Lettovský, L., and Smith, B. (2002). The development of the airline operations control center. In: *Handbook of Airline Economics* (D. Jenkins, ed.), pp. 197–215, McGraw-Hill.

Cohn, A. and Barnhart, C. (2002). Improving crew scheduling by incorporating key maintenance routing decisions. *Operations Research*, 51:387–396.

Cordeau, J., Stojković, G., Soumis, F., and Desrosiers, J. (2001). Benders decomposition for simultaneous aircraft routing and crew scheduling. *Transportation Science*, 35:375–388.

Daskin, M. and Panayoyopoulos, N. (1989). A Langrangian relaxation approach to assigning aircraft to routes in hub and spoke networks. *Transportation Science*, 23:91–99.

Day, P. and Ryan, D. (1997). Flight attendant rostering for short-haul airline operations. *Operations Research*, 45:649–661.

Desaulniers, G., Desrosiers, J., Dumas, Y., Marc, S., Rioux, B., Solomon, M., and Soumis, F. (1997a). Crew pairing at Air France. *European Journal of Operational Research*, 97:245–259.

Desaulniers, G., Desrosiers, J., Dumas, Y., Solomon, M. M., and Soumis, F. (1997b). Daily aircraft routing and scheduling. *Management Science*, 43:841–855.

Desaulniers, G., Desrosiers, J., Ioachim, I., Solomon, M., and Soumis, F. (1998). A unified framework for deterministic time constrained vehicle routing and crew scheduling problems. In: *Fleet Management and Logistics* (T. Crainic and G. Laporte, eds.), pp. 57–93, Kluwer Publishing Company.

Desaulniers, G., Desrosiers, J., Lasry, A., and Solomon, M. (1999). Crew pairing for a regional carrier. In: *Computer-Aided Transit Scheduling*: *Lecture Notes in Economics and Mathematical Systems* (N. Wilson, ed.), pp. 19–41. Springer Verlag.

Desrochers, M. and Soumis, F. (1989). A column generation approach to the urban transit crew scheduling problem. *Transportation Science*, 23:1–13.

Desrosiers, J., Dumas, Y., Solomon, M., and Soumis, F. (1995). Time constrained routing and scheduling. In: *Handbook in Operations Research and Management Science, Network Routing* (M. Ball, T. Magnanti, C. Monma, and G. Nemhauser, eds.), pp. 35–139, Elsevier Science Publishers.

Ehrgott, M. and Ryan, D. (2001). Bicriteria robustness versus cost optimization in tour of duty planning at Air New Zealand. *Technical Report*, University of Auckland.

Erdmann, A., Nolte, A., Noltemeier, A., and Schrader, R. (2001). Modeling and solving the airline schedule generation problem. *Annals of Operations Research*, 107:117–142.

Etschmaier, M. and Mathaisel, D. (1985). Airline scheduling: An overview. *Transportation Science*, 19:127–138.

Feo, T. and Bard, J. (1989). Flight scheduling and maintenance base planning. *Management Science*, 35:1415–1432.

Fisher, M. (1981). The Lagrangian relaxation method for solving integer programming problems. *Management Science*, 27:1–18.

Fisher, M. (1985). An applications oriented guide to Lagrangian relaxation. *Interfaces*, 15:10–21.

Galia, R. and Hjorring, C. (2003). Modeling of complex costs and rules in a crew pairing column generator. *Technical Report* CRTR-0304, Carmen Systems.

Gamache, M. and Soumis, F. (1998). A method for optimally solving the rostering problem. In: *Operations Research in the Airline Industry* (G. Yu, ed.), pp. 124–157. Kluwer Academic Publishers.

Gamache, M., Soumis, F., Villeneuve, D., Desrosiers, J., and Gelinas, E. (1998). The preferential bidding system at Air Canada. *Transportation Science*, 32:246–255.

Gamache, M., Soumis, F., Marquis, G., and Desrosiers, J. (1999). A column generation approach for large scale aircrew rostering problems. *Operations Research*, 47:247–262.

Garvin, M. (2000). Service delivery system: A regional airline perspective. In: *Handbook of Airline Operations* (G. Butler and M. Keller, eds.), pp. 419–428, McGraw-Hill.

Geoffrion, A. (1974). Lagrangean relaxation for integer programming. *Mathematical Programming Study*, 2:82–114.

Gopalan, R. and Talluri, K. (1998). The aircraft maintenance routing problem. *Operations Research*, 46:260–271.

Hane, C., Barnhart, C., Johnson, E., Marsten, R., Nemhauser, G., and Sigismondi, G. (1995). The fleet assignment problem: Solving a large-scale integer program. *Mathematical Programming*, 70:211–232.

Jacobs, T., Johnson, E., and Smith, B. (1999). O& D FAM: Incorporating passenger flows into the fleeting process. In: *Thirty-Ninth Annual AGIFORS Symposium* (R. Darrow, ed.), New Orleans.

Jarrah, A. and Diamond, J. (1997). The problem of generating crew bidlines. *Interfaces*, 27:49–64.

Jarrah, A., Yu, G., Krishnamurthy, N., and Rakshit, A. (1993). A decision support framework for airline flight cancellations and delays. *Transportation Science*, 27:266–280.

Kang, L. and Clarke, J. (2003). Degradable airline scheduling. *Technical Report*, Massachusetts Institute of Technology.

Karow, M. (2003). Virtual Hubs: An Airline Schedule Recovery Concept And Model. Master's thesis, Massachusetts Institute of Technology.

Kharraziha, H., Ozana, M., and Spjuth, S. (2003). Large scale crew rostering. *Technical Report* CRTR-0305, Carmen Systems.

Klabjan, D., Johnson, E., Nemhauser, G., Gelman, E., and Ramaswamy, S. (2001a). Airline crew scheduling with regularity. *Transportation Science*, 35:359–374.

Klabjan, D., Johnson, E., Nemhauser, G., Gelman, E., and Ramaswamy, S. (2001b). Solving large airline crew scheduling problems: Random pairing generation and strong branching. *Computational Optimization and Applications*, 20:73–91.

Klabjan, D., Johnson, E., Nemhauser, G., Gelman, E., and Ramaswamy, S. (2002). Airline crew scheduling with time windows and plane count constraints. *Transportation Science*, 36:337–348.

Kniker, T. (1998). Itinerary-based airline fleet assignment. *Ph.D Thesis*, Massachusetts Institute of Technology.

Kohl, N. and Karisch, S. (2004). Airline crew rostering: Problem types, modeling, and optimization. *Annals of Operations Research*, 127:223–257.

Kwok, L. and Wu, L. (1996). Development of an expert system in cabin crew pattern generation. *International Journal of Expert Systems*, 9:445–464.

Lavoie, S., Minoux, M., and Odier, E. (1988). A new approach for crew pairing problems by column generation with an application to air transportation. *European Journal of Operational Research*, 35:45–58.

Lettovský, L. (1997). Airline operations recovery: An optimization approach. *Ph.D Thesis*, Georgia Institute of Technology.

Lettovský, L., Johnson, E., and Smith, B. (1999). Schedule generation model. In: *Thirty-Ninth Annual AGIFORS Symposium* (R. Darrow, ed.), New Orleans.

Lettovský, L., Johnson, E., and Nemhauser, G. (2000). Airline crew recovery. *Transportation Science*, 34:337–348.

Listes, O. and Dekker, R. (2003). A scenario aggregation based approach for determining a robust airline fleet composition. *Technical Report* EI 2002-17, Rotterdam School of Economics.

Lohatepanont, M. and Barnhart, C. (2002). Airline schedule planning: Integrated models and algorithms for schedule design and fleet assignment. *Technical Report*, Massachusetts Institute of Technology.

Løve, M., Sørensen, K., Larsen, J., and Clausen, J. (2002). Disruption management for an airline - rescheduling of aircraft. In: *Applications of Evolutionary Computing: EvoWorkshops 2002* (S. Cagnoni, J. Gottlieb, E. Hart, M. Middendorf, and G. Raidl, eds.), volume 2279, *Lecture Notes in Computer Science*, pp. 315–324, Springer-Verlag Heidelberg.

Lučić, P. and Teodorović, D. (1999). Simulated annealing for the multi-objective aircrew rostering problem. *Transportation Research Part A*, 33:19–45.

Makri, A. and Klabjan, D. (2004). A new pricing scheme for airline crew scheduling. *INFORMS Journal on Computing*, 16:56–67.

Marsten, R. (1994). Crew planning at Delta Airlines. Presentation in: XV *Mathematical Programming Symposium*, Ann Arbor.

Marsten, R., Subramanian, R., and Gibbons, L. (1996). Junior analyst extraordinare (JANE): Route development at Delta Air Lines. In: *AGIFORS Proceedings*.

Martin, R. (1999). Large Scale Linear and Integer Optimization: A Unified Approach, Kluwer Academic Publishers.

Mercier, A., Cordeau, J.-F., and Soumis, F. (2003). A computational study of Benders decomposition for the integrated aircraft routing and crew scheduling problem. *Les Cahiers du GERAD* G-2003-48, HEC, Montréal, Canada.

Minoux, M. (1984). Column generation techniques in combinatorial optimization: A new application to crew pairing problems. In: *Proceedings XXIVth AGIFORS Symposium*.

Minoux, M. (1986). Mathematical Programming: Theory and Algorithms, Wiley-Interscience.

Paoletti, B., Cappelletti, S., and Lenner, C. (2000). Operations research models for the optimization of aircraft rotation and routing in the integrated resources management process. In: *Handbook of Airline Operations* (G. Butler and M. Keller, eds.), pp. 285–308, McGraw-Hill.

Rexing, B., Barnhart, C., Kniker, T., Jarrah, A., and Krishnamurthy, N. (2000). Airline fleet assignment with time windows. *Transportation Science*, 34:1–20.

Rosenberger, J., Johnson, E., and Nemhauser, G. (2001). Rerouting aircraft for airline recovery. *Transportation Science*, 37:408–421.

Rosenberger, J., Johnson, E., and Nemhauser, G. (2004). A robust fleet assignment model with hub isolation and short cycles. *Transportation Science*, 38:357–368.

Rushmeier, R. and Kontogiorgis, S. (1997). Advances in the optimization of airline fleet assignment. *Transportation Science*, 31:159–169.

Ryan, D. and Foster, B. (1981). An integer programming approach to scheduling. In: *Computer Scheduling of Public Transport Urban Passenger Vehicle and Crew Scheduling* (A. Wren, ed.), pp. 269–280, Elsevier Science B.V.

Schaefer, A., Johnson, E., Kleywegt, A., and Nemhauser, G. (2000). Airline crew scheduling under uncertainty. *Technical Report* TLI-01-01, Georgia Institute of Technology.

Sohoni, M. and Johnson, E. (2002a). Operational airline reserve crew planning. *Technical Report*, The Logistics Institute, Georgia Institute of Technology.

Sohoni, M. and Johnson, E. (2002b). An optimization approach to pilot recurrent training schedule. *Technical Report*, The Logistics Institute, Georgia Institute of Technology.

Sohoni, M., Johnson, E., and Bailey, G. (2003). Long range reserve crew manpower planning. *Technical Report*, The Logistics Institute, Georgia Institute of Technology.

Song, M., Wei, G., and Yu, G. (1998). A decision support framework for crew management during airline irregular operations. In: *Operations Research in the Airline Industry* (G. Yu, ed.), pp. 260–286, Kluwer Academic Publishers.

Sriram, C. and Haghani, A. (2003). An optimization model for aircraft maintenance scheduling and re-assignment. *Transportation Research Part A*, 37:29–48.

Stojković, G., Soumis, M., and Desrosiers, J. (1998). The operational airline crew scheduling problem. *Transportation Science*, 32:232–245.

Stojković, M. and Soumis, F. (2001). An optimization model for the simultaneous operational flight and pilot scheduling problem. *Management Science*, 47: 1290–1305.

Stojković, M. and Soumis, F. (2003). The operational flight and multi-crew scheduling problem. *Les Cahiers du GERAD* G-2000-27, HEC, Montréal, Canada.

Talluri, K. (1996). Swapping applications in a daily airline fleet assignment. *Transportation Science*, 30: pp. 237–248.

Teodorović, D. and Stojković, G. (1990). Model for operational daily airline scheduling. *Transportation Planning Technology*, 14:273–285.

The Airline Group of the International Federation of Operational Research Societies (AGIFORS). http://www.agifors.org.

Thengvall, B., Bard, J., and Yu, G. (2000). Balancing user preferences for aircraft schedule recovery during irregular operations. *IIE Transactions*, 32:181–193.

Thengvall, B., Bard, J., and Yu, G. (2003). A bundle algorithm approach for the aircraft schedule recovery problem during hub closures. *Transportation Science*, 37:392–407.

van Ryzin, G. and Talluri, K. (2002). Revenue management. In: *Handbook of Transportation Science* (R. Hall, ed.), pp. 599–661, Kluwer Academic Publishers.

Vance, P., Atamtürk, A., Barnhart, C., Gelman, E., Johnson, E., Krishna, A., Mahidhara, D., Nemhauser, G., and Rebello, R. (1997). A heuristic branch-and-price approach for the airline crew pairing problem. *Technical Report* LEC-97-06, Georgia Institute of Technology.

Wei, G. and Yu, G. (1997). Optimization model and algorithm for crew management during airline irregular operations. *Journal of Combinatorial Optimization*, 1:305–321.

Yan, S. and Lin, C. (1997). Airline scheduling for the temporary closure of airports. *Transportation Science*, 31:72–82.

Yan, S. and Tseng, C. (2002). A passenger demand model for airline flight scheduling and fleet routing. *Computers and Operations Research*, 29:1559–1581.

Yan, S. and Wang, C. (2001). The planning of aircraft routes and flight frequencies in an airline network operations. *Journal of Advanced Transportation*, 35:33–46.

Yan, S. and Young, H. (1996). A decision support framework for multi-fleet routing and multi-stop flight scheduling. *Transportation Research Part A*, 30:379–398.

Yen, J. and Birge, J. (2000). A stochastic programming approach to the airline crew scheduling problem. *Technical Report*, University of Washington.

Chapter 7

ROBUST INVENTORY SHIP ROUTING BY COLUMN GENERATION

Marielle Christiansen
Bjørn Nygreen

Abstract   We consider a real integrated ship scheduling and inventory management problem. A fleet of ships transports a single product between production and consumption plants. The transporter has the responsibility for keeping the inventory level within its limits at all actual plants, and there should be no need to stop the production at any plants caused by missing transportation possibilities.

Due to uncertainties in sailing time, we introduce soft inventory constraints and artificial penalty costs to the underlying model. The model is solved by a column generation approach. By introducing some model adjustments, the problem decomposes into a routing and scheduling subproblem for each ship and an inventory management subproblem for each port. The columns in the master problem represent ship schedules and port call sequences.

1.   Introduction

In several sectors in the shipping industry, the transporter often has a twofold responsibility. In this segment large quantities are transported, and normally considerable inventories exist at each end of a sailing leg. This is the case for transportation of, for instance, cement, chemicals and natural liquid gas (LNG). In some situations, the transporter has both the responsibility for the transportation and the inventories at the sources and the destinations. We consider a planning problem where a single product is transported. The problem is based on a real planning problem for Norsk Hydro ASA, which is the main transporter of ammonia in Europe.

This planning problem involves the design of a set of minimum cost routes and schedules for a given fleet of heterogeneous bulk ships that service a set of ports one or several times during the planning period. The transporter owns plants located near ports. These plants produce the product at the sources called *loading ports* and consume the product at the destinations called *unloading ports*. Inventory storage capacities are given in all ports, and there exist information about the consumption or production rates of the transported product. The ships transport this product from loading ports to unloading ports and if the product is loaded and unloaded in time, neither production nor consumption will be interrupted. Since the transporter owns the plants and controls the inventories both at the sources and destinations, the inventory costs do not come into play. In contrast to most ship scheduling problems, the number of calls at a given port during the planning period is not predetermined, neither is the quantity to be loaded and unloaded in each port call. A more detailed presentation of the real problem can be found in Christiansen (1999).

The shipping industry is concerned with uncertainty in time consumption at the ports and at sea due to strikes and mechanical problems at the ports and bad weather at sea. There will be a stochastic realization of the final plan, but still we base our model on deterministic planning. However, stochastic conditions can be partly considered in various ways even in deterministic models, such that it is easier to realize the plan. We introduce a pair of soft inventory bounds within the hard inventory bounds to reduce the possibility of violating the inventory bounds at the plants. Thus, the soft inventory bounds can be violated at a penalty, but it is not possible to exceed the storage capacity or drop below the lower inventory bound. These penalty costs will be incorporated into the optimization model in order to try to force the solution away from its inventory bounds and get more robust schedules.

In our near cooperation with Hydro, we have experienced that the company regards the use of soft inventory constraints as positive in order to generate more robust routes. This means that the inventory levels at the ports are kept further away from their inventory bounds in the mathematical model. In some planning situations, it is possible to achieve more robust routes without a large increase in transportation costs and sometimes without an increase at all. Even when the company studied different routes manually, they operated with explicit soft inventory bounds and reported the inventories outside the soft inventory bounds.

The purpose of this chapter is to present the real planning problem and the column generation solution approach while focusing on how to generate robust plans.

In Section 2, we discuss some research reported in the literature on integrated ship scheduling and inventory management. Section 3, gives a description of the real integrated ship scheduling and inventory management problem with focus on the uncertainty aspects. An arc flow formulation of the problem is presented in Section 4. Then, in Section 5, the solution approach is described. A few computational results are given in Section 6. Finally, some conclusions follow in Section 7.

2.   Related research within integrated ship scheduling and inventory management

The studied ship planning problem is an example of an integrated time constrained routing and inventory management problem. Christiansen (1999) gives an arc flow formulation of the problem and shows how the problem can be transformed to a decomposed formulation by introducing some important model adjustments. Then the model can be solved successfully by a Dantzig-Wolfe decomposition (column generation) approach for real instances of the problem. Christiansen and Nygreen (1998a) present the overall solution method, while the same authors in Christiansen and Nygreen (1998b) give a detailed description of the resulting subproblems from the Dantzig-Wolfe decomposition. Another solution approach to the same ship planning problem was developed by Flatberg et al. (2000). They have used an iterative improvement heuristic combined with an LP solver to solve this problem. The solution method presented consists of two parts. Their heuristic is used to solve the combinatorial problem of finding the ship routes, and an LP model is used to find the starting time of service at each call and the loading or unloading quantity. Computational results for real instances of the planning problem are reported. However, no comparisons in running time or solution quality of the results in Flatberg et al. (2000); Christiansen and Nygreen (1998a) exist.

Generally, ship routing and scheduling problems have been shown some attention in the literature, see for instance the surveys by Ronen (1983, 1993); Christiansen, Fagerholt and Ronen (2004) or the chapter on maritime transportation by Christiansen et al. (2004).

However, there exist only a few other research studies reported in the literature on integrated ship scheduling and inventory management. Fox and Herden (1999) describe a MIP model to schedule ships from ammonia processing plants to the agricultural market in Australia. The ammonia processing plants corresponds to the unloading ports in the planning problem described in this chapter, and at these plants the ammonia is further processed into different fertilizer products. The objective is to minimize freight, discharge, and inventory holding costs while taking into account the inventory, minimum discharge tonnage, and ship capacity constraints. The multi-period model is solved by commercial optimization software. An inventory routing problem similar to the one described here, but with multiple products is presented by Ronen (2002). A MIP model for the problem is formulated. Two approaches for solving the shipments planning problem were used. The additional complexity introduced by considering multiple products required separation of the shipments planning stage from the ship scheduling stage. In addition, there, the time dimension is discrete (daily resolution), and ship voyages have a single loading and a single unloading port. Finally, combined ship scheduling and inventory management problems for the coal industry are studied by Shih (1997); Kao, Chen and Lyu (1993); Liu and Sherali (2000). However, integrated routing and inventory problems for other transportation modes are well discussed in the literature; and the recently published paper by Kleywegt, Nori and Savelsbergh (2004) gives a short and updated review on such problems.

As described, ship scheduling is associated with a high degree of uncertainty. However, few ship scheduling contributions take this issue explicitly into account. Christiansen and Fagerholt (2002) study a ship scheduling problem, where the ports are closed for service at night and during the weekends. In addition, the loading and/or unloading of cargoes may take several days. This means that a ship will stay idle much of the time in port, and the total time in port will depend on the ship's arrival time. The objective is to find robust schedules that are less likely to result in ships staying idle in ports during the weekend, and to impose penalty costs for arrivals at risky times (i.e. close to weekends). The computational results show that the robustness of the schedules is increased at the price of increased transportation costs.

Flexibility, as well as robustness, are two important properties in ship scheduling. The flexibility aspect is considered in Fagerholt (2001) by introducing soft time windows outside the originally hard ones to a ship scheduling model. The motivation for introducing soft time windows instead of hard ones is that by allowing controlled time window violations for some cargoes, it may be possible to obtain better schedules and significant reductions in transportation costs. To control the time window violations, inconvenience costs for servicing cargoes outside their time windows are imposed.

3. The real robust integrated ship scheduling and inventory management problem

We consider an integrated ship scheduling and inventory management problem that is based on a real planning problem for Norsk Hydro ASA. A single product is produced at sources, and we call the associated ports loading ports. Similarly, the product is consumed at certain destinations and the corresponding ports are called unloading ports. Inventory storage capacities are given in all ports, and the planners have information about the production and consumption rates of the transported product. We assume that these rates are constant during the planning period. In contrast to most ship scheduling problems, the number of calls at a given port during the planning period is not predetermined, neither is the quantity to be loaded or unloaded in each port call. The production or consumption rate and inventory information at each port, together with ship capacities and the location of the ports, determine the number of possible calls at each port, the time windows for start of service and the range of feasible load quantities for each port call.

If the product is loaded and unloaded in time at the sources and destinations, respectively, neither production nor consumption will be interrupted. The planning problem is therefore to find routes and schedules that minimize the transportation cost without interrupting production or consumption. The transporter owns the plants and controls the inventories both at the sources and the destinations, so the inventory costs do not come into play. The transporter operates a heterogeneous fleet of ships. The ships have different cost structure, load capacity and specific ship characteristics.

In addition, the shipper trades ammonia with other operators to better utilize the fleet and to ensure the ammonia balance at its own plants. This issue is taken into account in the implemented version of the model, but disregarded in this presentation.

Some ships are owned by the company and others are chartered. Time charter rates exist for all ships. In the short-term, it is of no interest to plan a change of the fleet size, so the time charter costs have no influence on the planning of optimal routes and schedules. Therefore, we are concerned with the operations of a given number of ships within the planning period.

EXAMPLE   Figure 7.1 a) shows an artificial, simplified case consisting of five ports and two ships in the beginning of the planning period (Time $= 0$). Each potential port call is indicated by a node (i, m), where i is the port number and m the call number in the port. The first call to a port has $m = 1$, the second call has $m = 2$, and so forth. We see

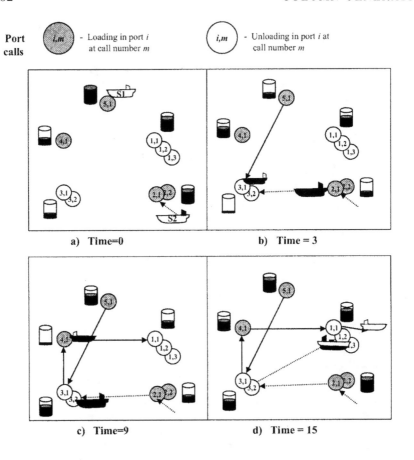

Figure 7.1. A solution for the ship planning problem with 5 ports and 2 ships.

that Port 1 can be called three times during the planning period. We have three loading ports and two unloading ports. Port 5 is the initial point for Ship 1. Ship 2 is empty at sea at the beginning of the planning period and starts service at port call $(2, 1)$ after some time. At Time $= 3$, we see from Figure 7.1 b) that Ship 1 has loaded up to its capacity at port call $(5, 1)$ before sailing to port call $(3, 1)$. Ship 2 loaded to its capacity at port call $(2, 1)$ before sailing toward port call $(3, 2)$. In the time interval $[3, 9]$, see Figures 7.1 b) and c), the load onboard Ship 1 has been unloaded at port call $(3, 1)$. Then the ship has continued to port call $(4, 1)$ to load up to the capacity of Ship 1. In the same time interval Ship 2 has sailed towards Port 3 and the ship has been waiting for some time in order to start unloading. In the time interval $[9, 15]$, see Figures 7.1 c) and d), Ship 1 sails towards Port 1 and unloads the

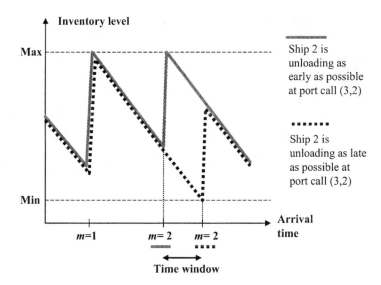

Figure 7.2.   The inventory level at Port 3 during the planning period.

load onboard at port call $(1,1)$. It leaves port call $(1,1)$ in the end of
the planning period. In the same time interval, Ship 2 has unloaded half
of its load at port call $(3,2)$ before it continued to port call $(1,2)$. Ship
2 unloads the rest of the quantity on board at port call $(1,2)$ after Time
$= 15$. For this ship, two unloading ports are called in succession. Port
call $(1,3)$ is not visited by any ship in the planning period.

Port 3 is called several times during the planning period. The solid,
grey line in Figure 7.2 shows the inventory level for Port 3 during the
planning period. Ship 2 unloads half of its load at port call $(3,2)$ as soon
as possible. Here it is important to ensure that the inventory level does
not exceed the maximum one when the unloading ends and that it is not
under minimum inventory level when the unloading starts. When the
inventory is within its bounds at arrival and departure, we know that it
is within its bounds at all other times.

Regardless of the rest of the planning problem and deviating from the
example in Figure 7.1, the broken line in Figure 7.2 illustrates another
extreme situation where Ship 2 starts the service at Port 3 as late as
possible. Here, the inventory level is not allowed to be under the minimum inventory level when the unloading starts. From these two extreme
scenarios for the inventory levels, we can derive the *feasible time window*
for port call $(3,2)$ given that the rest of the planning problem remains

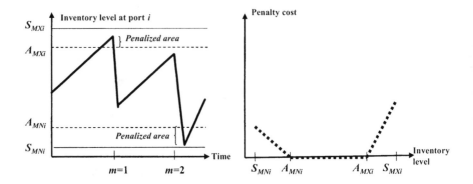

Figure 7.3. a) The inventory level and b) penalty costs for a loading port.

unchanged. In general the time windows for each port call can be derived from the inventory information.   □

As mentioned, the shipping industry is especially concerned with uncertainty in sailing times and time consumption at the ports due to
bad weather at sea and mechanical problems at the ports. To take the
uncertainty into account in a deterministic model, the minimum and
maximum inventory levels can be adjusted so that the absolute levels
cannot be reached in the planning model. The possibility of exceeding
the absolute levels in reality is reduced with increased adjustments in
the deterministic model.

Instead of adjusting the inventory bounds, we keep the old inventory
bounds and introduce another pair of soft bounds. At a given model
penalty it is possible to violate these soft inventory bounds, which we
call *alarm levels*. The constraints for not exceeding the alarm levels
become soft inventory constraints. However, we have still the hard inventory capacity constraints. In Figure 7.3 a), we see the inventory level
during the planning period for a production port. The hard inventory
(storage) interval is given by $[S_{MNi}, S_{MXi}]$, while the alarm interval is
$[A_{MNi}, A_{MXi}]$. From the figure, we see that at $m = 1$ the ship arrives
late such that the inventory level is above the upper alarm level and
will be penalized. Similarly, at $m = 2$ the ship arrives early and load
such a quantity that the inventory level is below the lower alarm level
at departure.

It is more critical to come too late to a port than too early. If you
come too late, the storage is full in a production port, and the company
has to stop the production. However, if you come too early and start

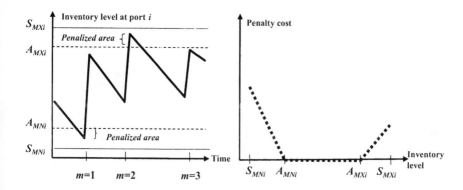

Figure 7.4.   a) The inventory level and b) penalty costs for an unloading port.

loading, the storage gets empty during the loading such that the optimal load quantity cannot be loaded. In such a situation, the actual ship must wait to start service until there is sufficient amount of product in storage. From Figure 7.3 b), we see that the artificial penalty costs for passing the lower alarm levels in a production port is less than exceeding the upper alarm levels. This comes from the fact that the capacity of the storages are more critical than the capacity of the ships.

As shown in Figure 7.3 b), we expect a linear penalty cost function, and the costs are increasing with the amount of violation. The unit penalty costs are adjusted for each port due to different production/consumption rates, inventory bounds and alarm levels. For instance, large unit penalty costs can be associated to ports with a large production rate, low storage capacity or alarm levels close to the inventory bounds, and small unit penalty costs to ports with some flexibility.

Figure 7.3 b) shows one possible cost function. Today, the company operates with alarm levels and associated costs in their regular planning, and the process of calibrating values on those costs has taken time. The penalty cost function and the alarm levels have to be determined in relation to each other. This means that changes of alarm levels would influence the penalty cost function, because the feasible interval of values for alarm levels changes. In addition, it is important that the penalty cost function is not too steep. That would simply imply that the hard inventory bounds are moved to the alarm levels.

In Figure 7.4, we see the inventory level during the planning period for an unloading (consumption) port and the associated artificial penalty costs as a function of the inventory level.

In Figure 7.4 a), we see that the unloading port is called three times during the planning period. In contrast to a production port, it is here important to ensure that the inventory level is not under minimum when unloading starts, and not above maximum when the unloading has finished.

For an unloading port, see Figure 7.4 b), the lower inventory level is most critical, and the penalty cost function has largest gradient at the lower alarm interval.

In reality, we have more flexibility than what has been modeled. In the model, we operate with fixed average speed. For a given practical situation, the transporter may be able to increase this speed to reach a port in time at an additional cost. This issue is not explicitly modelled, but the introduction of penalty costs approximately covers the increased costs.

4.   An arc flow formulation

The integrated ship scheduling and inventory management problem will be formulated as a deterministic cost minimization problem. The notation is based on the use of lower-case letters to represent subscripts and decision variables, and capital letters to represent constants. Capital letters are also used as literal subscripts to define mnemonic composite letters defining either variables or constants.

For simplicity, we disregard the initial and end conditions for the geographical position for each ship, the load onboard each ship and the inventory levels in the mathematical model. In this presentation of the model, we are not emphasizing an efficient model with regard to excluding redundant variables and constraints and reducing the variable bounds.

In the mathematical description of the problem, let $\mathcal{N}$ be the set of ports indexed by i and j. Further, let $\mathcal{V}$ be the set of available ships indexed by v. Each port can be called several times during the planning period, and $\mathcal{M}_i$ is the set of possible calls at port i indexed by m and n. The set of nodes in the flow network represents the set of possible port calls, and each such port call is specified by (i, m), $i \in \mathcal{N}$, $m \in \mathcal{M}_i$. Not all ships can visit all ports, so we specify flow networks for each ship v. Here, the set $\mathcal{A}_v$ contains all feasible arcs for ship v, which is a subset of $\{(i, m) \mid i \in \mathcal{N}, m \in \mathcal{M}_i\} \times \{(i, m) \mid i \in \mathcal{N}, m \in \mathcal{M}_i\}$.

The time required to load or unload one unit of a cargo at port i is given by T_{Qi}. T_{Sijv} represents the sailing time from port i to port j with ship v. Let $[T_{MNim}, T_{MXim}]$ denote the arrival time window associated with port call (i, m). This time window can be calculated based on other data in the model, such as the inventory conditions. See Figure 7.2 in Section 3. In addition, for some port calls the time windows might be explicitly given. In some ports, there is a minimum required time, T_{Bi}, between a departure of one ship and the arrival of the next ship, due to small port area or narrow channels from the port to the pilot station.

Each (i, m) a variable quantity interval given by $[Q_{MNim}, Q_{MXimv}]$, where Q_{MNim} is the minimum quantity to be (un)loaded at port call (i, m) given that the port is called, while Q_{MXimv} is the maximum quantity to be (un)loaded at port call (i, m) for ship v. The capacity of ship v is given by V_{CAPv}.

The levels of the inventory (or storage) have to be within a given interval at each port $[S_{MNi}, S_{MXi}]$. To increase the robustness of the generated plans, we introduce a lower alarm interval $[S_{MNi}, A_{MNi}]$ and an upper alarm interval $[A_{MXi}, S_{MXi}]$ with unit penalty costs of C_{Pi}^- and C_{Pi}^+ if the inventories are within these intervals when a ship arrives for (un)loading.

The production rate R_i is positive if port i is producing the product, and negative if port i is consuming the product. Further, constant I_i is equal to 1 if i is a loading port and -1 if i is an unloading port. From this information, the set $\mathcal{N}$ can be divided into a set of loading (pickup or production) ports, N_P, and a set of unloading (delivery or consumption) ports, N_D.

The total variable sailing cost, C_{Vijv}, that includes port, channel and fuel oil costs, corresponds to a sailing from port i to port j with ship v.

The following arc flow formulation involves two types of binary variables. The arc flow variable x_{imjnv}, $v \in \mathcal{V}$, $(i, m, j, n) \in \mathcal{A}_v$ equals 1 if ship v sails from port call (i, m) directly to port call (j, n) and 0 otherwise, and the slack variable w_{im}, $i \in \mathcal{N}, m \in \mathcal{M}_i$ is equal to 1 if no ship takes port call (i, m), and 0 otherwise. Also for other routing and scheduling problems, introduction of such slack variables has been favorable, e.g. Desaulniers et al. (1998).

In addition we use the following continuous variables. The time variable t_{im}, $i \in \mathcal{N}$, $m \in \mathcal{M}_i$ represents the time at which service begins at port call (i, m). Variable l_{imv}, $v \in \mathcal{V}$, $i \in \mathcal{N}$, $m \in \mathcal{M}_i$ gives the total load onboard ship v just after the service is completed at port call (i, m), while variable q_{Vimv}, $v \in \mathcal{V}$, $i \in \mathcal{N}$, $m \in \mathcal{M}_i$ represents the quantity loaded or unloaded at port call (i, m), when ship v visits (i, m). The variable $s_{im}(s_{Eim})$, $i \in \mathcal{N}$, $m \in \mathcal{M}_i$ represents the inventory level when service starts (ends) at port call (i, m), while $a_{im}^-(a_{im}^+)$ represents the amount the inventory level is below(above) the alarm inventory levels; $A_{MNi}(A_{MXi})$.

The objective and constraints in the arc flow formulation of the problem are as follows:

$$\min \sum_{v\in\mathcal{V}} \sum_{(i,m,j,n)\in\mathcal{A}_v} C_{Vijv} x_{imjnv} + \sum_{i\in\mathcal{N}} \sum_{m\in\mathcal{M}_i} (C_{Pi}^- a_{im}^- + C_{Pi}^+ a_{im}^+), \quad (7.1)$$

Subject to:
- Constraints regarding port visits and ship routes:

$$\sum_{v\in\mathcal{V}} \sum_{j\in\mathcal{N}} \sum_{n\in\mathcal{M}_j} x_{imjnv} + w_{im} = 1, \quad \forall i \in \mathcal{N}, \ m \in \mathcal{M}_i, \quad (7.2)$$

$$\left[\begin{array}{c} \sum_{j\in\mathcal{N}} \sum_{n\in\mathcal{M}_j} x_{jnimv} \\ -\sum_{j\in\mathcal{N}} \sum_{n\in\mathcal{M}_j} x_{imjnv} \end{array} \right] = 0, \quad \forall v \in \mathcal{V}, \ i \in \mathcal{N}, \ m \in \mathcal{M}_i, \quad (7.3)$$

$$w_{im} - w_{i(m-1)} \geq 0, \quad \forall i \in \mathcal{N}, \ m \in \mathcal{M}_i \backslash\{1\}, \quad (7.4)$$

- Constraints regarding the arrival times:

$$x_{imjnv} \left(\begin{array}{c} t_{im} + T_{Qi} q_{Vimv} \\ +T_{Sijv} - t_{jn} \end{array} \right) \leq 0, \quad \forall v \in \mathcal{V}, \ (i,m,j,n) \in \mathcal{A}_v, \quad (7.5)$$

$$T_{MNi} \leq t_{im} \leq T_{MXi}, \quad \forall i \in \mathcal{N}, \ m \in \mathcal{M}_i, \quad (7.6)$$

$$\left[\begin{array}{c} t_{im} - t_{i(m-1)} + T_{Bi} w_{im} \\ -\sum_{v\in\mathcal{V}} T_{Qi} q_{Vi(m-1)v} \end{array} \right] \geq T_{Bi}, \quad \forall i \in \mathcal{N}, \ m \in \mathcal{M}_i, \quad (7.7)$$

- Constraints regarding the quantities on the ships:

$$x_{imjnv}(l_{imv} + I_j q_{Vjnv} - l_{jnv}) = 0, \quad \forall v \in \mathcal{V}, \ (i,m,j,n) \in \mathcal{A}_v, \quad (7.8)$$

$$l_{imv} - \sum_{j\in\mathcal{N}} \sum_{n\in\mathcal{M}_j} V_{CAPv} x_{imjnv} \leq 0, \quad \forall v \in \mathcal{V}, \ i \in \mathcal{N}, m \in \mathcal{M}_i, \quad (7.9)$$

$$q_{Vimv} - \sum_{j\in\mathcal{N}} \sum_{n\in\mathcal{M}_j} Q_{MXimv} x_{imjnv} \leq 0, \quad \forall v \in \mathcal{V}, \ i \in \mathcal{N}, \ m \in \mathcal{M}_i, \quad (7.10)$$

$$\sum_{v\in\mathcal{V}} q_{Vimv} + Q_{MNim} w_{im} \geq Q_{MNim}, \quad \forall i \in \mathcal{N}, \ m \in \mathcal{M}_i, \quad (7.11)$$

- Constraints regarding the inventory conditions in the ports:

$$s_{i(m-1)} + \left[\begin{array}{c} -\sum_{v\in\mathcal{V}} I_i q_{Vi(m-1)v} \\ +R_i(t_{im} - t_{i(m-1)}) \end{array} \right] - s_{im} = 0, \quad \forall i \in \mathcal{N}, \ m \in \mathcal{M}_i, \quad (7.12)$$

$$s_{Eim} + \sum_{v\in\mathcal{V}} (I_i - R_i T_{Qi}) q_{Vimv} - s_{im} = 0, \quad \forall i \in \mathcal{N}, \ m \in \mathcal{M}_i, \quad (7.13)$$

$$s_{im} - a_{im}^+ \leq A_{MXi}, \quad \forall i \in \mathcal{N}_P, \ m \in \mathcal{M}_i, \quad (7.14)$$

$$s_{Eim} - a_{im}^+ \leq A_{MXi}, \quad \forall i \in \mathcal{N}_D, \ m \in \mathcal{M}_i, \quad (7.15)$$

$$s_{Eim} + a_{im}^- \geq A_{MNi}, \quad \forall i \in \mathcal{N}_P, \ m \in \mathcal{M}_i, \tag{7.16}$$

$$s_{im} + a_{im}^- \geq A_{MNi}, \quad \forall i \in \mathcal{N}_D, \ m \in \mathcal{M}_i, \tag{7.17}$$

$$a_{im}^- \leq A_{MNi} - S_{MNi}, \quad \forall i \in \mathcal{N}, \ m \in \mathcal{M}_i, \tag{7.18}$$

$$a_{im}^+ \leq S_{MXi} - A_{MXi}, \quad \forall i \in \mathcal{N}, \ m \in \mathcal{M}_i, \tag{7.19}$$

- Sign and integrality constraints:

$$x_{imjnv} \in \{0,1\}, \quad \forall v \in \mathcal{V}, \ (i,m,j,n) \in \mathcal{A}_v, \tag{7.20}$$

$$w_{im} \in \{0,1\}, \quad \forall i \in \mathcal{N}, \ m \in \mathcal{M}_i, \tag{7.21}$$

$$t_{im}, s_{im}, s_{Eim}, a_{im}^-, a_{im}^+ \geq 0, \quad \forall i \in \mathcal{N}, \ m \in \mathcal{M}_i, \tag{7.22}$$

$$l_{imv}, q_{Vimv} \geq 0, \quad \forall v \in \mathcal{V}, \ i \in \mathcal{N}, \ m \in \mathcal{M}_i. \tag{7.23}$$

The objective function (7.1) minimizes the total costs. Constraints (7.2) ensure that each port call is visited at most once. Constraints (7.3) describe the flow on the sailing route used by ship v. If no ship is visiting port call (i,m), we say that this port call is visited by a "dummy ship". The highest call numbers will be assigned to dummy ships in constraints (7.4), if not all port calls are visited in a specified port. Constraints (7.5) take into account the timing on the route. The time windows are given by constraints (7.6). If no ship is visiting port call (i,m), we will get an artificial start time within the time windows for a dummy ship. These artificial start times are used in the inventory balances. Constraints (7.7) prevent service overlap in the ports and ensure the order of real calls in the same port. A ship must complete its service before the next ship starts its service in the same port. The relationship between the binary flow variables and the ship load at each port call are described by constraints (7.8). Constraints (7.9) give the ship capacity at the port calls. Constraints (7.10) and (7.11) are the load bound constraints. From constraints (7.12), we find the inventory level at any port call (i,m) from the inventory level upon arrival at the port in the previous call $(i, m-1)$, adjusted for the (un)loaded quantity at the port call and the production/consumption between the two calls. From constraints (7.13) we calculate the departure inventories from the arrival inventories, adjusted for the (un)loaded quantities and the production or consumption from arrival to departure.

The general inventory limit and alarm quantity constraints at each port call are given in (7.14)–(7.19). Finally, the formulation involves binary and sign requirements (7.20)–(7.23) on the variables.

In this formulation, we have some non-linear constraints. See how these constraints are linearized in Christiansen (1999).

5.   A column generation approach

Due to the complexity of the model (7.1)–(7.23), only small sized data instances can be solved directly to optimality by use of standard commercial optimization software.

Therefore, a column generation approach is developed to solve the problem. First, in Section 5.1 we give a reformulation such that the model decomposes in an appropriate way. Then the mathematical formulation of the master problem in the Dantzig-Wolfe decomposition approach is presented in Section 5.2. Columns are generated and included in the master problem iteratively and the column generation is described in general terms in Section 5.3. These columns are generated by solving two types of subproblems; a ship routing and scheduling problem for each ship and an inventory management problem for each port. In Section 5.4, we describe the solution process of the subproblems. We focus on the inventory management subproblems to describe the effects on the algorithm when taking the uncertainty into account. Finally, in Section 5.5, we discuss the integer requirements and briefly comments on the algorithm for solving the integer part of the problem. Due to the generation of columns in the Dantzig-Wolfe decomposition approach, it is often called a column generation approach.

5.1   Model reformulation

If we try to decompose the model (7.1)–(7.23) directly, it does not separate due to the starting time t_{im} and the load quantity q_{Vimv} variables. These variables are needed in both types of subproblem; the ship scheduling and the inventory management subproblems. This issue is resolved by introducing new time and quantity variables, such that we get variables for each (i, m, v)-combination (t_{Vimv} and q_{Vimv}) and each port call (t_{im} and q_{im}) and introducing coupling constraints to the problem as follows:

$$(1 - w_{im}) \left(t_{im} - \sum_{v \in V} t_{Vimv} \right) = 0, \quad \forall i \in \mathcal{N},\ m \in M_i, \qquad (7.24)$$

$$q_{im} - \sum_{v \in V} q_{Vimv} = 0, \quad \forall i \in \mathcal{N},\ m \in M_i. \qquad (7.25)$$

We need the first parenthesis in (7.24) since the model (7.1)–(7.23) gives positive time values, t_{im}, for dummy port calls, and we want all t_{Vimv} equal to zero for such calls. At most one t_{Vimv} can be positive for each (i, m), so the upper bounds of (7.6) can be reformulated as:

$$t_{Vimv} - \sum_{j \in \mathcal{N}} \sum_{n \in \mathcal{M}_j} T_{MXim} x_{imjnv} \leq 0,$$
$$\forall v \in \mathcal{V}, \quad i \in \mathcal{N}, \; m \in \mathcal{M}_i, \quad (7.26)$$

We also need sign constraints for the newly introduced variables:

$$t_{Vimv} \geq 0, \quad \forall v \in \mathcal{V}, \; i \in \mathcal{N}, m \in \mathcal{M}_i, \quad\quad (7.27)$$
$$q_{im} \geq 0, \quad \forall i \in \mathcal{N}, \; m \in \mathcal{M}_i. \quad\quad (7.28)$$

With these new variables we are able to rewrite some of the previous constraints. In (7.5) we transform t_{im} and t_{jn} to t_{Vimv} and t_{Vjnv}, while we transform $\sum_{v \in \mathcal{V}} q_{Vimv}$ to q_{im} in (7.7), (7.11)–(7.13).

After this reformulation, the constraint set can be split into three groups. This makes it possible to solve the planning problem by a column generation approach. The first constraint group consists of (7.2), (7.24) and (7.25). These constraints are the common constraints where we synchronize the ship dependent variables and the port variables. These constraints constitute the master problem. The second constraint group consists of the ship routing constraints (7.3), (7.5), (7.8)–(7.10) and (7.26). None of the ship constraints include interaction between the ships, so these constraints constitute separate routing problems for each ship where the time windows and load on board the ship are considered. The port inventory constraints describe the inventory management for each port because there exists no interaction between the ports, and they are based on the last group of constraints (7.4), (7.6)–(7.7) and (7.11)–(7.19).

5.2   The master problem

According to the column generation approach, we use variables in the master problem corresponding to ship schedules and port call sequences instead of using the original variables from the arc-flow formulation. This transformation is well described in Christiansen (1999).

Let $\mathcal{R}_v$ be the set of ship schedules, indexed by r for ship v. Schedule r includes information about the geographical route, where X_{imjnvr} is set equal to 1 if the corresponding variable, x_{imjnv}, in the arc flow model has the value 1 for schedule r. In addition, the following information is given for each (v, r) combination: The number of visits, 0 or 1, at port call (i, m), A_{imvr}, the load quantity of each port call, Q_{Vimvr}, and the starting time of each port call T_{Vimvr}. No quantity and starting time information is given for "dummy calls". The transportation cost for sailing schedule r by ship v is C_{MVvr}.

A possible way to serve a port without any references to particular ships, is called a port call sequence. $\mathcal{S}_i$ constitutes the set of sequences for port i indexed by s. The value of W_{ims} is 1 if port call (i, m) in sequence s is served by a dummy ship and 0 otherwise. The values of Q_{Pims} and T_{Pims} represent the load quantity and starting time for port call (i, m) in sequence s, respectively. These two last values are set to zero for port calls with dummy ships. This enables us to remove $(1 - w_{im})$ in (7.24). The sum of the penalty costs for exceeding the alarm levels in port i by sequence s is C_{MPis}.

Now let variable y_{vr}, $v \in \mathcal{V}$, $r \in \mathcal{R}_v$ be 1 if ship v selects schedule r, and let variable z_{is}, $i \in \mathcal{N}$, $s \in \mathcal{S}_i$ be 1 if port i is served by sequence s. If all schedules and all sequences were known, we could solve our planning problem by solving the following master problem (7.29)–(7.37) instead of the arc flow formulation (7.1)–(7.23).

$$\min \sum_{v \in \mathcal{V}} \sum_{r \in \mathcal{R}_v} C_{MVvr} y_{vr} + \sum_{i \in \mathcal{N}} \sum_{s \in \mathcal{S}_i} C_{MPis} z_{is}, \qquad (7.29)$$

Subject to:

$$\sum_{v \in \mathcal{V}} \sum_{r \in \mathcal{R}_v} A_{imvr} y_{vr} + \sum_{s \in \mathcal{S}_i} W_{ims} z_{is} = 1, \qquad \forall i \in \mathcal{N}, \ m \in \mathcal{M}_i, \quad (7.30)$$

$$\sum_{v \in \mathcal{V}} \sum_{r \in \mathcal{R}_v} Q_{Vimvr} y_{vr} - \sum_{s \in \mathcal{S}_i} Q_{Pims} z_{is} = 0, \quad \forall i \in \mathcal{N}, \ m \in \mathcal{M}_i, \quad (7.31)$$

$$\sum_{v \in \mathcal{V}} \sum_{r \in \mathcal{R}_v} T_{Vimvr} y_{vr} - \sum_{s \in \mathcal{S}_i} T_{Pims} z_{is} = 0, \quad \forall i \in \mathcal{N}, \ m \in \mathcal{M}_i, \quad (7.32)$$

$$\sum_{r \in \mathcal{R}_v} y_{vr} = 1, \quad \forall v \in \mathcal{V}, \qquad (7.33)$$

$$\sum_{s \in \mathcal{S}_i} z_{is} = 1, \quad \forall i \in \mathcal{N}, \qquad (7.34)$$

$$y_{vr} \geq 0, \quad \forall v \in \mathcal{V}, \ r \in \mathcal{R}_v, \qquad (7.35)$$

$$z_{is} \geq 0, \quad \forall i \in \mathcal{N}, \ s \in \mathcal{S}_i, \qquad (7.36)$$

$$\sum_{r \in \mathcal{R}_v} X_{imjnvr} y_{vr} \in \{0, 1\}, \quad \forall v \in \mathcal{V}, \ (i, m, j, n) \in \mathcal{A}_v. \qquad (7.37)$$

The objective function (7.29) minimizes the transportation costs and the artificial inventory penalty costs for exceeding the alarm levels. Unlike usual vehicle routing problems solved by a column generation approach, the master problem includes additional coupling constraints for the load quantities and starting times to synchronize the ship schedule and port inventory aspects in addition to the usual visit constraints (7.30). These coupling constraints are given in (7.31) and (7.32). The convexity rows for the ships and ports are given in constraints (7.33) and

(7.34). The integer requirements are defined by (7.37) and correspond to declaring the original flow variables as binary variables.

5.3 Column generation

If all feasible ship schedules and port call sequences are known, then we get the optimal solution by solving the master problem. However, for real instances of the ship planning problem it is time consuming to generate all these schedules and sequences, and the number of such schedules and sequences would result in too many columns when solving the models. Instead, we solve the LP-relaxation of the restricted master problem which only differs from the continuous original master problem by having fewer variables. First, an initial restricted master problem is solved. Then some new columns are added to the restricted master problem. These columns correspond to ship schedules and port call sequences with least reduced costs in the solution of the master problem. This means that the dual values from the solution of the restricted master problem are transferred to the subproblems. The subproblems are solved, and ship schedules and port call sequences are generated. The restricted master problem is reoptimized with the added new columns, resulting in new dual values. This procedure continues until no columns with negative reduced costs exist, and no improvements can be made. At that point all the feasible solutions in the original master problem have been implicitly evaluated. A continuous optimal solution is then attained for both the original and the restricted master problem.

In the column generation we need symbols and values for the dual variables. The following dual variables $D_{Vim}, D_{Qim}, D_{Tim}, D_{Yv}$ and D_{Zi} are defined for constraints (7.30)–(7.34). Now we are able to write the reduced costs for y_{vr} and z_{is} as follows:

$$\overline{C}_{MVvr} = C_{MVvr} - \sum_{i \in \mathcal{N}} \sum_{m \in \mathcal{M}_i} A_{imvr} D_{Vim} - \sum_{i \in \mathcal{N}} \sum_{m \in \mathcal{M}_i} Q_{Vimvr} D_{Qim}$$
$$- \sum_{i \in \mathcal{N}} \sum_{m \in \mathcal{M}_i} T_{Vimvr} D_{Tim} - D_{Yv}, \qquad (7.38)$$

$$\overline{C}_{MPis} = C_{MPis} - \sum_{m \in \mathcal{M}_i} W_{ims} D_{Vim} + \sum_{m \in \mathcal{M}_i} Q_{Pims} D_{Qim}$$
$$+ \sum_{m \in \mathcal{M}_i} T_{Pims} D_{Tim} - D_{Zi}. \qquad (7.39)$$

5.4   The subproblems

The subproblems are formulated as shortest path problems and solved by specific dynamic programming algorithms on generated networks for each ship and each port.

In the subproblems, we discretize the possible load quantity intervals to obtain an easier structure of the subproblems. This means that instead of declaring the load quantity as a continuous variable, we just allow these variables to have a few discrete values. In the ship networks we use a separate node for each possible load quantity, while we in the port networks use a separate node for each possible cumulative load quantity. Still, the arrival time is declared as a continuous variable in the node network. Compared to the networks for the shortest path problem with time windows, see Desrosiers et al. (1995), with a node for each call, we get networks where each call is represented by several nodes. The number of calls to a port and the load quantity at each call are still unknown, and the inventory constraints remain active in the model. See Christiansen and Nygreen (1998a) for a justification of discretizing the possible load quantity interval for the real ship planning problem.

In this section, we want to show how the soft inventory constraints affect the subproblems. As mentioned, just the port subproblem is influenced by the soft inventory constraints, so we just describe this type of subproblem here. A detailed description of the ship subproblem can be found in Christiansen and Nygreen (1998b).

5.4.1   The inventory management problem for port i.

The inventory management problem (or port subproblem) can be separated into a problem for each port due to the fact that no constraint covers more than one port, and that the costs related to the ports are a sum of costs for each port. So we study the inventory management problem for *one* port i. The formulations for a loading port and an unloading port have a number of small differences. We will here only consider a loading (production) port, so we therefore assume that $i \in N_P$ whenever the formulation for a loading port differs from an unloading port. Before we continue the discussion of the port subproblems, we rewrite the objective function for port i.

Since the port call sequences with load and time information are not given in advance, we can represent the minimum reduced cost for port i, (7.39), in terms of the original variables, w_{im}, q_{im} and t_{im}. Each time the subproblem is solved, the dual variables have fixed values, so we write the objective function for the port subproblem as:

$$\min \sum_{m \in \mathcal{M}_i} \left[\begin{array}{c} C^-_{Pi} a^-_{im} + C^+_{Pi} a^+_{im} \\ -D_{Vim} w_{im} + D_{Qim} q_{im} + D_{Tim} t_{im} \end{array} \right]. \qquad (7.40)$$

If we compare (7.39) and (7.40), we see that the optimal value of (7.40) differs from the reduced cost of the corresponding port call sequence with $-D_{Zi}$. Since this term does not vary with different solutions for port i, it is excluded in the optimization. The resulting sequence is incorporated as a new column in the master problem if the value of (7.40) is less than D_{Zi}.

Before we continue, we need to take the mentioned discretization of the cumulative load quantities into account. Since the q_{im}–variables only take values in a discrete set, the same is true for the cumulative (un)loading in a port. Let $\mathcal{P}_{im}$ indexed by p or q be the set of subscripts referring to a possible cumulative (un)loaded quantity before call number m in port i, and let the corresponding quantity be Q_{imp}. From the initial inventory, the production/consumption rate and inventory capacity it is possible to calculate upper and lower bounds for the cumulative (un)loading during the planning period.

In the shortest path network for port i, we use nodes indexed by the port number, i, the call number, m, and an index, p, for the cumulative quantity (un)loaded before the call. In addition to nodes for different cumulative quantities before each call, we also need nodes, *end-nodes*, for all possible cumulative quantities during the planning period. These end-nodes are drawn as squares in Figure 7.5. The arcs that end in the end-nodes are called *end-arcs*.

Figure 7.5 shows a port network with a minimum of two and a maximum of three calls. A possible sequence may consist of nodes 1, 2, 9, 16 loading 15, 40 and 40 units at the three calls, while a sequence with two calls consists of the nodes 1, 5 and 13.

In the shortest path network where each node contains information about the cumulative amount (un)loaded in the corresponding port before the service in this node, the amount (un)loaded in a node is given on the arc out of the node.

The following constraints, with q_{im} instead of $\sum_{v \in \mathcal{V}} q_{Vimv}$, from the arc flow formulation describe the feasible region of the port subproblems: (7.4), (7.6)–(7.7) and (7.11)–(7.19).

From Figure 7.5, we see that constraints (7.4) and (7.11) are taken care of in the network. To be able to distinguish between different nodes for the same call number, we redefine some of the variables: t_{Pimp} (used for t_{im}) is the time for start of service in node (i, m, p) and q_{Pimpnq} (used for q_{im}) is the amount (un)loaded in node (i, m, p) when the next node

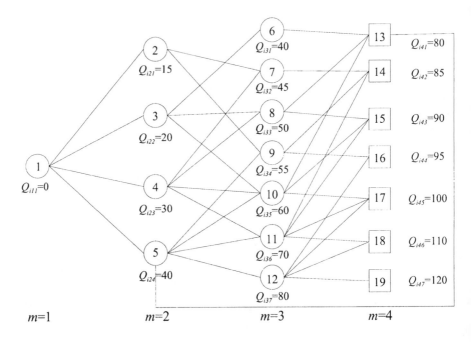

Figure 7.5.   Port network.

to visit is (i, n, q). The alarm levels at arrival in a node are called a^-_{imp} (used for a^-_{im}) and a^+_{imp} (used for a^+_{im}).

We use the equalities in (7.13) to eliminate the variable s_{Eim} from the formulation. Since the total amount (un)loaded before we arrive in a node is known, we are able to express the inventory level at arrival in a node, s_{Pim} (used for s_{im}), in the the following way, when the level in the beginning of the planning period is Q_{STi}:

$$s_{Pimp} = Q_{STi} - I_i Q_{imp} + R_i t_{Pimp}, \quad \forall m \in \mathcal{M}_i, \ p \in \mathcal{P}_{im}. \qquad (7.41)$$

All q_{Pimpnq}-variables are equal to the difference between the cumulative quantities in the nodes corresponding to (i, m, q) and (i, n, p).

$$q_{Pimpnq} = Q_{inq} - Q_{imp}, \quad \begin{aligned} &\forall m \in \mathcal{M}_i, \ p \in \mathcal{P}_{im}, \\ &\forall n \in \mathcal{M}_i, \ q \in \mathcal{P}_{in}. \end{aligned} \qquad (7.42)$$

After substituting these new variables into the formulation and also using (7.41)–(7.42) to eliminate s_{Pimp} and q_{Pimpnq}, we are able to rewrite (7.7) and (7.6) as:

$$\begin{bmatrix} t_{Pimq} \\ -t_{Pi(m-1)p} \end{bmatrix} \geq \begin{bmatrix} T_{Bi} + T_{Qi} Q_{imq} \\ -T_{Qi} Q_{i(m-1)p} \end{bmatrix}, \quad \forall m \in \mathcal{M}_i, \ q, p \in \mathcal{P}_{im}, \qquad (7.43)$$

$$T_{MNi} \leq t_{Pimp} \leq T_{MXi}, \quad \forall m \in \mathcal{M}_i, \ p \in \mathcal{P}_{im}. \tag{7.44}$$

The remaining constraints can after some manipulation be written as:

$$t_{Pimp} \leq \frac{1}{R_i}(S_{MXi} - Q_{STi} + I_i Q_{imp}), \quad \forall m \in \mathcal{M}_i, \ p \in \mathcal{P}_{im}, \tag{7.45}$$

$$t_{Pimp} \geq \frac{1}{R_i} \begin{bmatrix} S_{MNi} - Q_{STi} + I_i Q_{imp} + \\ (I_i - R_i T_{Qi})(Q_{inq} - Q_{imp}) \end{bmatrix},$$
$$\forall m \in \mathcal{M}_i, \ q, p \in \mathcal{P}_{im}, \tag{7.46}$$

$$a^-_{Pimp} = \max \begin{bmatrix} 0, A_{MNi} - Q_{STi} \\ +I_i Q_{imp} - R_i t_{Pimp} \end{bmatrix}, \quad \forall m \in \mathcal{M}_i, \ p \in \mathcal{P}_{im}, \tag{7.47}$$

$$a^+_{Pimp} = \max \begin{bmatrix} 0, Q_{STi} - A_{MXi} \\ -I_i Q_{imp} + R_i t_{Pimp} \end{bmatrix}, \quad \forall m \in \mathcal{M}_i, \ p \in \mathcal{P}_{im}. \tag{7.48}$$

The constraints we have left for this shortest path problem after the network has been constructed are therefore the following ones: Minimum time between nodes, (7.43), time windows for arriving in a node, (7.44) and (7.45), and earliest arrival time in a node depending on the leaving arc, (7.46).

We rewrite the objective (7.40) once more, partly with the new variables but without any use of new summation signs, before we write the non-linear expressions (7.47)–(7.48) into the new objective.

$$\min \sum_{m \in \mathcal{M}_i} \begin{bmatrix} C^-_{Pi} a^-_{Pimp} + C^+_{Pi} a^+_{Pimp} - D_{Vim} w_{im} \\ +D_{Qim} q_{Pimpnq} + D_{Tim} t_{Pimp} \end{bmatrix}. \tag{7.49}$$

We now let the binary variable x_{Pimpnq} be 1 if the arc from (i, m, p) to (i, n, q) is used in the shortest path. The "call" number representing the end of the planning period is numbered $n = |\mathcal{M}_i| + 1$. End-arcs that go from nodes with $m < |\mathcal{M}_i|$ to $n = |\mathcal{M}_i| + 1$ represent situations where $w_{ik} = 1$ for $m < k < n$. By use of x_{Pimpnq}, (7.42) and (7.47)–(7.48), we can eliminate w_{im}, q_{Pimpnq}, a^-_{Pimp} and a^+_{Pimp} from the objective function and get:

$$\min \sum_{m \in \mathcal{M}_i} \sum_{p \in \mathcal{P}_{im}} \sum_{n > m} \sum_{q \in \mathcal{P}_{in}} C^-_{Pi} \max \begin{bmatrix} 0, A_{MNi} - Q_{STi} + \\ I_i Q_{imp} - R_i t_{Pimp} \end{bmatrix} x_{Pimpnq} \tag{7.50}$$

$$+ \sum_{m \in \mathcal{M}_i} \sum_{p \in \mathcal{P}_{im}} \sum_{n > m} \sum_{q \in \mathcal{P}_{in}} C^+_{Pi} \max \begin{bmatrix} 0, Q_{STi} - A_{MXi} - \\ I_i Q_{imp} + R_i t_{Pimp} \end{bmatrix} x_{Pimpnq} \tag{7.51}$$

$$+ \sum_{m \in \mathcal{M}_i} \sum_{p \in \mathcal{P}_{im}} \sum_{n > m} \sum_{q \in \mathcal{P}_{in}} D_{Tim} t_{Pimp} x_{Pimpnq} \tag{7.52}$$

$$+ \sum_{m \in \mathcal{M}_i} \sum_{p \in \mathcal{P}_{im}} \sum_{n > m} \sum_{q \in \mathcal{P}_{in}} D_{Qim}(Q_{inq} - Q_{imp}) x_{Pimpnq} \tag{7.53}$$

$$+ \sum_{m \in \mathcal{M}_i} \sum_{p \in \mathcal{P}_{im}} \sum_{q \in \mathcal{P}_{in}} \sum_{n=|\mathcal{M}_i|+1} \left(- \sum_{k=m+1}^{n-1} D_{Vik} \right) x_{Pimpnq}. \tag{7.54}$$

The final objective function for the port subproblem, (7.50)–(7.54), is written with five terms. This function has to be minimized subject to the time constraints (7.43)–(7.46). The three first terms (7.50)–(7.52) of the objective depend on the nodes and the arrival times in the nodes. The next two terms (7.53)–(7.54) depend on the arcs used out of the nodes. The coefficients in the three last terms (7.52)–(7.54) depend on the dual variables in the master problem, so they will change from one master iteration to the next. Constraints (7.46) are used in each port subproblem and they are special in the sense that one part of the arrival time window in a node depends on the arc used out of the node. This makes the port shortest path problems slightly harder to solve. However, since all the port networks are acyclic, these shortest path problems are easy to solve.

Since the terms (7.50)–(7.52) depend linearly or piecewise linearly on the continuous start time variable, t_{Pimp}, the dynamic programming formulas become piecewise linear functions of time, t_{Pimp}. By minimizing over the arcs into each node, we obtain a piecewise linear cost function with possible steps. The cost function is nonincreasing because we use the costs of beginning within time t_{Pimp}. Introduction of inventory penalty costs increases the number of linear pieces in the cost function, so the computational time is expected to increase slightly when considering alarm levels. See Christiansen and Nygreen (1998b) for a detailed discussion of the dynamic programming solution method for the case without alarm levels and penalty costs.

5.5   Integer requirements and the integer solution approach

The generation of columns for the master problem continues until no negative reduced costs found in the subproblems exist, and no improvements can be made. A continuous optimal solution is then attained.

For each ship, the solution is regarded as integer feasible if all partly used routes ($y_{vr} > 0$) for a given ship visit the same set of port calls in the same order. This means that the underlying flow variables must be binary, and constant X_{imjnvr} gives information about the number of times ship v sails directly from port call (i, m) to port call (j, n) on route

r. These binary requirements to the underlying flow variables are given in (7.37).

We see that the final solution may consist of several positive y-values for the same ship, representing the same geographical route by having the same values (0 or 1) on the X_{imjnvr}-constants, but they may differ in start time and/or load quantity. The convex combination of columns for each ship, by using the y-weights, gives the start time and the load quantity of each port call. For each ship, all partial used routes satisfying (7.37) give the same transportation costs, so the convex combination of the costs has the same costs. The calculation of the transportation costs becomes correct.

The binary requirements (7.37) also imply that all positive port variables give the same number of calls in each port. We can therefore treat z_{is} as an ordinary continuous variable in respect of integer number of calls in the ports. A convex combination of columns for each port, by using the z-weights, gives the start time, load quantity and penalty costs. The resulting start time and load quantity involve no problems, but let us consider the penalty costs once more.

The total penalty costs for a port i is the sum of the calculated penalty costs for the partial used sequences $(z_{is} > 0)$; $\sum_{s \in \mathcal{S}_i} C_{MPis} z_{is}$. Each C_{MPis} consists of the calculated penalty costs for each call in sequence s for port i. This means that we calculate the penalty for each sequence and use a convex combinations of those. In order to be correct, we should first take a convex combination of the arrival times for each call and then use these times in (7.47)–(7.48) to calculate the alarm quantity and penalty.

In treating the z_{is} variables as continuous, we overestimate the penalty costs. Since these costs are artificial and here used to force the solution away from the inventory bounds, we accept the approximate calculation of the penalty costs.

So by introducing soft inventory constraints, we include no new integer requirements to the model. The LP-relaxed solution approach described in this section is embedded in a branch-and-bound search. In each branch-and-bound node, the existing feasible columns remain in the matrix and new columns are generated until no improvements of the continuous solution are possible. The integer requirements can be achieved by branching on the underlying structure. We change the subproblems, and by this the networks, dynamically such that we exclude solutions not permitted because of the last branching. Branching entities are combinations of arcs and nodes in the networks and the time window widths. The different branching entities are further described in Christiansen and Nygreen (1998a).

Table 7.1.   A few computational results

Instance	I1a	I1b	I2a	I2b	I3a	I3b	I4a	I4b
Relative TW-width	8		15		28		40	
Soft inv.constraints	No	Yes	No	Yes	No	Yes	No	Yes
IP-value	54.9	64.3	54.5	56.7	53.9	54.5	53.5	54.0
Transportation cost	54.9	54.9	54.5	54.9	53.9	54.1	53.5	54.0
Penalty cost	(22.7)	9.2	(8.1)	1.7	(18.1)	0.2	(18.8)	0.0
Over estimation	-	0.2	-	0.1	-	0.2	-	0.0
BB-nodes	1	5	3	15	35	15	157	99
# port columns	143	221	195	522	1521	1004	7135	5002
# ship columns	133	219	205	787	1976	1241	9169	6067

6.   Computational results

Here, we present a few results from a real planning problem in Europe. The problem is solved by the column generation approach described in Section 5 by combining a commercial subroutine library for mathematical programming with user written code. The library has been used to solve the master problem while the shortest path algorithms and the branch-and-bound algorithm are user-written.

In Table 7.1, we give results for a case of the planning problem for Northern Europe. The case consists of 3 ships and 11 ports. Some ports can be called only once, while others can at most be called five times during the planning period. The length of the planning period in this case is one month. Here, we want to focus on the structural effects of introducing soft inventory constraints into the problem, so we limit ourselves to present one case from the real planning problem.

In order to examine how the widths of the time windows influence the use of penalties, we run the case with four different sets of time window widths. The instance with widest time window width reflects the real planning problem best, while the others have artificially tight time windows. We give the average time window width in percentage of the planning period length. These percentages express the widths after they have been reduced in a preprocessing phase. We present results with average time window widths from 8% to 40%.

Each of the instances in Table 7.1 is run with and without soft inventory constraints, and we report the objective values, the number of nodes in the branch-and-bound tree and the number of generated ship and port columns for both objective types.

In Table 7.1, we give information about the integer objective values in row "IP-value." These values decrease with increasing time window width, as cheaper routes may be found. In addition, we see that the objectives increase by including the soft inventory constraints. However, this cost increase becomes less with increased time window width, because the feasible region has increased and there is more flexibility built into the model.

Further on, we split the objective value into (1) the transportation costs for the ships, (2) the correctly calculated, artificial penalty costs for violating the soft inventory constraints and (3) the over estimation of the penalty costs for the I?b-instances. The sum of the two last cost rows give the approximate penalty costs used in the solution approach. The calculated penalty costs are given in parentheses for the instances (I?a) run without soft inventory constraints, as well. In general more expensive route pattern is found, when we include the soft inventory constraints. The penalty costs decreases with increased time windows for the instances with soft inventory constraints. For the I4b-instance no penalty costs occur at all. We see that the difference in the correct cost calculation and the cost calculation chosen is limited for all except for one instance; I3b. For that instance the penalty costs and the over estimation is equal. However, the over estimation is limited compared to the calculated penalty costs for the instances where the soft inventory constraints are not included in the optimization.

For all, except instance I1a, we had to call branch-and-bound. The same branching strategy is used in solving all the instances. In the table, we specify the number of nodes explored in the branch-and-bound tree until the search is completed. From the number of nodes we see that there are no evident increase in computational burden by including the soft inventory constraints.

7.   Concluding remarks

This ship planning problem is an integrated multi pickup and delivery problem with time windows and a multi inventory management problem.

Time constrained routing problems solved by a Dantzig-Wolfe decomposition approach, typically, decompose into a subproblem for each vehicle. In contrast, this ship planning problem decomposes into a subproblem for each ship and a subproblem for each port, and the master problem includes additional coupling constraints to synchronize the ship schedules and port inventory aspects.

The main objective in this planning situation is to keep the inventory level within its bounds at all actual ports, so there should be no need to stop the production at any ports due to full or empty inventories caused by missing transportation possibilities. The shipping industry is concerned with uncertainties in sailing time and time consumption at port. To reduce the possibility of violating the inventory constraints in practice, we introduce soft inventory constraints. This means that we introduce artificial penalties to the underlying model for violating the soft inventory constraints. In this way, we try to force the solution away from its inventory bounds and get more robust routes.

We saw that the inventory constraints could be transformed into time windows for start of service at a port call, when we introduced networks for the ship routing subproblem and the port inventory management subproblem. Similarly, soft time windows replace the soft inventory bound.

The soft time windows can easily be incorporated into the solution approach. In the port subproblem, we search for port call sequences with least reduced cost. Before introducing the soft time windows, these cost functions were found to be piecewise linear and non-increasing over the start time, due to continuously declared start times and the dual information from the time coupling constraints. For each port call, the penalty cost function is also piecewise linear over the start time, with up to three pieces. The structure of the existing cost function and the new penalty cost contribution is similar, so the soft time windows can easily be incorporated into the subproblem.

The master problem gets additional cost terms in the objective function representing the artificial penalty costs. Appropriate penalty costs are associated with each port call sequence, and a convex combination of port call sequences for each port gives the start time, load quantity and penalty costs for each port call. The penalty costs are calculated in an approximate way to avoid the introduction of new integer requirements. Technically, there is no problem introducing new integer requirements, but the computational time would increase. The computational results show that the over estimation of penalty costs is limited compared to the calculated penalty costs for the instances where the soft inventory constraints are not included in the optimization. Therefore, we regard the approximate penalty costs as acceptable.

When we introduce the soft inventory constraints into the model, we see from the computational results that more expensive route patterns are found to reduce the penalty costs as much as possible. For illustration purposes, we run some instances of the planning problem with artificial tight time windows. The instances with widest time window width describes the real planning problem best. We saw that the penalty costs decreases with increased time window, because the feasible region has increased and there is more flexibility built into the model.

References

Christiansen, M. (1999). Decomposition of a combined inventory and time constrained ship routing problem. *Transportation Science*, 33(1):3–16.

Christiansen, M. and Fagerholt, K. (2002). Robust ship scheduling with multiple time windows. *Naval Research Logistics*, 49(6):611–625.

Christiansen, M., Fagerholt, K., Nygreen, B., and Ronen, D. (2004). Maritime Transportation. Forthcoming in: *Handbooks in Operations Research and Management Science, Transportation* (C. Barnhart and G. Laporte, eds.), North-Holland, Amsterdam.

Christiansen, M., Fagerholt, K., and Ronen, D. (2004). Ship routing and scheduling: status and perspectives. *Transportation Science*, 38(1):1–18.

Christiansen, M. and Nygreen, B. (1998a). A method for solving ship routing problems with inventory constraints. *Annals of Operations Research*, 81:357–378.

Christiansen, M. and Nygreen. B. (1998b). Modelling path flows for a combined ship routing and inventory management problem. *Annals of Operations Research*, 82:391–412.

Desaulniers, G., Desrosiers, J., Ioachim, I., Solomon, M.M., and Soumis, F. (1998). A Unified Framework for Deterministic Time Constrained Vehicle Routing and Crew Scheduling Problems. In: *Fleet Management and Logistics* (T. Crainic and G. Laporte, eds), pp. 57–93, Kluwer Academic Publishers.

Desrosiers, J., Dumas, Y., Solomon, M.M., and Soumis, F. (1995). Time Constrained Routing and Scheduling. In: *Handbooks in Operations Research and Management Science, Network Routing* (M.O. Ball, T.L. Magnanti, C.L. Monma, and G.L. Nemhauser, eds), volume 8, pp. 35–139, North-Holland, Amsterdam.

Fagerholt, K. (2001). Ship scheduling with soft time windows—An optimization based approach. *European Journal of Operational Research*, 131:559–571.

Flatberg, T., Haavardtun, H., Kloster, O., and Løkketangen, A. (2000). Combining exact and heuristic methods for solving a Vessel routing problem with inventory constraints and time windows. *Ricerca Operativa*, 29(91):55–68.

Fox, M. and Herden, D. (1999). Ship Scheduling of Fertilizer Products. *OR Insight*, April–June.

Ioachim, I., Gélinas, S., Soumis, F., and Desrosiers, J. (1998). A dynamic programming algorithm for the shortest path problem with time windows and linear node costs. *Networks*, 31(3):193–204.

Kao, C., Chen, C.Y., and Lyu, J. (1993). Determination of optimal shipping policy by inventory theory. *International Journal of Systems Science*, 24(7):1265–1273.

Kleywegt, A.J., Nori, V.S., and Savelsbergh, M.W.P. (2004). Dynamic programming approximations for a stochastic inventory routing problem. *Transportation Science*, 38(1):42–70.

Liu, C.-M., and Sherali, H.D. (2000). A coal shipping and blending problem for an electric utility company. *OMEGA*, 28:433–444.

Ronen D. (1983). Cargo ships routing and scheduling: Survey of models and problems. *European Journal of Operational Research*, 12:119–126.

Ronen D. (1993). Ship scheduling: The last decade. *European Journal of Operational Research*, 71:325–333.

Ronen D (2002). Marine inventory routing: Shipments planning. *Journal of the Operational Research Society*, 53:108–114.

Shih, L.-H. (1997). Planning of fuel coal imports using a mixed integer programming method. *International Journal of Production Economics*, 51:243–249.

Chapter 8

SHIP SCHEDULING WITH RECURRING VISITS AND VISIT SEPARATION REQUIREMENTS

Mikkel M. Sigurd
Nina L. Ulstein
Bjørn Nygreen
David M. Ryan

Abstract   This chapter discusses an application of advanced planning support in designing a sea-transport system. The system is designed for Norwegian companies who depend on sea-transport between Norway and Central Europe. They want to achieve faster and more frequent transport by combining tonnage. This requires the possible construction of up to 15 new ships with potential investments of approximately 150 mill US dollars. The problem is a variant of the general pickup and delivery problem with multiple time windows. In addition, it includes requirements for recurring visits, separation between visits and limits on transport lead-time. It is solved by a heuristic branch-and-price algorithm.

1.   Introduction

Increased pressure on road networks and increasing transport requirements make companies look for new transport solutions. This spurred an initiative to create a new liner shipping service. The initiative came from a group of Norwegian companies who need transport between locations on the Norwegian coastline and between Norway and The European Union. While few producers on the Norwegian coast have sufficient load to support a cost efficient, high frequency sea-transport service, they can reduce costs and decrease transport lead-time by combining their loads on common ships. They agreed upon a tender (transport offer) which was proposed to a number of shipping companies. The tender specifies the number of cargos per week and time constraints for pickup and delivery. It also states the requirements regarding ship-types and loading and unloading techniques. For rapid handling, all goods must be transported in containers. Finally the tender specifies the yearly payment each company will make to be part of this transportation system. Today there are neither ships nor harbour facilities to support the proposed solution. Thus, major investments are necessary. Estimates indicate that investments in ships alone, can amount to about 150 mill US dollars. We present a model which calculates an optimal solution to the requirements in the tender. The model includes selection of an optimal fleet composition, ship routing and visit-schedules. The problem is formulated as a set partitioning model and solved by a heuristic branch-and-price algorithm. The next section presents the system requirements in more detail. In Section 3 the problem is compared to other fleet design and routing problems. Our choice of master and pricing problem is presented in Sections 4 and 5. The branching strategy is described in Section 6. Section 7 presents results, while Section 8 conclude with some remarks on the model choice and on the results.

2.   Problem description

In this section we will first take a closer look at the ship requirements, and then describe requirements pertaining to customers transport demand.

To achieve fast transport, it is necessary to limit both the travel time and the loading and unloading time. Faster ships can substantially decrease the travel time. While traditional cargo ships travel at about 16 knots, cargo ships can be designed to travel at up to 25 knots. With this speed a ship can travel from Trondheim to Rotterdam in 35 hours. This represents a reduction in travel time of about 20 hours compared to traditional cargo ships. Although higher speed increases variable costs, as fuel consumption for ships increase exponentially with speed, this may be outweighed by a reduction in the number of required ships, reduced inventory costs and the need to satisfy customers' lead-time requirements. Combining tonnage leads to an increased number of port visits. To limit the loading and unloading time, ships need to use a roll-on roll-off technology. This means that cargo is rolled onto the ships by trucks and not lifted by cranes. The existing fleet of ships serving the North-Sea region cannot adopt this technology. Therefore the system requires construction of new ships. The shipping companies have in collaboration with the customers proposed a number of candidate ship-types. It is possible to construct any number of each candidate ship-type. The candidate ships vary in cost, capacity and speed. Some ships have properties which

Table 8.1. Alternative visit-patterns for a customer with three visits per week and at least one day in-between visits.

nr	Mon	Tue	Wed	Thu	Fri	Sat	Sun
1	X	O	X	O	X	O	O
2	X	O	X	O	O	X	O
3	X	O	O	X	O	X	O
4	O	X	O	X	O	X	O
5	O	X	O	X	O	O	X
6	O	X	O	O	X	O	X
7	O	O	X	O	X	O	X

prevent them from visiting particular harbours. The constructed ships will be used in full by the system. The fixed weekly cost of a ship covers crew costs, financial costs and maintenance costs. The financial cost of a ship equals the depreciation cost from constructing the ship. The variable cost depends mainly on the fuel consumption, which is calculated as a function of the travel distance and speed.

The tender includes transport of 68 cargos per week between 21 harbours, 20 in Norway and one in Rotterdam. The total transport volume is approximately 2000 containers weekly. All customers specify a pickup port and a delivery port, a weekly load and a frequency. The frequency states the number of shipments per week. The weekly load is distributed evenly among the shipments. For each shipment, there can be single or multiple *time-windows* for pickup and for delivery. If, for example a cargo can be collected between Monday and Wednesday but only within the opening hours of the port, there will be three time windows, one for each day. The maximum *lead-time* from pickup to delivery limits the time from when a cargo is picked up until it is delivered. Lead-time requirements apply to perishable goods such as fish and to goods where customers require rapid delivery. Customers with multiple visits per week, can demand a minimum time between visits or limit the number of visits during a given number of days. If a customer requires at least one day between visits in the pickup port and a ship visits on Monday, then visits on Sunday and Tuesday are forbidden. Table 8.1 shows the seven feasible *visit-patterns* for a customer with three cargos and at least one day in-between service. If the customer instead requests not more than two visits per three days, there are 21 additional feasible visit-patterns. It is possible to enumerate all feasible visit-patterns for customers with separation requirements for the visits. If desired, this

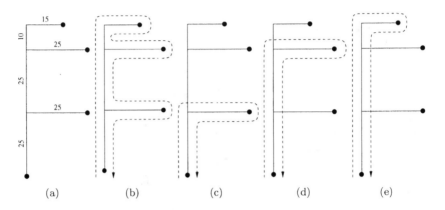

Figure 8.1. An example showing how extending the planning period from one week to two weeks can improve the solution.

can also be done for customers without requirements for separation of visits.

To facilitate planning, companies want a weekly recurring *visit-schedule*, similar to a bus-schedule. The visit-schedule must state the day when each visit is made, but does not restrict the time of day for making the visit. This implies that the same visit-pattern will be repeated each week for each customer. The visit-schedule is not given a priori, so the shipping companies must decide on an optimal visit-schedule which complies with the customers requirements. Let a route denote a sequence of pickup and delivery visits on given days made by a particular ship-type. A recurring visit-schedule can be met by a set of weekly recurring routes, as cyclic routes with the duration of one week will visit the same customer(s) on the same day each week. Alternatively, ships can use cyclic routes with a longer duration than one week. Then the requirement for weekly recurring visit-patterns at each customer must be fulfilled by combining routes. For example, if the route length is two weeks, two ships can alternate on visiting the same customers every second week. Figure 8.1 illustrates how it is possible to reduce costs by 50% by allowing two-week routes instead of one-week routes. Figure 8.1(a) depicts a problem instance with four harbours. The most southerly harbour is Rotterdam and the other three harbours are on the Norwegian coastline. The three Norwegian harbours must be visited once a week to pick up cargo bound for Rotterdam. The numbers on the edges indicates the length of the edge in hours of sailing. Thus the route shown in 8.1(b) takes 250 hours ≈ 10.5 days and it covers all visits. Thus two ships sailing this route, starting one week apart will cover all visits in every week. On the other hand, if the maximum route length is set to one-week, the three routes shown in Figure 8.1(c)–8.1(e) are required to cover all visits, which means using an additional ship.

The duration of the routes will be a multiple of the length of the recurring visit-schedule, which is one week. Ship owners suggest a sensible duration for routes should be two weeks. This is therefore used in the further description. With a two week planning period, the planning period includes two weekly visit-schedules. After two weeks, the plan is repeated. The ships travel on two week cyclic routes and visit the depot at least once during this time. The ships can be anywhere on their route at the start of the planning period.

Based on the tender and on the candidate ship-types, the mathematical model produces:

- a fleet of ships,

- a published and fixed visit-schedule,

- a recurring route for each ship.

The model takes into account requirements for:

- separation of visits to the same customer,

- time-windows (multiple) for pickup and delivery,

- visits on same days each week,

- lead-time from pickup to delivery.

The properties of the routes are:

- start and end at the same harbour,

- maximum route length (in weeks),

- time for each visit,

- load always less than ship capacity,

- lead-time from pickup to delivery must be met,

- time window constraints must be met,

- port/ship compatibility must be met.

Unlike bus scheduling, the ships do not return back to the "garage" at regular intervals. However, one harbour is special and will be used as a depot. The southernmost harbour, Rotterdam, receives and sends large quantities of goods from and to Norwegian harbours. Since Rotterdam is relatively far away from Norway, we can safely assume that optimal routes never bring goods destined for Norway, to Rotterdam and back to Norway. Thus, after unloading and before reloading, ships will be empty in this harbour. Because of the large portion of cargo destined for Rotterdam, all ships will travel from Norway to Rotterdam at least once during a planning period. Hence, Rotterdam is used as the depot for all ships. Ships can leave from the depot at any time during the week. The ships are allowed to visit the depot again within the route, but they need to return to the depot and discharge all cargo within the maximum route duration. It is possible to include additional depots, in particular, if the assumption that ships will always be empty in the depot still holds.

The model assumes deterministic transport demand. This is based on data from each of the collaborating companies. The sensitivity of the solution to changes in demand is further discussed in Section 8.

3.   Similar problems in the literature

3.1   Sea-transport

Ship scheduling and fleet planning often involves decisions which represent large monetary values. Constructing or acquiring a ship costs millions of US dollars and daily operation costs amounts to thousands of dollars. Improved utilization and fleet planning can lead to great benefits. This should motivate the use of decision support. According to Ronen (1993), optimization based decision support was not often applied in the shipping industry before 1993. This lack of interest, (Ronen, 1983), explains as a result of strong traditions for other planning methods as well as a range of operational factors in which sea transport problems differs from vehicle routing problems. In a more recent review Christiansen, Fagerholt and Ronen (2004) report on an increase in the number of studies for maritime transport planning. As most of the studies consider industrial and tramp shipping, the literature on liner shipping problems is still sparse. This does not reflect the development in global capacity in liner shipping, which was nearly doubled from 1991 to 1995.

Most liner shipping problems consider fleet size and mix in addition to fleet deployment. Cho and Perakis (1996) present models for routing and fleet design in a liner shipping problem. They present a LP model for a problem with a fixed fleet of ships and a set of candidate routes. This becomes an IP problem when the number of ships is allowed to vary. The IP problem minimizes the costs of satisfying customers demand by assigning ships to routes. The candidate routes are suggested by the planners. In a later application Powell and Perakis (1997) expand on the issue of fleet deployment and include penalties for days when ships are idle. Fagerholt (1999) describes another model for a liner shipping system. Because of limited total route duration, all routes are generated a priori. Then, from a set of predefined alternative ships, the least costly ship is assigned to each route and set partitioning is used to select among the routes. The limitation to all three approaches is that they only work on relatively constrained problems. In contrast to the former studies, Rana and Vickson (1991) present a profit maximization model which constructs routes of favourable visits and selects routes from this set. They apply a Lagrange decomposition approach to solve the problem. By relaxing the demand constraints, the problem is decomposed into one problem for each ship.

Christiansen (1999) describes another decomposition approach to solve an industrial shipping problem with transport of only one bulk product, but it also involves managing inventory held at the ports. Therefore, it includes decisions on both load quantity and routes. The solution approach uses Dantzig-Wolfe decomposition. Suggestions for routes and load quantities are generated in a subproblem. The master problem selects routes that minimize cost while controlling inventory levels. This problem is solved by a heuristic branch-and-bound algorithm with generation of new columns when needed.

3.2   Related problems

Similar problems to the liner shipping problem are found in freight transport, train scheduling and in the airline industry. Such problems are often referred to as service network design problems, see Crainic and Laporte (1997).

The demand requirements with sets of legal visit-patterns at each customer are similar to those given in periodic VRP (PVRP) problems. Cordeau, Gendreau and Laporte (1997) present the currently known best heuristic for the PVRP. However, as their method uses the fact that all routes in the PVRP last for only one day and also do not have pickup and delivery, their method is not directly relevant to our problem.

4.   Mathematical model formulation

The model constructs a fleet of ships, a visit-schedule and routes. A Dantzig-Wolfe decomposition is applied to formulate a set partitioning problem which is in turn solved by a branch-and-price algorithm. Desrosiers et al. (1995) reports that this approach has been successfully applied to numerous routing and scheduling problems while Barnhart et

al. (1998) discuss how this method can be used to solve various classes
of integer models.

The decomposition gives a *master problem* and a *pricing problem*. The
master problem selects, from a set of candidate routes, routes which
minimize the cost of satisfying customers requirements. With a huge
number of possible routes only a subset of routes can be included in the
master problem. New routes are generated in the pricing problem and
added to the master problem. Dual values from the master problem are
used to modify the objective function of the pricing problem to encourage
new routes for poorly serviced cargoes. With delayed column generation,
new routes are generated throughout the branching process. When no
improving routes can be found the algorithm terminates. As further
described in Section 5, heuristic methods are used to solve the pricing
problem. This gives a heuristic branch-and-price algorithm.

The system must service a given number of cargos. Each cargo has an
origin and a destination node. There may be time window constraints
on pickup and delivery and possibly lead-time constraints for delivery
of the cargo. Since new ships will be constructed, ship characteristics
are not fixed. In theory, both speed and capacity of the ships could be
modelled as continuous variables. However, shipping companies prefer
some standardization. Therefore speed and capacity are modelled as
stepwise functions. This results in a finite number of ship-types. There
is no limit on the number of ships of each type.

4.1   The master problem

The master problem selects routes from a subset of all routes. Recall that a route is a sequence of pickup and delivery visits on given
days made by a particular ship-type. There are two ways to model the
composition of routes for the planning problem. One approach is to
construct routes which last for the duration of the planning period in
the pricing problem. This way, only complete routes are selected in the
master problem and selecting a route also involves using a ship. Alternatively, a collection of shorter routes can be proposed to the master
problem. Then a sequence of shorter routes for each ship are selected
in the master problem. This requires additional constraints on the daily
utilization of unique ships. Both approaches have been tested. The
approach with complete routes gave better results. Therefore only this
approach is further described.

The plan for one planning period is repeated at the end of the period.
With complete routes which last for the duration of the planning period,
a ship travels exactly one route. A ship can be anywhere on its route

at the beginning of the planning period. In other words, a route which start from the depot on day d in the planning period will wrap around and finish on day $d-1$ in the planning period. Let S denote the set of ship-types s, and $\mathcal{R}^s$ denote the set of candidate routes r for ship-type s. With complete routes, the cost of using a ship can be included in the cost for the route, C_r^s, and calculated in the pricing problem.

Let the variable z_r^s be one if route r, with ship-type s, is used. There are no restrictions on how many times each ship-type can be used, so different routes which use the same ship-type can be selected simultaneously. However, a particular route r can only be used once, as each visit can only be made once, so $z_r^s \in \{0,1\}$. The master problem selects routes z_r^s to minimize system costs, while ensuring that all requests are served. The cost minimization objective is given in (8.1).

$$\min \sum_{s \in S} \sum_{r \in R^s} C_r^s z_r^s. \tag{8.1}$$

Each customer requires one or more shipments per week between its pickup port and its delivery port. The cargos which belong to one customer are identical in terms of load, lead-time requirements, origin and destination. Let $\mathcal{G}$ denote the set of all ports, and g denote a particular pickup or delivery port. Note that ports are customer-specific so that different ports may correspond to the same geographical location. The term "harbour" refers to a particular geographical location. A visit-pattern gives the legal days d of visit at port g in compliance with the separation requirements for the port. Each port g must be visited according to one of its feasible visit-patterns. Since customers want weekly recurring visits a one-week visit-pattern is repeated for each week in the planning period. With a one week recurring visit-schedule and a two week planning period, each visit-pattern consist of 14 days. Let $\mathcal{D}$ denote the set of all days in the planning period. Within these 14 days, a weekly pattern is repeated two times. Let $\mathcal{P}^g$ denote the set of feasible patterns p for port g. The first week $d = \{1, \ldots, 7\}$ of a visit-pattern with visits on Tuesday, Thursday and Saturday, can be expressed by: $\{\overline{M}_{g1}^p, \ldots, \overline{M}_{g7}^p\} = [0\ 1\ 0\ 1\ 0\ 1\ 0]$. The variable u_g^p is one when pattern p is used for port g. B_{gdr}^s is one when route r with ship s visits port g on day d. Restriction (8.2) requires that if a route visits port g on day d, then a visit-pattern p with a feasible visit on day d must be selected.

$$\sum_{s \in S} \sum_{r \in R^s} B_{gdr}^s z_r^s - \sum_{p \in \mathcal{P}^g} \overline{M}_{gd}^p u_g^p = 0 \quad \forall g \in \mathcal{G}, \qquad d \in \mathcal{D}. \tag{8.2}$$

These pattern matching constraints can be reformulated as partitioning constraints by replacing the service pattern $\overline{M}_{gd}^p$, by the "inverse".

The "inverse" representation is: $\{M^p_{g1}, \ldots, M^p_{g7}\} = [1\ 0\ 1\ 0\ 1\ 0\ 1]$. In addition, restriction (8.4) ensures that exactly one pattern is used for each customer. This requirement also ensures that all cargos are served exactly once. The master problem can now be expressed as:

$$\min \sum_{s \in \mathcal{S}} \sum_{r \in R^s} C^s_r z^s_r \tag{8.3}$$

subject to

$$\sum_{p \in \mathcal{P}^g} u^p_g = 1 \quad \forall g \in \mathcal{G}, \tag{8.4}$$

$$\sum_{s \in \mathcal{S}} \sum_{r \in R^s} B^s_{gdr} z^s_r + \sum_{p \in \mathcal{P}^g} M^p_{gd} u^p_g = 1 \quad \forall g \in \mathcal{G}, \quad d \in \mathcal{D}, \tag{8.5}$$

$$z^s_r \in \{0, 1\} \quad \forall s \in \mathcal{S}, \quad r \in R^s, \tag{8.6}$$

$$u^p_g \in \{0, 1\} \quad \forall g \in \mathcal{G}, \quad p \in \mathcal{P}. \tag{8.7}$$

The combination of routes selected in the master problem must visit each port on days which comply with a feasible visit-pattern for the port. Selecting a visit-pattern fulfills both the requirement for weekly recurring visits and for separation of visits. In addition, it ensures that all cargos from/to the port are collected or delivered. Since pickup and delivery restrictions are fulfilled in all routes which are proposed by the pricing problem, it is sufficient with a pattern restriction in either the pickup port or the delivery port for each customer to ensure that all shipments are made. In this system the customers want both regular departures and arrivals of goods and therefore most customers specify separation requests in both pickup and delivery ports.

Constraints (8.4) and (8.5) have dual variables π_g and δ_{gd} respectively. Since constraint (8.4) does not involve the route variable z^s_r, only the δ_{gd} duals influence costs in the pricing problem.

5.   The pricing problem

When solving the master problem using delayed column generation, columns must be added during the branching process. Finding or *generating* suitable columns to add to the master problem is known as the pricing problem. Only columns with negative reduced costs are added to the master problem, since only these may go into the basis.

The reduced cost of a route equals the violation of the corresponding constraint in the dual linear program. Thus, the reduced cost $\bar{c}^{\pi,\delta}_{s,r}$ of route $r \in R^s$ is:

$$\bar{c}^{\pi,\delta}_{s,r} = C^s_r - \sum_{g \in \mathcal{G}} \sum_{d \in \mathcal{D}} B^s_{gdr} \delta_{gd}.$$

Adding a negative reduced cost column corresponds to adding a violated inequality in the dual linear program. If no negative reduced cost column exists, the optimal solution of the restricted master problem is also optimal for the complete master problem. The pricing problem consists of finding one or more legal routes with negative reduced cost.

The pricing problem is represented in a directed graph with edge weights. For each pickup (or delivery) port g and for each day d, create a node i, where $i = f(g, d)$. Furthermore, add a depot node 0 and add edges from the depot node to all pickup nodes and from all delivery nodes to the depot node. Also add edges between all feasible node pairs. A node pair (i, j) is feasible if there is sufficient time to visit i before visiting j given the time windows of i and j. The cost C_r^s of a route r sailed by ship s, consists of a fixed cost and a sailing cost. The fixed cost is independent of how we use the ship. The sailing cost, however, is proportional to the distance traveled by the ship. Both costs depend on the ship-type used.

The sailing cost can be assigned to edges going into the visits. Given a dual vector $\bar{\delta}$, the value of dual δ_{gd} can be subtracted from the edge going into node $i = f(g, d)$. In this graph, the reduced cost of a route equals the length of the path in the graph consisting of a sequence of nodes corresponding to the visits on the route in that order.

To allow routes to wrap around the to the beginning of the planning period, append a copy of the network, less the depot node, to the end of the horizon. Add arcs from all delivery nodes in this network to the original depot node. Then solve one subproblem for each ship-type and for each day in the planning period.

Since the reduced cost of a route is equal to the length of the path in the above graph, the pricing problem can be solved as a shortest path problem with additional constraints that disqualify illegal routes. This problem is commonly known as a *resource constrained shortest path problem* (RCSP). In our case the additional constraints will make sure that all routes comply with the capacity, pickup and delivery, time-windows, lead-time and visit-pattern requirements which are described in Section 2.

Several methods have been proposed to solve the resource constrained shortest path problem, see Mehlhorn and Ziegelmann (2000). However, when dealing with non-additive "difficult" constraints like "pickup before delivery", "time windows", and "visit-patterns" only dynamic programming algorithms have been used to solve the problem. Dumas, Desrosiers and Soumis (1991) were the first to propose a dynamic programming algorithm for the pickup and delivery problem with time windows used in a column generation setting like this.

We solve the resource constrained shortest path problem with a heuristic two Phase algorithm. First, we use the fact that most cargos go between Rotterdam and ports on the Norwegian coastline. Because of this, we can assume that all ships will visit Rotterdam at least once in a planning period, either to pickup cargo, to deliver cargo or both. As explained in Section 2, since Rotterdam is relatively far away from the Norwegian ports we can assume that a ship unloads all cargo whenever it visits Rotterdam and does not carry cargo between two Norwegian ports through Rotterdam.

Following our assumptions, a route is comprised by a number of tours starting and ending with an empty ship in Rotterdam. Phase I of the algorithm will generate these tours and Phase II will collect the tours into complete routes. The fixed cost of using a ship of type s, is added in Phase II. A heuristic dynamic programming algorithm is used to solve Phase I and a k-shortest path algorithm with a feasibility check is used to solve Phase II. The two phases are described in detail below.

5.1   Creating tours

Creating good legal tours is the main part of the pricing problem. With a planning period of two weeks the problem involves 272 visits on 10 different ship-types. This is more than similar algorithms have been able to solve. Running an exact dynamic programming algorithm on problem instances of this size is too time-consuming because of the combinatorial complexity. A heuristic dynamic programming algorithm is designed in order to make the problem practically solvable.

The basic idea of the algorithm is similar to previously proposed dynamic programming algorithms for solving resource constrained shortest path problems, see Dumas, Desrosiers and Soumis (1991). We start with an empty ship in Rotterdam and extend the route to each of the ports. For every visit, we create a new label which contains all relevant information on the route so far. We check that candidate visits do not make the route illegal, i.e. we check that:

- We do not deliver a cargo that has not yet been picked up.

- We do not visit Rotterdam with any cargos bound for other ports.

- The ship has available capacity for picking up the cargo.

- The visit is made within an open time window. If there is a later time window, we wait.

- The visit is consistent with a visit-pattern for this node, especially if the same visit has been made previously on this route.

- We still have time to deliver all onboard cargos within their lead-time and within the planning period. (This corresponds to solving a travelling salesman problem with time windows by complete enumeration for the earliest cargos).

If any of the constraints are violated, the route is not extended to make the visit. As stated, the number of states created in the dynamic programming algorithm has to be pruned further in order to run the algorithm in practice. Thus, a number of heuristic constraints that the routes must satisfy are added. For each visit we check that:

- The waiting time is no more than 12 hours. We enforce this rule since good routes will not have long periods of waiting in ports.

- All cargos on the ship bound for a particular harbour are delivered when a ship visits the harbour. We enforce this rule since we expect that good routes will unload the cargos as soon as possible to make room for picking up more cargo.

- The route should not change direction more than five times on the Norwegian coast line. We enforce this rule since we expect that good routes will not make too many detours.

- All deliveries can be made without changing directions, if the route has already changed direction five times.

These additional constraints help prune away unpromising routes while running the dynamic programming algorithm. However, many routes will comply with these constraints, so additional pruning is needed. For this purpose the algorithm is constrained to a *limited subsequence* of visits for every state we create. For a given state, we list all possible visits we can make next. Each visit is assigned a score based on the route so far and on the best continuation of the route. Instead of creating new states for all possible next visits, we pick the three best scoring pickups and the three best scoring deliveries and create states only for those. The score equals the reduced cost of the visit in question plus the minimum reduced cost of delivering the cargos on board after this visit. Thus, the best subsequences of a given visit may change in every pricing iteration since the dual variables have changed. Also, the best subsequences of a visit may be different depending on which day we make the visit, since we have a different dual variable for every visit for every day of the planning period.

We use *dominance* to further reduce the number of dynamic programming states. If two states are placed on the same node and have made the same number of pickup and deliveries and one state has arrived there

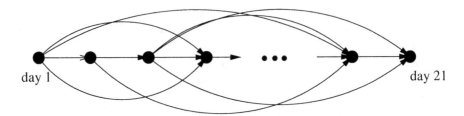

Figure 8.2. An auxiliary graph representing tours.

before the other with a lower reduced cost and one has been at least as far north as the second state, the second state is removed. This is done to remove unpromising routes that are similar, but inferior, to other routes at an early stage. Whenever the ship visits the depot, it has created a complete, legal tour. If the tour has negative reduced cost, it is saved in our pool of tours. The tours are not added directly as columns in the master problem, since they are typically much shorter than the planning period. Instead, tours from the tour pool are combined to create full length routes, which are added to the master problem.

5.2   Combining tours

Given a pool of tours created by our dynamic programming algorithm described above, the pricing problem is reduced to combining one or more tours into legal, full length routes. This is done by running a k-shortest path algorithm on an auxiliary graph, which is defined as follows:

- For every day in the planning period, we add a node in the auxiliary graph. The nodes constitute a time-line.

- For every node, we add an edge from the node to the next node on the time line. The edge weights are 0 on these edges. We do not add an edge going out of the last node.

- For every tour found in Phase 1, we add an edge from the node corresponding to the start time to the node corresponding to the end time of the tour. The edge weights are calculated in every column generation iteration as the reduced cost of the tours.

The graph layout is shown in Figure 8.2. The weights on the edges representing the tours equal the reduced cost of the tours. Minimum reduced cost combinations of tours can be found by a shortest path algorithm on the graph in Figure 8.2 with node i as source and node $i + 21$ as target. To handle tours that cross the end of the planning period (e.g. a tour that starts on day 18 and ends on day 4), another 21 nodes are added to represent day 1 to 21 in the next planning period.

Combining tours with conflicting visit-patterns gives infeasible routes. Conflicting visit-patterns occur if tours visit the same customer on days which do not match any legal visit-pattern. A constrained shortest path algorithm can be used to find a legal minimum cost combination of tours. Since we wish to return more than one combination in every pricing iteration, a constrained k-shortest path algorithm can be used to solve the pricing problem.

Traditionally, this problem has been solved by a dynamic programming algorithm, where states are expanded by combining tours, whilst checking that the combination of tours are legal with respect to the visit-patterns. However, checking whether two tours have contradicting visit-patterns is rather time consuming. Furthermore, most tour combinations will be legal. For these reasons, the time-consuming check is not performed in the k-shortest path algorithm. Instead we first run an unconstrained k-shortest path algorithm as proposed by Carlyle and Wood (2003) to find a number of shortest paths which may or may not be legal. This algorithm first identifies the shortest path. Then it returns all paths which are within a factor of $1 + \varepsilon$ of the shortest path for some specified ε. This is done by enumerating subpaths while checking if candidate subpaths can be extended to paths of acceptable length. Afterwards, a rather expensive visit-pattern check examines the routes, starting with the cheapest route and continuing until the desired number of legal routes has been found. Whenever an illegal combination of tours appears, this combination is inserted into a data structure which allows fast lookups. If the same tour combination appears in a later pricing iteration, it can be discarded without running the expensive check. In fact, the fast lookup can also be done in the k-shortest path algorithm, to avoid combinations, which have already been found illegal. A tree data structure with a sorted sequence of tour-numbers as keys is used to provide the fast lookups.

The running time of the k-shortest path algorithm depends on the number of paths to return. Therefore, only a small number of paths are returned early in the convergence. If there are too few legal paths within these paths, the number of paths to return is increased and the algorithm is re-run. In practice, the k-shortest path algorithm by Carlyle and Wood (2003), combined with a post-check and a fast lookup data structure performs very efficiently.

Both the Phase I algorithm which creates tours and the Phase II algorithm which combines the tours are run for every ship-type and every day in every pricing iteration. The fixed cost of using a particular ship-type is added to the cost of each route in Phase II. Once a tour has been added to the tour pool, it is never removed. Hence, all generated tours are available for the tour combining algorithm in later pricing iterations.

6. Finding integer solutions

The optimal solutions to the linear relaxation of the master problem will generally not be integer. This means that in the optimal solution to the relaxed problem, more than one route variable at fractional value cover the transport of some cargo. Constraints (8.5) which include route and pattern variables may also permit the transport of some cargo by pattern variables at fractional values. The pattern variables will be integer in solutions where all the route variables are integer. If all pattern variables are integer, the resulting problem is a general pickup and delivery problem with time-windows. While it is hard to rank patterns, problem knowledge can be used to identify desirable routes. Therefore, the branching scheme works on the route variables, driving them towards integer values. This will simultaneously give integer pattern variables.

6.1 Branching on pairs of visits

We apply the branching strategy proposed by Ryan and Foster (1981) which has proven successful in a number of set partitioning applications. Recall that a row in the constraint matrix correspond to a particular visit on a given day. Following the Ryan–Foster branching strategy, it is possible to choose two rows k and l in the matrix such that the sum of route variables where the corresponding visits occur consecutively is strictly between 0 and 1. This corresponds to choosing two visits (n_1, d_1) and (n_2, d_2), where a visit is defined by a pickup (or delivery) n_i of a cargo group and a day d_j. Given two visits, on the first branch, demand that the pickup (or delivery) n_1 on day d_1 is followed by the pickup (or delivery) n_2 on day d_2. On the second branch, demand that the two visits are not made consecutively on those specific days.

The master problem changes on both branches, since we must remove variables which violate the branching restriction. On one branch we must remove all route variables which make one of the two visits n_1 on d_1 or n_2 on d_2, but not both. On the other branch we must remove all route variables which make both visits n_1 and n_2 consecutively on the specified days. The variables are not permanently removed, but rendered inactive, so that they can be easily put into the master problem again in other branches of the branch-and-bound tree. To prevent generating and adding violating variables to the master problem, the pricing problem needs to be changed. In any node in the branch-and-bound tree, the pricing problem must reflect the restrictions of the branch. Phase II of the pricing algorithm works on a pool of tours, represented by edges in a directed acyclic graph. To avoid generating routes violating the branching constraints, all tours which violate the constraints are removed. Again, the tours are not really removed, but merely hidden so that they can be easily restored in other branches of the branch-and-bound tree. We do not branch on consecutive depot visits. All branching decisions are imposed within the tours which means that the branching decisions do not restrict the combination of tours which is done in Phase II of the pricing algorithm. However, it is necessary to prevent generating new tours which violate branching constraints. Hence, the dynamic programming algorithm, which constitutes Phase I of the pricing algorithm, checks if any branching constraints are violated when building the tours, discarding the violating tours.

6.2 Using problem information in branching

Although the above branching strategy will work for any choice of visits (n_1, d_1), (n_2, d_2), the performance of the branching strategy is much improved by selecting a sensible pair of consecutive visits. This is guided by knowledge of the problem and by information from the fractional solutions to the master problem. For every pair of rows in our constraint matrix (i.e. for every pair of days and every pair of pickup/deliveries) a score is calculated based on this information. In a depth first strategy we branch on the pair of rows with best score.

As mentioned, the score is based on knowledge of the problem instance and information from the fractional solution at hand. The information from the problem instance tells us whether two consecutive visits fit well together. For example, if the visits are far apart in either time or geography, the visits receive a low score, which implies that the visits do not fit well together. More specifically, for a visit in node n_1 on day d_1 followed by a visit in n_2 on day d_2 the following factors are used in the score calculation:

- A penalty proportional to the distance between n_1 and n_2.

- A penalty proportional to the time difference $d_1 - d_2$ less the average travel time from n_1 to n_2.

- A penalty if the *direction* of cargos does not fit. For example, if n_1 is a pickup of a cargo bound for a harbour north of n_1 and n_2 lies south of n_1, the ship will need to go south to n_2 and then go north later on. Such detours are penalized. There are 14 cases similar to the above where the directions of two cargos do not fit and where making the visits consecutively gives a detour. In all cases the penalty is proportional to the length of the detour.

The choice of visits to branch on should also be guided by the fractional solution at hand, since routes with high fractional values in an optimal LP-solution are likely to be good routes in an IP-solution as well. The sum of fractions is calculated for every day and every consecutive pair of visits. The sum of fractions for a particular visit pair is the sum of the values of the route variables that make both these visits on these days. The sum of fractions is added to the score, so that a high sum of fractions results in a higher score. This is based on the assumption that routes chosen in the optimal solution to the relaxed master problem are similar to the routes in an optimal solution to the IP master problem.

As stated, the pair of visits with the highest score is selected for branching. This pair constitutes visits which we think are likely to be made consecutively on a route. On the first branch these visits must be made consecutively on two specific days and on the other branch they can not be made consecutively. Since we pick pairs of visits which fit well together, we will follow the first branch most often, which is the stronger of the two since it fixes the two visits together.

7.   Computational results

The heuristic branch-and-price algorithm described in Section 6 has been implemented and tested on test instances based on data provided by the collaborating companies. As could be expected, our results show that there are substantial savings in making two-week planning periods compared to one-week planning periods. The best solution found using a one-week planning period requires 15 ships and has an objective value of 20.28, whereas the best solution found using a two-week planning period only requires 13 ships and has an objective value of 17.67 per week. This constitutes a saving of 14.8%. In Table 8.2 we show some details of the performance of the branch-and-price algorithm on the two test instances.

The two week planning period is advantageous compared to the one week planning period because it allows a greater flexibility in the length of the tours. Where all the tours in the one week planning period are between 5 and 7 days long, the tours of the two week planning period are between 4 and 13 days long. Hence in the two week planning period each ship make either a single long tour, two medium tours or a short tour and a longer tour. In the one week planning period all ships make medium length tours.

	1 week instance	2 week instance
No. of constraints	624	1170
No. of columns	1624	10128
No. of B&B-nodes	45	107
No. of pricing itr.	124	565
CPU time (sec.)	440	40 576
Best solution found		
No. of ships	15	13
Cost pr. week	20.28	17.67

Table 8.2. The table shows details of running the branch-and-price heuristic on the test instances based on data provided by the collaborating companies.

8. Concluding remarks

We have chosen to solve the problem using column generation. Another approach would be to enumerate all possible combinations of feasible visit-patterns in all ports. This would give a finite number of alternative visit-schedules. For each visit-schedule one could solve a pickup and delivery problem with time windows and with lead-time as the only additional requirement. Unfortunately, the number of visit-schedules renders this approach impractical. The problem has 78 ports each with an average of 7.1 alternative patterns. This gives about $78^{7.1}$ (10^{13}) alternative visit-schedules or problems to solve. Various heuristic approaches could also be used. However, most would have difficulties incorporating the separation of visits, recurring visits and lead-time constraints. Since these restrictions require considering all routes simultaneously.

Demand is modelled as fixed. This is realistic, as the companies will have to pay for the requested capacity irrespective of whether or not they use it. Likewise the shipping company must guarantee a minimum transport capacity to each partner. It is likely that unused capacity will be traded among the partners and also sold to customers outside the system. This way there can be room for some variations in the load. The shipping companies need to consider whether they assume a net increase in total transport load and thus want to use ships with surplus capacity.

Acknowledgments Companies who have contributed to this work are Elkem, Norsk Hydro, Norske Skog and Statoil. The work has also involved the research institution MARINTEK.

References

Barnhart C., Johnson E.L., Nemhauser G.L., Savelsberg M.W.P., and Vance P.H. (1998). Branch-and-price: Column generation for solving huge integer programs. *Operations Research*, 46:316–329.

Carlyle W.M. and Wood R.K. (2003). Near-Shortest and K-Shortest Simple Paths. *Technical Report*, Department of Operations Research, Naval Postgraduate School, Monterey, CA 93943, USA.

Cho S.-C. and Perakis A.N. (1996). Optimal liner fleet routing strategies. *Maritime Policy & Management*, 23:249–259.

Crainic, T.G. and Laporte G. (1997). Planning models for freight transportation. *European Journal of Operational Research*, 97:409–438.

Christiansen, M. (1999). Decomposition of a combined inventory and time constrained ship routing problem. *Transportation Science*, 33:3–16.

Christiansen, M., Fagerholt, K. and Ronen, D. (2004). Ship routing and scheduling: Status and perspectives. *Transportation Science*, 38:1–18.

Cordeau, J.-F. Gendreau, M., and Laporte, G. (1997). A tabu search heuristic for periodic and multi-depot vehicle routing problems. *Networks*, 30:105–119.

Desrosiers, J., Dumas, Y., Solomon, M., and Soumis, F. (1995). Time constrained routing and scheduling. In: *Network Routing* (M.O. Ball, T.L. Magnanti, C.L. Monma, and G.L. Nemhauser, eds.), volume 8, Handbooks in Operations Research and Management Science, pp. 35–139, North-Holland, Amsterdam.

Dumas, Y., Desrosiers, J., and Soumis, F. (1991). The pickup and delivery problem with time windows. *European Journal of Operational Research*, 54:7–22.

Fagerholt, K. (1999). Optimal fleet design in a ship routing problem. *International Transactions in Operations Research*, 6:453–464.

Laporte, G. and Osman, I.H. (1995). Routing problems: A bibliography. *Annals of Operations Research*, 61:227–262.

Mehlhorn, K. and Ziegelmann, M. (2000). Resource Constrained Shortest Paths. *Proc. 8th European Symposium on Algorithms* (ESA2000), pp. 326–337, LNCS 1879 Springer, Berlin.

Powell, B.J. and Perakis, A.N. (1997). Fleet deployment optimization for liner shipping: an integer programming model. *Maritime Policy & Management*, 24:183–192.

Rana, K. and Vickson, R.G. (1991). Routing ships using Lagrangean relaxation and decomposition. *Transportation Science*, 25:201–214.

Ronen, D. (1983). Cargo ships routing and scheduling: Survey of models and problems. *European Journal of Operational Research*, 12:119–126.

Ronen, D. (1993). Ship scheduling: The last decade. *European Journal of Operational Research*, 71:325–333.

Ryan, D.M. and Foster, B.A. (1981). An integer programming approach to scheduling. In: *Computer Scheduling of Public Transport* (A. Wren, ed.), North-Holland Publishing Company.

Chapter 9

COMBINING COLUMN GENERATION AND LAGRANGIAN RELAXATION

Dennis Huisman
Raf Jans
Marc Peeters
Albert P.M. Wagelmans

Abstract   Although the possibility to combine column generation and Lagrangian relaxation has been known for quite some time, it has only recently been exploited in algorithms. In this paper, we discuss ways of combining these techniques. We focus on solving the LP relaxation of the Dantzig-Wolfe master problem. In a first approach we apply Lagrangian relaxation directly to this extended formulation, i.e. no simplex method is used. In a second one, we use Lagrangian relaxation to generate new columns, that is Lagrangian relaxation is applied to the compact formulation. We will illustrate the ideas behind these algorithms with an application in lot-sizing. To show the wide applicability of these techniques, we also discuss applications in integrated vehicle and crew scheduling, plant location and cutting stock problems.

1.   Introduction

In this chapter we consider (mixed) integer programming problems in minimization form. Obviously, lower bounds for such problems can be computed through a straightforward calculation of the LP relaxation. Dantzig-Wolfe decomposition and Lagrangian relaxation are alternative methods for obtaining tighter lower bounds. The key idea of Dantzig-Wolfe decomposition (Dantzig and Wolfe, 1960) is to reformulate the problem by substituting the original variables with a convex combination of the extreme points and a linear combination of the extreme rays of the polyhedron corresponding to a substructure of the formulation. Throughout the paper, we will assume that this polyhedron is bounded. Therefore, only the extreme points are needed. This substitution results in the master or extended formulation, which contains the linking constraints from the original compact formulation and additional convexity constraints. When solving the LP relaxation of the master problem, column generation is used to deal with the large number of variables. Starting with a restricted master problem which contains only a small subset of all columns, we generate the other columns when they are needed. This is done by solving a so called pricing problem in which one or more variables with negative reduced costs are determined. After each execution of the pricing procedure, we calculate the optimal value of the LP relaxation of the restricted master problem, $\overline{v}_{RDW}$. This provides an upper bound on the optimal value of the Dantzig-Wolfe relaxation, $\overline{v}_{DW}$, which itself is a lower bound for the optimal IP value v_P. When a simplex algorithm is used to solve the restricted master problem, we obtain optimal values of the dual variables corresponding to the linking and convexity constraints. These values are used in the pricing problem to check if we can generate new columns with negative reduced cost. If we find such columns, we add them to the relaxed master problem and reoptimize, otherwise we have found the optimal value of the Dantzig-Wolfe relaxation $\overline{v}_{DW}$. This value will usually be tighter than $\overline{v}_P$, the value of the LP relaxation of the original compact formulation.

In Lagrangian relaxation, the complicating constraints are dualized into the objective function. Given a specific vector of positive multipliers l, the Lagrangian relaxation problem always gives a lower bound, $\overline{v}_{LR}(l)$, on the optimal IP value v_P. The Lagrangian dual problem consists of finding the maximum lower bound: $\overline{v}_{LD} = \max_{l \geq 0} \overline{v}_{LR}(l)$. Typically, the latter problem is solved using an iterative procedure, where in subsequent iterations, the Lagrangian multiplier vector l is updated and we solve a new Lagrangian problem with these updated multipliers. In this chapter we focus on the subgradient method (Fisher, 1985, e.g.) for approximating the optimal multipliers, although more advanced methods such as the bundle method (Lemaréchal, Nemirovskii and Nesterov, 1995, e.g.) or the volume algorithm (Barahona and Anbil, 2000) exist.

There exists a strong relationship between Dantzig-Wolfe decomposition and Lagrangian relaxation. It is well known that when the Lagrangian relaxation is obtained by dualizing exactly those constraints that are the linking constraints in the Dantzig-Wolfe reformulation, the optimal values of the Lagrangian dual, $\overline{v}_{LD}$, and the LP relaxation of the Dantzig-Wolfe reformulation, $\overline{v}_{DW}$, are the same. In fact, one formulation is the dual of the other (Geoffrion, 1974; Fisher, 1981). Furthermore, the optimal dual variables λ for the linking constraints in the master problem correspond to optimal multipliers l for the dualized constraints in the Lagrangian relaxation (Magnanti, Shapiro and Wagner,

1976). Moreover, the subproblem that we need to solve in the column generation procedure is the same as the one we have to solve for the Lagrangian relaxation except for a constant in the objective function. In the column generation procedure, the values for the dual variables are obtained by solving the LP relaxation of the restricted master problem, whereas in the Lagrangian relaxation, the Lagrangian multipliers are updated by subgradient optimization.

Both approaches have advantages and disadvantages. Lagrangian relaxation provides a lower bound on the optimal IP value v_P at each iteration of the subgradient algorithm, but no primal solution is available. In many applications, the dual information is used in a heuristic fashion to obtain a primal solution. On the other hand, column generation directly provides a primal solution at each iteration, which can be used to construct feasible solutions for the MIP in a rounding heuristic. Further, the Lagrangian lower bound can be computed without much difficulty at each step of the column generation process. There are also differences in the computational implementation and convergence behaviour. The subgradient algorithm is usually stopped after a fixed number of iterations, without the guarantee of having found the optimal value $\bar{v}_{LD}$ (Fisher, 1985). However, the subgradient optimization for updating the Lagrangian multipliers is computationally inexpensive and easy to implement. The simplex optimization of the master problem, on the other hand, is computationally expensive and a tailing-off effect, i.e. slow convergence towards the optimum in the final phase of the algorithm, is generally observed (Barnhart et al., 1998; Vanderbeck and Wolsey, 1996). The use of problem specific information can guide the choice of the Lagrangian multipliers and can lead to a faster convergence, whereas we do not have the same freedom in the column generation approach where the master problem provides the values of the dual variables.

In this chapter we will discuss how the relationship between Dantzig-Wolfe decomposition and Lagrangian relaxation can be exploited to develop improved algorithms combining the strengths of both methods. We discuss two ways in which the two techniques can be combined efficiently. To be more specific, Lagrangian relaxation can be applied to the master problem to approximate the optimal values of the dual variables or it can be used on the original compact formulation of the problem to generate good columns. However, notice that we will only discuss column generation within the framework of DW decomposition, but it can also be considered as a general LP pricing technique. For the combination of column generation and Lagrangian relaxation within this framework, we refer to Löbel (1998); Fischetti and Toth (1997). In order to explain the general principles within the framework of DW decomposition in Section 2, we use the example of capacitated lot-sizing. In Sections 3–5, other applications and their specific implementation issues are discussed.

2.   Theoretical framework and basic approaches

2.1   Preliminaries

We will illustrate the basic approaches for combining column generation and Lagrangian relaxation using the Capacitated Lot-Sizing Problem (CLSP). In this problem we determine the timing and level of production for several items on a single machine with limited capacity over a discrete and finite horizon. For a more comprehensive description, we refer to Kleindorfer and Newson (1975) or Trigeiro, Thomas and Mc-Clain (1989). Let P be the set of products $\{1, \ldots, n\}$ with index i and T the set of time periods $\{1, \ldots, m\}$ with index t. We have the following parameters: d_{it} is the demand of product i in period t; sc_i, vc_i and hc_i are the set up cost, variable production cost and holding cost for product i, respectively; vt_i is the variable production time for product i and cap_t is the capacity in period t. There are three decision variables: x_{it} is the amount of production of product i in period t; s_{it} is the inventory level of product i at the end of period t; $y_{it} = 1$ if there is a set up for product i in period t, $y_{it} = 0$ otherwise. The mathematical formulation of the CLSP is then as follows:

$$\min \sum_{i \in P} \sum_{t \in T} (sc_i y_{it} + vc_i x_{it} + hc_i s_{it}) \tag{9.1}$$

$$\text{subject to } s_{i,t-1} + x_{it} = d_{it} + s_{it} \quad \forall i \in P, \forall t \in T, \tag{9.2}$$

$$x_{it} \leq M y_{it} \quad \forall i \in P, \ \forall t \in T, \tag{9.3}$$

$$\sum_{i \in P} vt_i x_{it} \leq cap_t \quad \forall t \in T, \tag{9.4}$$

$$y_{it} \in \{0,1\}, \quad x_{it} \geq 0, \quad s_{it} \geq 0, \quad s_{i,0} = 0 \quad \forall i \in P, \ \forall t \in T. \tag{9.5}$$

The objective function (9.1) minimizes the total costs, consisting of the set up cost, the variable production cost and the inventory holding cost. Constraints (9.2) are the inventory balancing constraints: Inventory left over from the previous period plus current production can be used to satisfy current demand or build up more inventory. Constraints (9.3) are the set up forcing constraints: If there is any positive production in period t, a set up is enforced. In order to make the formulation stronger, the 'big M' is usually set to the minimum of the sum of the remaining demand over the horizon and the total production which is possible with the available capacity. Next, there is a constraint on the available capacity in each period (9.4). Finally, there are the non-negativity and integrality constraints (9.5). We let v_{LS} and $\overline{v}_{LS}$ denote the optimal objective value for problem (9.1)–(9.5) and its LP relaxation, respectively.

Decomposition approaches for this problem hinge on the observation that when we disregard the capacity constraints (9.4), the problem decomposes into an uncapacitated lot-sizing problem for each item i. Let S^i be the set of feasible solution for subproblem i: $S^i = \{(x_{it}, y_{it}, s_{it}) \mid$ (9.2), (9.3), (9.5)$\}$ and $S = \bigcup_{i \in P} S^i$. In the Dantzig-Wolfe decomposition, we keep the capacity constraints in the master problem and add a convexity constraint for each item (Manne, 1958; Dzielinski and Gomory, 1965)). The new columns represent a production plan for a specific item over the full time horizon. Let Q_i be the set of all extreme point production plans for item i; z_{ij} is the new variable representing production plan j for item i; c_{ij} is the total cost of set up, production and inventory for production plan j for item i and r_{ijt} is the capacity usage of the production in period t according to plan j for item i. The LP relaxation of a restricted master problem then looks as follows:

$$\overline{v}_{RDWLS} = \min \sum_{i \in P} \sum_{j \in \widetilde{Q}_i} c_{ij} z_{ij} \tag{9.6}$$

$$\text{subject to} \sum_{i \in P} \sum_{j \in \widetilde{Q}_i} r_{ijt} z_{ij} \leq cap_t \quad \forall t \in T, \tag{9.7}$$

$$\sum_{j \in \widetilde{Q}_i} z_{ij} = 1 \quad \forall i \in P, \tag{9.8}$$

$$z_{ij} \geq 0 \quad \forall i \in P, \ \forall j \in \widetilde{Q}_i. \tag{9.9}$$

where $\widetilde{Q}_i$ is a subset of Q_i. Additional columns (variables) are generated when they are needed, using the information of the optimal dual variables λ_t (≤ 0) and π_i of the capacity and convexity constraints, respectively. In the pricing problem, we check for each item i if we can generate a new column by solving the following subproblem:

$$rc_i^*(\lambda, \pi) = \min_{(x,y,s) \in S^i} \sum_{t \in T} (sc_i y_{it} + vc_i x_{it} + hc_i s_{it}) - \sum_{t \in T} vt_i x_{it} \lambda_t - \pi_i. \tag{9.10}$$

If such a column with negative reduced cost is found, we add it to the restricted master problem, reoptimize this problem and perform another pricing iteration; otherwise we have found the optimal Dantzig-Wolfe bound, $\overline{v}_{DWLS}$.

In Lagrangian relaxation, the capacity constraints (9.4) are dualized in the objective function with non-positive multipliers $l = \{l_1, l_2, \ldots, l_m\}$:

$$\overline{v}_{LRLS}(l) = \min_{(x,y,s)\in S} \sum_{i\in P} \sum_{t\in T} (sc_i y_{it} + vc_i x_{it} + hc_i s_{it})$$

$$+ \sum_{t\in T} l_t \left(cap_t - \sum_{i\in P} vt_i x_{it} \right). \quad (9.11)$$

Note that we use here non-positive Lagrangian multipliers in order to show the similarity with the non-positive dual variables λ. The Lagrangian problem also decomposes into single item uncapacitated lot-sizing problems. For each item i we have the following subproblem:

$$\overline{v}_{LRLS,i}(l) = \min_{(x,y,s)\in S^i} \sum_{t\in T} (sc_i y_{it} + vc_i x_{it} + hc_i s_{it}) - \sum_{t\in T} vt_i x_{it} l_t. \quad (9.12)$$

We see that the subproblem of calculating the minimum reduced cost (9.10) in the Dantzig-Wolfe decomposition and the subproblem in the Lagrangian relaxation (9.12) are identical, except for a constant in the objective function. The solution of the Lagrangian dual problem gives the maximum lower bound $\overline{v}_{LDLS} = \max_{l\leq 0} \overline{v}_{LRLS}(l)$. In iterative steps, the multipliers are updated in order to attain this Lagrangian dual bound. Let $x^* = (x^*_{11}, x^*_{12}, \ldots, x^*_{1m}, \ldots, x^*_{n1}, x^*_{n2}, \ldots, x^*_{nm})$ be the optimal production quantities for the Lagrangian problem (9.11) with multipliers l^k at iteration k, then the following standard subgradient update formulas (Fisher, 1981) result in a new vector of multipliers l^{k+1}:

$$l^{k+1}_t = \min \left(0, l^k_t + \beta_k \left(cap_t - \sum_{i\in P} vt_i x^*_{it} \right) \right) \quad t = 1,\ldots,m, \quad (9.13)$$

$$\beta_k = \alpha \frac{\left(ub - \overline{v}_{LRLS}(l^k) \right)}{\sum_{t\in T} (cap_t - \sum_{i\in P} vt_i x^*_{it})^2}. \quad (9.14)$$

Equation (9.14) determines the step-size, where $0 < \alpha \leq 2$ and the value ub is an upper bound on v_{LS}.

During column generation, the value of the restricted master problem $\overline{v}_{RDWLS}$ provides an upper bound on the optimal Dantzig-Wolfe relaxation value $\overline{v}_{DWLS}$. However, a lower bound can be easily calculated as well. Let $rc^*_i(\lambda, \pi)$ be the minimum reduced cost for subproblem i with the current optimal dual variables λ and π, then

$$\sum_{i\in P} rc^*_i(\lambda, \pi) + \overline{v}_{RDWLS} \leq \overline{v}_{DWLS} \leq \overline{v}_{RDWLS}. \quad (9.15)$$

This lower bound is actually equal to the Lagrangian lower bound using the current optimal dual variables λ as multipliers:

$$
\begin{aligned}
\overline{v}_{LRLS}(\lambda) &= \sum_{i \in P} \overline{v}_{LRLS,i}(\lambda) + \sum_{t \in T} \lambda_t cap_t \\
&= \sum_{i \in P} \overline{v}_{LRLS,i}(\lambda) - \sum_{i \in P} \pi_i + \sum_{i \in P} \pi_i + \sum_{t \in T} \lambda_t cap_t \\
&= \sum_{i \in P} rc_i^*(\lambda, \pi) + \overline{v}_{RDWLS},
\end{aligned}
$$

where in the final step, equivalence between $\sum_{i \in P} \pi_i + \sum_{t \in T} \lambda_t cap_t$ and $\overline{v}_{RDWLS}$ follows from LP duality. This lower bound was already proposed by Lasdon and Terjung (1971) who used column generation to solve a large production scheduling problem. It has also been discussed for other specific problems such as discrete lot-sizing and scheduling (Jans and Degraeve, 2004), machine scheduling (Van den Akker, Hurkens and Savelsbergh, 2000), vehicle routing (Sol, 1994), a multicommodity network-flow problem (Holmberg and Yuan, 2003) and the cutting stock problem (Vanderbeck, 1999). A general discussion can be found in Wolsey (1998); Martin (1999). Vanderbeck and Wolsey (1996) provide a slight strengthening of this bound. The bound can be used for early termination of the column generation procedure, reducing the tailing-off effect. For IP problems with an integer objective function value, we can also stop if the value of this lower bound rounded up is equal to the value of the restricted master problem rounded up.

2.2   Using Lagrangian relaxation on the extended formulation

Instead of using the simplex algorithm to obtain the optimal dual variables of the (restricted) master problem, one can also use Lagrangian relaxation to approximate these values. Cattrysse et al. (1993); Jans and Degraeve (2004) apply this technique for solving a variant of the capacitated lot-sizing problem. A similar integration of Dantzig-Wolfe decomposition and Lagrangian relaxation is also used for the generalized assignment problem (Cattrysse, Salomon and Van Wassenhove, 1994), and integrated vehicle and crew scheduling which is the topic of Section 3.

In order to approximately solve the LP relaxation of the restricted master problem (9.6)–(9.9), we dualize the capacity constraint (9.7) into the objective function (9.6) with non-positive multipliers l_t:

$$\overline{v}_{LR-RDW}(l) = \min \sum_{i \in P} \sum_{j \in \tilde{Q}_i} c_{ij} z_{ij}$$

$$+ \sum_{t \in T} l_t \left(cap_t - \sum_{i \in P} \sum_{j \in \tilde{Q}_i} r_{ijt} z_{ij} \right) \tag{9.16}$$

$$\tag{9.17}$$

$$\text{subject to } \sum_{j \in \tilde{Q}_i} z_{ij} = 1 \quad \forall i \in P, \tag{9.18}$$

$$z_{ij} \geq 0 \quad \forall i \in P, \ \forall j \in \tilde{Q}_i. \tag{9.19}$$

The problem decomposes into subproblems per item that are easy to solve, because taking the column with the lowest total cost for each item results in the optimal solution. The optimal Lagrangian multipliers are iteratively approximated via a standard subgradient optimization procedure. At the end of a subgradient phase, the Lagrangian multipliers l_t are an approximation of the optimal dual variables λ_t. Next, the optimal dual variable π_i of the convexity constraint for item i can be approximated by the value p_i as follows:

$$p_i = \min_{j \in \tilde{Q}_i} \left(c_{ij} - \sum_{t \in T} l_t r_{ijt} \right). \tag{9.20}$$

The Lagrangian multipliers l_t and p_i can be used to generate new columns in the pricing subproblem (9.10). The new columns are added to the restricted master problem and in a subsequent step the optimal dual variables λ and π for the updated restricted master problem are again approximated by Lagrangian relaxation.

Given the Lagrangian multipliers l_t and p_i, we can still compute a lower bound:

$$\sum_{i \in P} rc_i^*(l, p) + \overline{v}_{LR-RDW}(l) \leq \overline{v}_{DWLS}. \tag{9.21}$$

This can again be proven by starting from the Lagrangian relaxation $\overline{v}_{LRLS}(l)$ (9.11), which gives a valid lower bound for any $l \leq 0$:

$$\overline{v}_{LRLS}(l) = \sum_{i \in P} \overline{v}_{LRLS,i}(l) - \sum_{i \in P} p_i + \sum_{i \in P} p_i + \sum_{t \in T} l_t cap_t$$

$$= \sum_{i \in P} rc_i^*(l, p) + \sum_{i \in P} p_i + \sum_{t \in T} l_t cap_t$$

$$= \sum_{i \in P} rc_i^*(l, p) + \sum_{i \in P} \min_{j \in \tilde{Q}_i} \left(c_{ij} - \sum_{t \in T} l_t r_{ijt} \right) + \sum_{t \in T} l_t cap_t$$

$$= \sum_{i \in P} rc_i^*(l, p) + \overline{v}_{LR-RDW}(l).$$

What are the advantages of approximating the optimal dual variables by Lagrangian relaxation instead of computing them exactly with a simplex algorithm? Bixby et al. (1992); Barnhart et al. (1998) note that in case of alternative dual solutions, column generation algorithms seem to work better with dual variables produced by interior point methods than with dual variables computed with simplex algorithms. The latter give a vertex of the face of solutions whereas interior point algorithms give a point in the center of the face, providing a better representation of it. From that perspective, Lagrangian multipliers may also provide a better representation and speed up convergence. Computational experiments from Jans and Degraeve (2004) indicate that using Lagrangian multipliers indeed speeds up convergence and decreases the problem of degeneracy. Lagrangian relaxation has the additional advantage that during the subgradient phase possibly feasible solutions are generated. The subgradient updating is also fast and easy to implement. Finally, this procedure eliminates the need for a commercial LP optimizer.

2.3   Using Lagrangian relaxation on the compact formulation

This approach is based on the observation that when the Lagrangian relaxation is obtained by dualizing exactly those constraints that are the linking constraints in the Dantzig-Wolfe reformulation, the same subproblem results. Consequently, the solutions generated by the Lagrangian subproblems can also be added as new columns to the master problem. This was first proposed by Barahona and Jensen (1998) for a plant location problem and by Degraeve and Peeters (2003) for the cutting stock problem. These applications are discussed in Sections 4 and 5, respectively. It has also been applied successfully to the capacitated lot-sizing problem (Degraeve and Jans, 2003), that is used again to illustrate the technique. The procedure essentially consists of a nested double loop. In the outer loop, optimal dual variables for the restricted master problem (9.6)-(9.9) are obtained by the simplex method. In the inner loop, the Lagrangian subproblem of the compact formulation (9.11) is solved during several iterations, each time with dual variables which are updated with a subgradient optimization procedure. A generic procedure is depicted in Figure 9.1.

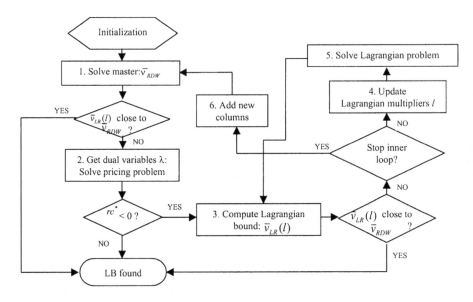

Figure 9.1. Outline of algorithm.

After initialization, the LP relaxation of the restricted master problem (9.6)–(9.9) is solved (Box 1). Next the optimal dual variables λ and π are passed to the pricing problem (9.10), which is then solved to find a new column (Box 2). If the reduced cost is non-negative for each subproblem, then the Dantzig-Wolfe bound $\overline{v}_{DW}$ is found. Otherwise, the inner loop starts (Box 3), where in the first iteration the Lagrangian bound $\overline{v}_{LR}(l)$ (9.11) is computed, using the optimal dual variables of the restricted master problem. This bound is then compared with the objective value of the restricted master problem $\overline{v}_{RDW}$. For a pure integer programming problem with integer coefficients in the objective function, the procedure terminates if both values rounded up are equal, and the Dantzig-Wolfe bound equals $\lceil \overline{v}_{RDW} \rceil = \lceil \overline{v}_{LR}(l) \rceil$. For a mixed integer programming problem, the algorithm may be terminated, if the difference between both values is smaller than a pre-specified percentage. Other stopping criteria could also be checked. For instance, Barahona and Jensen (1998) stop the inner loop after a fixed number of iterations. If no stopping criteria are satisfied, then the Lagrangian multipliers are updated using subgradient optimization (Box 4). The value ub in (9.14) is an upper bound on $\overline{v}_{LD}$, and therefore, ub can be set equal to the LP bound of the last solved restricted master problem $\overline{v}_{RDW}$, since $\overline{v}_{RDW} \geq \overline{v}_{DW} = \overline{v}_{LD}$. Next the algorithm proceeds with solving a new Lagrangian problem, with the updated multipliers (Box 5). The Lagrangian bound is computed again and the inner loop continues, until a stopping criterion is met. Next, we switch back to the outer loop. We add to the restricted master problem the columns with negative reduced costs (obtained in Box 2) and the ones generated in the inner loop if they are not yet present (Box 6).

The main advantage of this procedure is that the LP relaxation of the master problem does not need to be solved each time to get new dual variables necessary for pricing out a new column. Solving the LP relaxation to optimality is computationally much more expensive than performing an iteration of the subgradient optimization procedure. At each subgradient iteration, a new column is found and these columns are expected to be "good" because the Lagrangian multipliers prices converge towards the optimal dual variables of the LP relaxation of the restricted master problem. A second advantage is that we can stop the column generation short of proving LP optimality of the master problem, because the Lagrangian relaxation provides lower bounds on the optimal LP value. Barahona and Jensen (1998) mention this fact as the main motivation for performing a number of subgradient iterations between two consecutive outer loop iterations. This procedure tries to combine the speed of subgradient optimization with the exactness of the Dantzig-Wolfe algorithm. In addition, the procedure provides a primal solution on which branching decisions or rounding heuristics can be based, which is not the case if only subgradient optimization is used. Computational results from Degraeve and Jans (2003) indicate that this method speeds up the column generation procedure. With this hybrid method, it takes about half the time to find the lower bound compared to the traditional method.

3.   Application 1: Integrated vehicle and crew scheduling

In this section we discuss the application of a combined column generation/Lagrangian relaxation algorithm to the integrated vehicle and crew scheduling problem. Vehicle and crew scheduling are two of the most important planning problems in a bus company. After a short problem description, we present a formulation for the integrated problem (in case of multiple-depots) to which we apply the approach outlined in Subsection 2.2. Some interesting, recent references on the integrated problem are Freling (1997); Haase, Desaulniers and Desrosiers (2001); Freling, Huisman and Wagelmans (2003) for the single-depot case, and Gaffi and Nonato (1999); Huisman, Freling and Wagelmans (2003) for the multiple-depot case.

3.1   Problem description

The *multiple-depot vehicle and crew scheduling problem* (MD–VCSP) can be defined as follows. Given a set of *trips* within a fixed planning horizon, it minimizes the total sum of vehicle and crew costs such that both the vehicle and the crew schedule are feasible and mutually compatible. Each trip has fixed starting and ending times, and can be assigned to a vehicle and a crew member from a certain set of depots. Furthermore, the travelling times between all pairs of locations are known. A vehicle schedule is feasible if (1) all trips are assigned to exactly one vehicle, and (2) each trip is assigned to a vehicle from a depot that is allowed to drive this trip. From a vehicle schedule it follows which trips have to be performed by the same vehicle and this defines so-called vehicle *blocks*. The blocks are subdivided at *relief points*, defined by location and time, where and when a change of driver may occur and drivers can enjoy their break. A *task* is defined by two consecutive relief points and represents the minimum portion of work that can be assigned to a crew. These tasks have to be assigned to crew members. The tasks that are assigned to the same crew member define a crew *duty*. Together the duties constitute a crew schedule. Such a schedule is feasible if (1) each task is assigned to one duty, and (2) each duty is a sequence of tasks that can be performed by a single crew, both from a physical and a legal point of view. In particular, each duty must satisfy several complicating constraints corresponding to work load regulations for crews. Typical examples of such constraints are maximum working time without a break, minimum break duration, maximum total working time, and maximum duration.

3.2   Mathematical formulation

Let $N = \{1, 2, \ldots, n\}$ be the set of trips, numbered according to increasing starting time. Define D as the set of depots and let s^d and t^d both represent depot d. Furthermore, for the crew we distinguish between two types of tasks, viz., *trip tasks* corresponding to trips, and *dh-tasks* corresponding to deadheading. A *deadhead* is defined as a period that a vehicle is moving in time or space without passengers. E^d is the set of deadheads between two trips i and j.

We define the vehicle scheduling network $G^d = (V^d, A^d)$, which is an acyclic directed network with nodes $V^d = N^d \cup \{s^d, t^d\}$, and arcs $A^d = E^d \cup (s^d \times N^d) \cup (N^d \times t^d)$. Note that N^d is the subset of N that can be serviced by depot d, since it is not necessary that all trips can be served from each depot. Let c_{ij}^d be the vehicle cost of arc $(i, j) \in A^d$.

Furthermore, let K^d denote the set of duties corresponding to depot d and f_k^d denote the crew cost of duty $k \in K^d$, respectively. Moreover, $K^d(i)$ denotes the set of duties covering the trip task corresponding to trip $i \in N^d$, which means that we assume that a trip corresponds to exactly one task. $K^d(i,j)$ denotes the set of duties covering the dh-tasks corresponding to deadhead $(i,j) \in A^d$. Decision variable y_{ij}^d indicates whether an arc (i,j) is used and assigned to depot d or not, while x_k^d indicates whether duty k corresponding to depot d is selected in the solution or not. The multiple-depot vehicle and crew scheduling problem (MD–VCSP) can be formulated as follows.

$$\min \sum_{d \in D} \sum_{(i,j) \in A^d} c_{ij}^d y_{ij}^d + \sum_{d \in D} \sum_{k \in K^d} f_k^d x_k^d \tag{9.22}$$

$$\text{subject to} \sum_{d \in D} \sum_{j:\,(i,j) \in A^d} y_{ij}^d = 1 \quad \forall i \in N, \tag{9.23}$$

$$\sum_{d \in D} \sum_{i:\,(i,j) \in A^d} y_{ij}^d = 1 \quad \forall j \in N, \tag{9.24}$$

$$\sum_{i:\,(i,j) \in A^d} y_{ij}^d - \sum_{i:\,(j,i) \in A^d} y_{ji}^d = 0 \quad \forall d \in D,\ \forall j \in N^d, \tag{9.25}$$

$$\sum_{k \in K^d(i)} x_k^d - \sum_{j:\,(i,j) \in A^d} y_{ij}^d = 0 \quad \forall d \in D,\ \forall i \in N^d, \tag{9.26}$$

$$\sum_{k \in K^d(i,j)} x_k^d - y_{ij}^d = 0 \quad \forall d \in D,\ \forall(i,j) \in A^d, \tag{9.27}$$

$$x_k^d \in \{0,1\} \quad \forall d \in D,\ \forall k \in K^d, \tag{9.28}$$

$$y_{ij}^d \in \{0,1\} \quad \forall d \in D,\ \forall(i,j) \in A^d. \tag{9.29}$$

The objective is to minimize the sum of vehicle and crew costs. The first three sets of constraints, (9.23)–(9.25), correspond to the formulation of the vehicle scheduling problem. Notice that in this formulation constraints (9.24) are redundant. However, it is useful to have these constraints when we relax constraints (9.25), as will be done in the algorithm. Constraints (9.26) assure that each trip task will be covered by a duty from a depot if and only if the corresponding trip is assigned to this depot. Furthermore, constraints (9.27) guarantee the link between vehicles and crews. That is, a vehicle performs deadhead (i,j) if and only if the corresponding dh-task is assigned to a driver from the same depot.

Notice that this formulation is already an extended one. We would obtain a similar formulation, if we would apply Dantzig-Wolfe decomposition on a compact formulation of this problem. Desrosiers et al. (1995) show how such a transformation can be applied on the multicommodity flow problem with resource constraints, which has as special case all kind of vehicle and crew scheduling problems.

3.3 Algorithm

Below we first give a schematic overview of a combined column generation/Lagrangian relaxation algorithm to solve the MD–VCSP. Afterwards, we discuss the steps related to Lagrangian relaxation (1, 2 and 4) in more detail. For details about the other steps, we refer to Huisman, Freling and Wagelmans (2003).

STEP 1   Find an initial feasible solution and take as initial set of columns the duties in that solution.

STEP 2   Solve a Lagrangian dual problem with the current set of columns approximately, i.e. perform some subgradient optimization steps to update the multipliers. This gives a lower bound for the current restricted master problem.

STEP 3   Modify multipliers to prevent that columns are generated twice.

STEP 4   Generate columns (duties) with negative reduced cost and update the set of columns.

STEP 5   Compute an estimate of a lower bound for the (full) master problem. If the gap between this estimate and the lower bound found in Step 2 is small enough (or another termination criterion is satisfied), go to Step 6; otherwise, return to Step 1.

STEP 6   Construct feasible solutions by applying a Lagrangian heuristic.

To approximate the optimal value of the restricted master problem in Step 1, we use the relaxation of model MD–VCSP, where the equality signs in the constraints (9.25)–(9.27) are first replaced by "greater-than-or-equal" signs. These constraints are subsequently relaxed in a Lagrangian way. That is, we associate non-negative Lagrangian multipliers κ_j^d, λ_i^d, μ_{ij}^d with constraints (9.25), (9.26), (9.27), respectively. Then the optimal solution of the remaining Lagrangian subproblem can be obtained by inspection for the x variables and by solving a large *single-depot vehicle scheduling problem* (SDVSP) for the y variables.

The values of the Lagrangian multipliers obtained after applying a subgradient algorithm can be used to generate new columns. However, to assure that all columns in the current restricted master problem have non-negative reduced costs such that the corresponding duties will not be generated again in the pricing problem, we use an additional procedure (Step 3) to update the Lagrangian multipliers after solving the Lagrangian relaxation. This can be done with a greedy heuristic, that modifies these multipliers in such a way that columns in the current restricted master problem $\widetilde{K}^d$ have non-negative reduced costs and the value of the Lagrangian function does not decrease. We denote $\bar{f}_k^d$ as the reduced cost of column $k \in K^d$, which is equal to

$$
f_k^d - \sum_{i \in N(k,d)} \lambda_i^d - \sum_{(i,j) \in A(k,d)} \mu_{ij}^d, \tag{9.30}
$$

where $N(k,d)$ and $A(k,d)$ are the set of trip tasks and dh-tasks in duty k from depot d, respectively. The heuristic is described below (see also Freling, 1997; Carraresi, Girardi and Nonato, 1995):

for each column $k \in \widetilde{K}^d$ with $\bar{f}_k^d < 0$;

$$
\delta := \frac{\bar{f}_k^d}{|N(k,d)| + |A(k,d)|};
$$

for each trip task $i \in N(k,d)$: $\lambda_i^d := \lambda_i^d + \delta$;

for each dh-task $(i,j) \in A(k,d)$: $\mu_{ij}^d := \mu_{ij}^d + \delta$;

update the reduced costs for all columns $l \in \widetilde{K}^d$ and $l > k$.

Finally, we will discuss Step 4, where we compute an estimate of a lower bound for the master problem given a lower bound for the current restricted master problem. The latter bound, denoted by $\Phi'(\kappa, \lambda, \mu)$, is obtained in Step 1. Then the expression:

$$
\Phi'(\kappa, \lambda, \mu) + \sum_{d \in D} \sum_{k \in K^d \setminus \widetilde{K}^d} \min(\bar{f}_k^d, 0) \tag{9.31}
$$

is a lower bound for the (full) master problem for each vector (κ, λ, μ). This can be proven in a similar way as in Subsection 2.2. Therefore, we will skip this proof here.

Notice, however, that we do not calculate this lower bound in each iteration, since for generating new columns it is not necessary to calculate the reduced costs for all of them. Therefore, we estimate this bound in each iteration by taking only into account the reduced costs of the columns that we actually add to the master problem. This estimate can be used to stop the column generation part of the algorithm earlier without exactly obtaining a lower bound.

Table 9.1. Computational results MD–VCSP.

# trips	80	100	160	200
# iter.	17.4	25.2	36.8	39.5
cpu m.	154.7	403.9	982.8	1641.5
cpu p.	148.7	510.7	3529.8	4769.5
cpu t.	317.5	942.3	4721.3	6675.0
# found	10	10	4	2
gap (%)	5.37	5.31	5.75	6.52

3.4   Some results

The algorithm presented in the previous subsection has been used to solve several problem instances arising from real-world applications as well as randomly generated instances. In Table 9.1 we summarize some of the results for randomly generated instances with two depots (see Huisman, Freling and Wagelmans, 2003). We report the average number of iterations of the column generation algorithm, and the average computation times for the master problem (cpu m.) and pricing problem (cpu p.), respectively. Furthermore, we give the total average computation time for computing the lower bound (cpu t.). These averages are computed over the instances for which a lower bound is found within 3 hours of cpu time on a Pentium III 450MHz personal computer (128MB RAM). Therefore, we also report the number of instances (out of 10) for which we actually found a lower bound. In the remainder of the table, we report the average gaps between the lower and upper bounds. Notice that all computation times are mentioned in seconds.

In Table 9.1, we only provide results for instances up to 200 trips, since for larger instances we were not able to compute a lower bound within 3 hours computation time. The average gaps between the feasible solutions and the lower bound are about 5% for those instances. However, for large instances we can still use the suggested algorithm to compute feasible solutions by terminating the lower bound phase after a maximum computation time and then continue with Step 5. In practice, this is already quite satisfactory. Therefore, these types of algorithms can be used to solve practical problem instances in an integrated way.

4.   Application 2: Plant location

Barahona and Jensen (1998) apply the procedure described in Subsection 2.3 to a plant location problem with minimum inventory. Given a set N of customers, each requiring a set of parts $D_i \subset P, i \in N$, where P denotes the set of all parts, and a set of M possible locations, the objective is to minimize the total costs such that every customer is served, a bound on the total number of warehouses is not exceeded and a service criterion is met. The total costs consist of a fixed costs f_j, for $j \in M$, if a warehouse is opened at location j, a transportation cost c_{ij} if customer i is served from warehouse j, and an inventory cost h_{jk}, if part k is stored in warehouse j. A part must be stored in a warehouse if a customer, requiring that part, is assigned to the warehouse. The service criterion implies that a given percentage of the total demand must be delivered within a certain time limit. Let y_j be 1, if warehouse j is opened, and 0 otherwise, let x_{ij} be 1 if customer i is assigned to warehouse j, and 0 otherwise, and let z_{jk} be 1, if part k must be stored in warehouse j, and 0 otherwise. Then the model can be stated as follows.

$$\min \sum_{j \in M} f_j y_j + \sum_{i \in N} \sum_{j \in M} c_{ij} x_{ij} + \sum_{j \in M} \sum_{k \in P} h_{jk} z_{jk} \qquad (9.32)$$

$$\text{subject to } \sum_{j \in M} x_{ij} = 1 \quad \forall i \in N, \qquad (9.33)$$

$$\sum_{i \in N} \sum_{j \in M} d_{ij} x_{ij} \geq t, \qquad (9.34)$$

$$\sum_{j \in M} y_j \leq L, \qquad (9.35)$$

$$x_{ij} \leq y_j \quad \forall i \in N, \forall j \in M, \qquad (9.36)$$

$$x_{ij} \leq z_{jk} \quad \forall i \in N, \forall j \in M, \forall k \in D_i, \qquad (9.37)$$

$$x_{ij}, y_j, z_{jk} \in \{0, 1\} \quad \forall i \in N, \forall j \in M, \forall k \in P. \qquad (9.38)$$

The objective (9.32) is to minimize the total costs, i.e. the sum of fixed costs for opening warehouses, transportation and inventory costs. Constraints (9.33) impose that every customer must be assigned to one location. Constraint (9.34) is the service criterion, i.e. suppose that the company would like that 95% of the demand can be served within two hours, then t equals 95% of the total demand and d_{ij} is equal to the demand of customer i, if the travel time between i and j is less than two hours, and 0 otherwise. Constraint (9.35) implies that at most L locations can be opened. Constraints (9.36) and (9.37) define the relations between the variables, i.e. a customer can only be assigned to a warehouse, if the warehouse is open (9.36), and, if customer i is assigned to a warehouse, then all parts D_i of customer i must be present in the warehouse (9.37).

The Dantzig-Wolfe reformulation consists of implicitly considering every possible assignment of customers to locations. Hence, the objective function and constraints of (the LP relaxation of) the master problem correspond to (9.32)-(9.35) and the original variables are replaced by a convex combination of the extreme points of the polytope defined by (9.36)-(9.38). Barahona and Jensen (1998) show that the pricing problem is equivalent to a minimum cut problem. They observed that the convergence of the Dantzig-Wolfe algorithm is very slow for this problem and that the lower bound obtained by adding the reduced cost of the columns that price out to the value of the current restricted master problem, is very poor in the first iterations of the Dantzig-Wolfe algorithm and improves only slowly. After solving the LP relaxation of the current restricted master problem, they perform a fixed number of subgradient iterations on the original problem to improve the bound, using the master problem's optimal dual variables as starting values for the subgradient procedure. Next, all columns are added to the LP relaxation of the restricted master problem, which is then re-optimized. If the new optimal objective value and the Lagrangian lower bound are close to each other, then a heuristic is applied to obtain an integer solution. They are able to obtain good solutions for problems with about 200 locations, 200 parts and 200 customers within about one hour of computation time on a RS6000-410, using OSL (IBM Corp., 1995) to solve the LPs.

5.   Application 3: Cutting stock

Degraeve and Peeters (2000) use a combination of the simplex method and subgradient optimization to speed up the convergence of the column generation algorithm of Gilmore and Gomory (1961) for the one-dimensional cutting stock problem (CSP). This procedure is used to compute the LP relaxation at every node of the branch-and-price tree of the algorithm described in Degraeve and Peeters (2003). The CSP can be defined as follows. Given an unlimited stock of a raw material type of length c and a set of n items with widths $w_1, \ldots, w_n$ and demands $d_1, \ldots, d_n$, cut as few raw material types as possible, such that the demand is satisfied and the total width of the items cut from a raw material type does not exceed its length c. Let P be the set of all feasible cutting patterns, or

$$P = \left\{ p \in \mathbb{Z}_+^n : \sum_{i=1}^{n} w_i p_i \leq c \right\}. \tag{9.39}$$

Let z_p be the number of times pattern p is selected in the solution, then the Gilmore and Gomory formulation can be stated as follows:

$$\min \sum_{p \in P} z_p \tag{9.40}$$

$$\text{subject to } \sum_{p \in P} p_i z_p \geq d_i \quad \forall i \in 1, \ldots, n, \tag{9.41}$$

$$z_p \in \{0, 1, 2, \ldots\} \quad \forall p \in P. \tag{9.42}$$

The objective function (9.40) minimizes the total number of cut raw material, whereas constraints (9.41) are the demand constraints and constraints (9.42) the integrality and non-negativity restrictions. The LP relaxation of (9.40)–(9.42) can be solved by column generation, where the pricing problem is a bounded knapsack problem, if one does not allow that the number of items present in a cutting pattern exceeds the demand, i.e. $p_i \leq d_i$.

Using the procedure described in Subsection 2.3, Degraeve and Peeters (2000) are able to achieve a substantial reduction in required CPU time to solve the LP relaxation of (9.40)–(9.42). Like Barahona and Jensen (1998), they use a limit on the number of subgradient iterations in the inner loop of Figure 9.1, but, in addition, the inner loop is interrupted, if a new column has non-negative reduced cost, or if the Lagrangian bound rounded up equals the master problem's objective value rounded up, as explained earlier in Figure 9.1. If this last condition holds, the Dantzig-Wolfe lower bound is found. Otherwise, all different columns generated in the inner loop are added to the restricted master problem. First it is checked if the value of the best Lagrangian lower bound rounded up is equal to the value of the new restricted master problem rounded up. Then, the algorithm can be terminated, otherwise the next iteration of the outer loop continues.

Table 9.2 presents the results of the computation times for cutting stock instances with 50, 75 and 100 items for 4 different width intervals given in the first row, in which the item widths are uniformly distributed. The demand is uniformly distributed with an average of 50 and the raw material length equals 10000. The experiments were run on a Dell Pentium Pro 200Mhz PC (Dell Dimension XPS Pro 200n) using the Windows95 operating system, the computation times are averages over 20 randomly drawn instances and given in seconds. The LPs are solved using the industrial LINDO optimization library version 5.3 (Schrage, 1995). The columns labelled "DW" present the traditional Dantzig-Wolfe algorithm and the columns labelled "CP" present the results of the combined procedure of Figure 9.1. We observe that the

Table 9.2. Computational results, Cutting Stock Problem.

int	[1,2500]		[1,5000]		[1,7500]		[1,10000]	
n	DW	CP	DW	CP	DW	CP	DW	CP
50	0.44	0.21	1.47	0.52	0.67	0.46	0.14	0.10
75	1.14	0.47	4.82	1.12	4.26	1.14	0.53	0.27
100	3.19	0.84	15.96	2.05	14.78	3.99	1.65	0.73

reduction in CPU time is higher, when the number of items is higher, and can be as high as a factor 8.

6.   Conclusion

We discussed two ways to combine Lagrangian relaxation and column generation. Since this combination has not been used quite often, there are many interesting research questions open. For example, should we use another method to approximate the Lagrangian dual, e.g. a multiplier adjustment method? Furthermore, when implementing such algorithms one has to make decisions with respect to issues such as column management.

In the first method, we used Lagrangian relaxation to solve the extended formulation. Therefore, no simplex method was necessary anymore, which has several advantages. First of all, it decreases the problem of degeneracy and speeds up the convergence. Furthermore, master problems with a larger number of constraints are most often faster solved with Lagrangian relaxation than with a LP solver. We showed this by solving the multiple-depot vehicle and crew scheduling problem.

In the second method, Lagrangian relaxation was used to generate new columns. It is an effective method to speed up convergence of the Dantzig-Wolfe column generation algorithm. The method seems to be quite robust, since it gives good results on three totally different problems, and this without much fine-tuning of the parameters. Several issues can be further investigated. For example, how many subgradient iterations do we allow in the inner loop of Figure 9.1? This is also related to the number of columns that we want to add in an inner loop: All new columns, the ones with negative reduced cost or only the ones with the most negative reduced cost? Adding more columns leads possibly to a faster convergence, but larger restricted master problems are also more difficult to solve. Do we initialize the multipliers in the Lagrangian relaxation part with the best Lagrangian multipliers of the previous step, with the optimal dual variables provided by the simplex algorithm for the current restricted master problem, or some combination? Clearly, there are ample opportunities for research into the effective combination of column generation and Lagrangian relaxation.

References

Barahona, F. and Anbil, R. (2000). The volume algorithm: Producing primal solutions with a subgradient method. *Mathematical Programming Series A*, 87:385–399.

Barahona, F. and Jensen, D. (1998). Plant location with minimum inventory. *Mathematical Programming*, 83:101–112.

Barnhart, C., Johnson, E.L., Nemhauser, G.L., Savelsbergh, M.W.P., and Vance, P.H. (1998). Branch-and-price: Column generation for solving huge integer programs. *Operations Research*, 46:316–329.

Bixby, R.E., Gregory, J.W., Lustig, I.J., Marsten, R.E., and Shanno, D.F.(1992). Very large-scale linear programming: A case study in combining interior point and simplex methods. *Operations Research*, 40:885–897.

Carraresi, P., Girardi, L., and Nonato, M. (1995). Network models, Lagrangean relaxation and subgradients bundle approach in crew scheduling problems. In: *Computer-Aided Transit Scheduling, Proceedings of the Sixth International Workshop* (J.R. Daduna, I. Branco, and J.M. P. Paixão, eds.), pp. 188–212. Springer Verlag.

Cattrysse, D.G., Salomon, M., Kuik, R., and Van Wassenhove, L.N. (1993). A dual ascent and column generation heuristic for the discrete lotsizing and scheduling problem with setup times. *Management Science*, 39:477–486.

Cattrysse, D.G., Salomon, M., and Van Wassenhove, L.N.(1994). A set partitioning heuristic for the generalized assignment problem. *European Journal of Operational Research*, 72:167–174.

Dantzig, G.B. and Wolfe, P. (1960). Decomposition principle for linear programming. *Operations Research*, 8:101–111.

Degraeve, Z. and Jans, R. (2003). A new Dantzig-Wolfe reformulation and Branch-and-Price algorithm for the capacitated lot sizing problem with set up times. *Technical Report* ERS-2003-010-LIS, ERIM, Erasmus University Rotterdam, the Netherlands.

Degraeve, Z. and Peeters, M. (2000). *Solving the linear programming relaxation of cutting and packing problems: A hybrid simplex method/subgradient optimization procedure.* Technical Report OR0017, Departement Toegepaste Economische Wetenschappen, K.U. Leuven, Belgium.

Degraeve, Z. and Peeters, M. (2003). Optimal integer solutions to industrial cutting-stock problems: Part 2. Benchmark results. *INFORMS Journal on Computing*, 15:58–81.

Desrosiers, J., Dumas, Y., Solomon, M.M., and Soumis, F. (1995). Time Constrained Routing and Scheduling. In: *Network Routing* (M.O. Ball, T.L. Magnanti, C.L. Monma, and G.L. Nemhauser, eds.), Handbooks in Operations Research and Management Science, Volume 8, pp. 35–139. North-Holland.

Dzielinski, B.P. and Gomory, R.E. (1965). Optimal programming of lot sizes, inventory and labour allocations. *Management Science*, 11:874–890.

Fischetti, M. and Toth, P. (1997). A polyhedral approach to the asymmetric traveling salesman problem. *Management Science*, 43:1520–1536.

Fisher, M.L. (1981). The Lagrangian relaxation method for solving integer programming problems. *Management Science*, 27:1–18.

Fisher, M.L.(1985). An applications oriented guide to Lagrangian relaxation. *Interfaces*, 15:10–21.

Freling, R. (1997). Models and techniques for integrating vehicle and crew scheduling. *Ph.D. Thesis*, Tinbergen Institute, Erasmus University Rotterdam.

Freling, R., Huisman, D., and Wagelmans, A.P.M. (2003). Models and algorithms for integration of vehicle and crew scheduling. *Journal of Scheduling*, 6:63–85.

Gaffi, A. and Nonato, M. (1999). An integrated approach to extra-urban crew and vehicle scheduling. In: *Computer-Aided Transit Scheduling* (N.H.M. Wilson, ed.), pp. 103–128. Springer Verlag.

Geoffrion, A.M. (1974). Lagrangean relaxation for integer programming. *Mathematical Programming Study*, 2:82–114.

Gilmore, P.C. and Gomory, R.E. (1961). A linear programming approach to the cutting stock problem. *Operations Research*, 9:849–859.

Haase, K., Desaulniers, G., and Desrosiers, J. (2001). Simultaneous vehicle and crew scheduling in urban mass transit systems. *Transportation Science*, 35:286–303.

Holmberg, K. and Yuan, D. (2003). A multicommodity network flow problem with side constraints on paths solved by column generation. *INFORMS Journal on Computing*, 15:42–57.

Huisman, D., Freling, R., and Wagelmans, A.P.M. (2003). *Multiple-depot integrated vehicle and crew scheduling*. Technical Report EI2003-02, Econometric Institute, Erasmus University Rotterdam, the Netherlands. Forthcoming in: *Transportation Science*.

IBM Corp. Optimization subroutine library: Guide and references. (1995).

Jans, R. and Degraeve, Z. (2004). An industrial extension of the discrete lot sizing and scheduling problem. *IIE Transactions*, 36:47–58.

Kleindorfer, P.R. and Newson, E.F.P. (1975). A lower bounding structure for lot-size scheduling problems. *Operations Research*, 23:299–311.

Lasdon, L.S. and Terjung, R.C.(1971). An efficient algorithm for multi-item scheduling. *Operations Research*, 19:946–969.

Lemaréchal, C., Nemirovskii, A.S., and Nesterov, Y.E. (1995). New variants of bundle methods. *Mathematical Programming*, 69:111–148.

Löbel, A. (1998). Vehicle scheduling in public transit and Lagrangian pricing. *Management Science*, 44:1637–1649.

Magnanti, T.L., Shapiro, J.F., and Wagner, M.H. (1976). Generalized linear programming solves the dual. *Management Science*, 22:1195–1203.

Manne, A.S. (1958). Programming of economic lot sizes. *Management Science*, 4:115–135.

Martin, R.K. (1999). *Large Scale Linear and Integer Optimization: A Unified Approach*. Kluwer Academic Publishers.

Schrage, L. (1995). LINDO: Optimization Software for Linear Programming. Lindo Systems Inc., Chicago IL.

Sol, M. (1994). *Column generation techniques for pickup and delivery problems*. Ph.D Thesis, Eindhoven University of Technology.

Trigeiro, W., Thomas, L.J., and McClain, J.O. (1989). Capacitated lot sizing with set-up times. *Management Science*, 35:353–366.

Van den Akker, J.M., Hurkens, C.A.J., and Savelsbergh, M.W.P. (2000). Time-indexed formulations for machine scheduling problems: Column generation. *INFORMS Journal on Computing*, 12:111–124.

Vanderbeck, F. (1999). Computational study of a column generation algorithm for bin packing and cutting stock problems. *Mathematical Programming Series A*, 86:565–594.

Vanderbeck, F. and Wolsey, L.A. (1996). An exact algorithm for IP column generation. *Operations Research Letters*, 19:151–159.

Wolsey, L.A. (1998). *Integer Programming*. Wiley.

Chapter 10

DANTZIG-WOLFE DECOMPOSITION FOR JOB SHOP SCHEDULING

Sylvie Gélinas
François Soumis

Abstract   This chapter presents a formulation for the job shop problem based on Dantzig-Wolfe decomposition with a subproblem for each machine. Each subproblem is a sequencing problem on a single machine with time windows. The formulation is used within an exact algorithm capable of solving problems with objectives C_{max}, T_{max}, as well as an objective consistent with the Just-In-Time principle. This objective involves an irregular cost function of operation completion times. Numerical results are presented for 2 to 10 machine problems involving up to 500 operations.

1.   Introduction

The job shop problem is a classical scheduling problem (French, 1982) that consists of scheduling n jobs on m machines. A single machine processes one job at a time and a job is processed by one machine at a time. Processing of one job on one machine is called an operation. The length of an operation is fixed; once begun, an operation may not be interrupted. The sequence of machines is known for each job. This sequence defines precedence constraints between the operations. The sequence of jobs on each machine must be determined so as to minimize a function of the operation completion times. This chapter considers problems in which jobs do not necessarily involve operations on all machines, but only on a subset of machines. This problem is occasionally referred to as the general job shop problem.

The problems considered in this chapter can be classified using the classical notation such as $(J|r_i, d_i|C_{max})$, $(J|r_i, d_i|T_{max})$ and $(J|r_i, d_i|JIT)$. The job shop problem is NP-hard in the strong sense (Rinnooy Kan, 1976; Garey and Johnson, 1979) and is one of the hardest problems to solve in practice. The most successful exact algorithms use a branch and bound procedure. See McMahon and Florian (1975); Lageweg, Lenstra and Rinnooy Kan (1977), Barker and McMahon (1985); Carlier and Pinson (1989), and Brucker, Jurisch and Sievers (1994).

Three branching schemes are generally used: conflict resolution in the disjunctive graph (Roy and Sussman, 1964), generation of active schedules (Giffler and Thompson, 1960), and time oriented branching (Marten and Shmoys, 1996). The conflict resolution scheme produces a complete schedule at each node of the branch tree; schedules thus produced may, however, violate the machine constraints. Two descendants are obtained by selecting two operations in conflict and imposing an order on them. In the generation of active schedules scheme, schedules are constructed sequentially from the root of the tree to a terminal node. A node of the branch tree is associated with a feasible schedule for a subset of operations. Barker and McMahon (1985), by contrast, propose a scheme that obtains a feasible schedule for all operations at each node of the branch tree. This procedure branches by rearranging operations in a critical block to yield an improved schedule. The head-tail adjustment proposed by Brinkkotter and Bruckner (2001) is also of this type.

A lower bound may be obtained by relaxing the machine constraints for all machines except one, to yield m scheduling problems on a single machine with time and precedence constraints $(n|r_i, d_i, \text{prec}|C_{\max})$. While this problem is NP-hard, even large instances of it may be solved efficiently (Carlier, 1982; McMahon and Florian, 1975). The maximum value found for the m problems gives a lower bound for the job shop problem. Lageweg, Lenstra and Rinnooy Kan (1977) discuss this bound, as well as many others obtained by relaxing one or more aspects of the scheduling problem on one machine. Balas, Lenstra and Vazacopoulos (1995) propose an improvement that considers delayed precedence constraints between operations. They used the resulted bound within the shifting bottleneck heuristic of Adams, Balas and Zawack (1988). Brucker and Jurisch (1993) obtain a bound derived from two-job scheduling problems. Problems are solved in polynomial time using a graphical method. Numerical results show that the bound obtained in this way is superior to that obtained from single-machine scheduling problems if the ratio between the number of machines and the number of jobs is large.

Other authors propose a relaxation of the problem based on a mathematical formulation. Fisher (1973) uses Lagrangian relaxation to solve problems with min-sum type objectives. He dialyzes the machine constraints and conserves precedence constraints. This method is tested on eight problems involving up to five jobs and four machines. Hoitomt et al. (1993) present an augmented Lagrangian approach for the weighted quadratic tardiness job shop problem. They obtained feasible solutions within 4% of their respective lower bounds for 125–145 job problems with 1 to 3 operations per job. Fisher et al. (1983) propose two aggregate-constraint formulations for the objective C_{max}, in which either the precedence or machine constraints are grouped into a linear combination. While these formulations yield a better bound than those obtained from single-machine scheduling problems or Lagrangian relaxation, the effort required discourages any search beyond the root of the search tree.

Currently available exact methods are not capable of solving large-scale problems. Carlier and Pinson (1989) solved the famous 10-machine, 10-job problem of Muth and Thompson (1963), more than 25 years after its publication. Applegate and Cook (1991) have solved some 150, 225-operation problems using many families of cutting phases and some good heuristics to find feasible solutions. However, some problems of 150, 200, 225-operations stayed unsolved. Brucker et al. (1994) have solved problems having up to 300 operations on a Sun 4/20 workstation. Optimal solutions of some of these problems require close to five days of CPU time.

In view of the difficulty of the job shop problem, algorithms should be developed to address the special structure of the problems encountered in industry. This chapter describes an efficient exact algorithm to solve problems with many jobs and few operations per job. Such problems appear more and more frequently in manufacturing, where increasingly versatile machines are capable of processing jobs with few changes per machine. The real problems from Pratt & Witney presented in Hoitomt et al. (1993) have this property (1 to 3 operations per job).

Most algorithms are designed to minimize the total length of operations (C_{max}) and are poorly adapted to other objectives. In practice, however, other objectives such as minimizing inventory costs or penalties arising from delivery delays are frequently more interesting. While the objective T_{max} has been examined in the context of the one-machine scheduling problem, it has received little attention with regard to the general job shop problem. Furthermore, the literature contains virtually no discussion on exact methods for irregular functions of completion times. This chapter discusses objectives such as the total length of operations (C_{max}), the maximum tardiness (T_{max}) and an objective consistent with the Just-In-Time principle.

The algorithm presented here is of the branch and bound variety. A lower bound is obtained using Dantzig-Wolfe decomposition, (Dantzig and Wolfe, 1960). Unlike Fisher, we use a primal approach and relax precedence constraints rather than machine constraints. A primal approach yields faster convergence; furthermore, precedence constraints play a reduced role in problems involving few operations per job. The solution to the master problem provides a lower bound for the job shop problem, which is incorporated into a branching scheme based on conflict resolution. This formulation can be adapted to several objectives, including irregular functions of completion times. This is not the case for the aggregate-constraint formulation of Fisher et al. (1983), which is only valid for minimizing $C_{\max}$.

The Dantzig-Wolfe decomposition was used for parallel machine scheduling problems by van den Akker et al. (1995) and Chen and Powell (1999a,b, 2003). This scheduling problem does not involve precedence constraints and is simpler than the job shop scheduling problem. The column generation algorithms proposed by these authors are straightforward translations of the algorithm for vehicle routing problems presented by Desrochers et al. (1992). The present chapter proposes a column generation algorithm for a more complex problem.

More references on methods using constraint programming, metaheuristics, neural networks, can be found in surveys on the job shop scheduling problem by Blazewicz et al. (1996) and Join and Meeron (1999). More recent work using constraint propagation was presented by Dorndorf et al. (2000, 2002).

2.   Mathematical formulation

We consider n jobs (index j), m machines (index i) and N operations (indices u, v). Let p_u be the time required for operation u, i_u the machine on which operation u is to be carried out, r_u the earliest time at which operation u may begin and d_u the latest time at which operation u may end.

Precedence relations are contained in the set

$$A = \{(u, v) | u \text{ and } v \text{ are successive operations of the same job,}$$
$$\text{and } u \text{ preceeds } v\}$$

and pairs of operations carried out on machine i are contained in set

$$B_i = \{\{u, v\} \mid u \neq v, \text{ and } i_u = i_v = i\}, \quad i = 1, \ldots, m.$$

The job shop problem is formulated using a min-max type objective function. We consider a function $g_u(C_u)$ of completion time C_u for operation u. We assume that the function $g_u(C_u)$ is piecewise linear, but not necessarily monotone. We define $g_{\max}$, a variable that takes the maximum value of the quantities $g_u(C_u)$ for all operations u.

We define the following variables,

C_u: completion time for operation u, $u = 1, \ldots, N$,

g_{max}: cost of the solution.

The job shop problem can be formulated as follows.

$$\min g_{max} \tag{10.1}$$

$$\text{s.t.} \qquad g_{max} \geq g_u(C_u) \quad u = 1, \ldots, N, \tag{10.2}$$

$$C_u \leq C_v - p_v \quad (u, v) \in A, \tag{10.3}$$

$$r_u + p_u \leq C_u \leq d_u \quad u = 1, \ldots N, \tag{10.4}$$

$$C_u \leq C_v - p_v \vee C_v \leq C_u - p_u \quad \{u, v\} \in B_i, \ i = 1, \ldots, m. \tag{10.5}$$

We minimize a min-max type function of the operation completion times (10.1)–(10.2) under precedence constraints (10.3), time constraints (10.4) and machine constraints (10.5). The machine constraints are disjunctive and require that two operations carried out on the same machine may not occur at the same time. These constraints make the problem non-linear and hence difficult to solve. The next section reformulates the job shop problem using Dantzig-Wolfe decomposition.

3.   Decomposition

We have formulated the job shop problem as a non-linear problem with disjunctive constraints. Relaxing the precedence constraints yields a problem that is separable by machine. Each problem corresponds to a single-machine scheduling problem whose solution provides a lower bound on the job shop problem. This idea is applied within the framework of Dantzig-Wolfe decomposition.

Constraints used to compute the objective (10.2), as well as the precedence constraints (10.3), are left in the master problem. The time constraints (10.4) and machine constraints (10.5) are transferred to the subproblems. Each subproblem generates schedules for one machine. The master problem selects a convex combination of the generated schedules that satisfies the precedence constraints. The Dantzig-Wolfe decomposition provides the optimal solution for a linear problem, and for the job shop problem, it provides a lower bound. This bound is better than that obtained by ignoring the precedence constraints, because of the exchange of information between the master problem and the subproblems.

3.1   Master problem

The following notation is used in the master problem formulation.

Ω_i: set of schedules that satisfy the time constraints (10.4) and machine constraints (10.5) for machine i, $i = 1, \ldots, m$,

C_u^h: completion time for operation u in schedule h, $u = 1, \ldots, N$, $h \in \Omega_{i_u}$,

y_h : variable associated with schedule h, $\forall h \in \Omega_i$, $i = 1, \ldots, m$.

The master problem is then:

$$\min g_{\max} \tag{10.6}$$

$$\text{s.t.} \quad g_{\max} \geq \sum_{h \in \Omega_{i_u}} y_h g_u(C_u^h) \quad u = 1, \ldots, N, \tag{10.7}$$

$$\sum_{h \in \Omega_{i_u}} y_h C_u^h \leq \sum_{h \in \Omega_{i_v}} y_h C_v^h - p_v \quad \forall (u,v) \in A, \tag{10.8}$$

$$\sum_{h \in \Omega_i} y_h = 1 \quad i = 1, \ldots, m, \tag{10.9}$$

$$y_h \geq 0, \quad \forall h \in \Omega_i, \quad i = 1, \ldots, m. \tag{10.10}$$

The master problem is formulated using variables associated with schedules for a single machine. Its objective is to find a convex combination of schedules for each machine such that the precedence constraints (10.8) are satisfied and the cost (10.6)–(10.7) is minimized. The convex combination constraints are given by (10.9) and (10.10).

The set of schedules for machine i is not convex, because of the machine constraints (which are disjunctive). This set contains a polytope for each ordering; each point in a polytope corresponds to a schedule for the ordering associated with the polytope. Polytopes are bounded, because of the time windows and have a finite number of extreme points

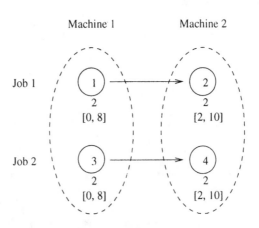

Figure 10.1.   A job shop problem.

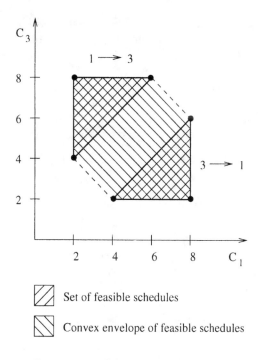

Figure 10.2.   Schedules for machine 1.

and no extreme rays. The master problem obtains schedules as convex combinations of points in the set Ω_i. These schedules belong to the convex envelope of Ω_i, but not necessarily to the set Ω_i itself. $\text{Conv}(\Omega_i)$, the convex envelope of Ω_i is the convex combination of a finite set of extreme points; the union of the finite number of orderings on machine i of the finite set of extreme points of the polytope for this ordering. Because $\Omega_i \subseteq \text{Conv}(\Omega_i)$, Dantzig-Wolfe decomposition provides a lower bound for the job shop problem.

A job shop problem with two jobs, two machines and four operations is illustrated in Figure 10.1. All operations last two units of time. Job processing may begin at time 0, and must be completed by time 10. Figure 10.2 illustrates all feasible schedules for machine 1, which make up two polytopes associated either with the ordering 1-3 or the ordering 3-1. Each point in the set corresponds to a schedule for these two operations. The convex envelope of this set contains schedules that may be selected in the solution of the Dantzig-Wolfe decomposition.

Since there is a large number of sequences for each machine, it is impossible to enumerate all of them in a job shop problem unless the number of operations on each machine is very small. There is an even greater number of extreme schedules, which represent extreme points of schedule polytopes for all sequences. The master problem considers only a subset of schedules for each machine. New schedules are generated by subproblems as necessary.

3.2   Subproblems

Subproblem S_i finds the schedule at minimum marginal cost for machine i. Dual variables for the master problem are denoted as follows:

$\gamma_u \geq 0$ $u = 1, \ldots, N$,   Constraints for computing the objective (10.7),

$\alpha_{uv} \geq 0$ $(u,v) \in A$,   Precedence constraints (10.8),

λ_i $i = 1, \ldots, m$.   Convexity constraints (10.9).

Column h^* of minimum reduced cost $\bar{c}_{h^*}$ for subproblem S_i is such that

$$\bar{c}_{h^*} = \min_{h \in \Omega_i} \left\{ \sum_{u|i_u=i} \gamma_u g_u(C_u^h) + \sum_{u|i_u=i} \sum_{v|(u,v)\in A} \alpha_{uv} C_u^h \right.$$
$$\left. - \sum_{u|i_u=i} \sum_{v|(v,u)\in A} \alpha_{vu} C_u^h - \lambda_i \right\}$$

$$= \min_{h \in \Omega_i} \left\{ \sum_{u|i_u=i} \gamma_u g_u(C_u^h) + \sum_{u|i_u=i} \left(\sum_{v|(u,v)\in A} \alpha_{uv} - \sum_{v|(v,u)\in A} \alpha_{vu} \right) C_u^h \right\} - \lambda_i$$

$$= \min_{h \in \Omega_i} \left\{ \sum_{u|i_u=i} \left(\gamma_u g_u(C_u^h) + w_u C_u^h \right) \right\} - \lambda_i$$

$$= \min_{h \in \Omega_i} \left\{ \sum_{u|i_u=i} g_u'(C_u^h) \right\} - \lambda_i$$

where $w_u = \sum_{v|(u,v)\in A} \alpha_{uv} - \sum_{v|(v,u)\in A} \alpha_{vu}$
and $g_u'(C_u^h) = \gamma_u g_u(C_u^h) + w_u C_u^h$, $u = 1, \ldots, N$.
Subproblem S_i is formulated as follows:

$$\min \sum_{u|i_u=i} g_u'(C_u) \tag{10.11}$$

s.t. $r_u + p_u \leq C_u \leq d_u$ $u \mid i_u = i$,   (10.12)

$C_u \leq C_v - p_v \lor C_v \leq C_u - p_u$ $\{u,v\} \in B_i$.   (10.13)

This subproblem is a sequencing problem on a single machine with time constraints and an objective of minimizing a piecewise linear function of the completion times: $n|r_u, d_u| \sum g_u'(C_u)$. The problem is difficult because the cost function is irregular (the weights w_u may be positive or negative and the function $g_u(C_u)$ is an irregular function). This problem is solved using a dynamic programming algorithm. A dynamic programming state is associated with a set of operations X and cost function $G_X(t)$. The function $G_X(t)$ gives the minimum cost of a feasible schedule that carries out all operations of X and ends at the latest at time t. The functions $G_X(t)$ are evaluated by stages. At stage k, functions are evaluated for sets having k operations, using function values for sets containing $(k-1)$ operations. Details of the algorithm may be found in Gélinas and Soumis (1997).

At each iteration of the Dantzig-Wolfe algorithm, the master problem is solved using the simplex algorithm. The solution provides the values of the dual variables, which are then used in the subproblems to obtain new schedules, that is, new columns for the master problem. Columns are added to the master problem if their marginal cost is negative, giving rise to a new iteration. The procedure terminates when each subproblem generates a column with nonnegative marginal cost. The solution to the master problem is then the optimal solution for all columns, whether or not they are considered explicitly.

3.3   Branching

Dantzig-Wolfe decomposition provides a lower bound for the job shop problem. Although the solution satisfies the precedence and time constraints, it may violate the machine constraints because $\Omega_i \subseteq \text{Conv}(\Omega_i)$. If all machine constraints are satisfied, the solution is optimal for the job shop problem. Otherwise, there are operations carried out concurrently on the same machine. In this case, a pair (u, v) of operations that conflict on one machine is selected and two new problems are created by imposing an order on these operations: either operation u is carried out before operation v, or operation v is carried out before operation u. The new problems are solved using Dantzig-Wolfe decomposition and this process continues until the branching tree has been thoroughly explored. The lower bound may be used to prune branches from the tree.

To respect the order imposed by the branching, precedence constraints are added between operations carried out on one machine. These constraints are easily handled in the subproblem solution. Dynamic programming states that do not satisfy constraints are not constructed. The subproblem then becomes a sequencing problem on a single machine with time and precedence constraints $\left(n \mid r_u, d_u, \text{prec} \mid \sum g'_u(C_u)\right)$.

An optimal schedule for the job shop problem of Figure 10.1, with the objective $C_{\max}$, is illustrated in Figure 10.3. The optimal schedule ends

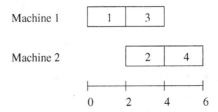

Figure 10.3.   Optimal solution for objective $C_{\max}$.

at time 6. For this example, the Dantzig-Wolfe solution combines two schedules for each machine

$$\begin{pmatrix} C_1 \\ C_3 \end{pmatrix} = 0.5 \begin{pmatrix} 2 \\ 4 \end{pmatrix} + 0.5 \begin{pmatrix} 4 \\ 2 \end{pmatrix} = \begin{pmatrix} 3 \\ 3 \end{pmatrix},$$

$$\begin{pmatrix} C_2 \\ C_4 \end{pmatrix} = 0.5 \begin{pmatrix} 4 \\ 6 \end{pmatrix} + 0.5 \begin{pmatrix} 6 \\ 4 \end{pmatrix} = \begin{pmatrix} 5 \\ 5 \end{pmatrix}.$$

This solution is infeasible because operations 1 and 3 are carried out concurrently on machine 1, and operations 2 and 4 are carried out concurrently on machine 2. An optimal, feasible solution is then obtained using the branching tree. Finally, note that, for this example, the Dantzig-Wolfe decomposition provides a lower bound of 5, while an approach ignoring the precedence constraints would provide a bound of 4.

3.4 $C_{\max}$ objective

The present formulation can be used to model problems with objectives of type $C_{\max}$ if we set $g_u(C_u) = C_u$. The master problem then becomes:

$$\min C_{\max} \tag{10.14}$$

s.t. $\quad C_{\max} \geq \sum_{h \in \Omega_{i_u}} y_h C_u^h \quad u = 1, \dots, N, \tag{10.15}$

$\phantom{s.t. \quad C_{\max} \geq \sum_{h \in \Omega_{i_u}}}$ constraints (10.8), (10.9) and (10.10).

Another formulation, providing a better bound in the linear relaxation of the Dantzig Wolfe decomposition, has been developed for this specific objective. In this formulation, the objective is a function of the weighted mean completion time for each operation. The order of operations is not the same in the generated schedules. Some schedules for machine i may have operation u as their final operation, while others may have operation $v \neq u$ as their final operation. The mean completion time for operations u and v may be far smaller than the mean completion time for all operations on machine i.

It would be better to express makespan constraints as a function of the completion time for each machine. To accomplish this, a fictitious operation is added for each machine, i.e., $u = N + 1, N + 2, \ldots, N + m$. We set $p_{N+i} = 0$, $r_{N+i} = \max_{u|i_u=i} r_u + p_u$, $d_{N+i} = \max_{u|i_u=i} d_u$, $i = 1, \ldots, m$ and require that the operation $N + i$ be the last carried out on machine i.

Constraints (10.15) of the master problem are then replaced by

$$C_{\max} \geq \sum_{h\in\Omega_{i_u}} y_h C_u^h \quad u = N + 1, \ldots, N + m. \tag{10.16}$$

A dual variable γ_i is defined for each machine. The objective of subproblem S_i is to minimize the weighted sum of the operation completion times $(n \mid r_u, d_u, \text{prec} \mid \sum w'_u C_u)$ where

$$w'_u = \begin{cases} w_u, & u \mid i_u = i \quad \text{and} \quad u \leq N, \\ \gamma_i, & u = N + i. \end{cases}$$

The cost function is also irregular in the operation completion times and the same dynamic programming algorithm described in Section 3.2 can be applied.

With this formulation, the Dantzig-Wolfe solution for the problem illustrated in Figure 10.1 has a cost of 6.

$$\begin{pmatrix} C_1 \\ C_3 \\ C_5 \end{pmatrix} = 0.5 \begin{pmatrix} 2 \\ 4 \\ 4 \end{pmatrix} + 0.5 \begin{pmatrix} 4 \\ 2 \\ 4 \end{pmatrix} = \begin{pmatrix} 3 \\ 3 \\ 4 \end{pmatrix},$$

$$\begin{pmatrix} C_2 \\ C_4 \\ C_6 \end{pmatrix} = 0.5 \begin{pmatrix} 4 \\ 6 \\ 6 \end{pmatrix} + 0.5 \begin{pmatrix} 6 \\ 4 \\ 6 \end{pmatrix} = \begin{pmatrix} 5 \\ 5 \\ 6 \end{pmatrix}.$$

This solution, which is optimal for the Dantzig-Wolfe decomposition, has the same cost as the optimal solution of the job shop problem.

4. Implementation issues

An exact algorithm for the job shop problem has been implemented using three types of objectives:

$C_{\max}$: We set $g_u(C_u) = C_u$ and modify the formulation as described in Section 3.4.

T_{max}: We set $g_u(C_u) = \max\{C_u - d'_u, 0\}$ where d'_u is the latest time at which operation u may terminate and not be late.

JIT (**Just-In-Time**): We set $g_u(C_u) = |C_u - d'_u|$ where d'_u is the desired termination time for operation u. A penalty is paid if operation u terminates before or after time d'_u. Different penalty costs can be used for earliness and tardiness without increasing the complexity of the solution approach.

The algorithm uses Dantzig-Wolfe decomposition and a branching strategy based on conflict resolution. It implements several exact and heuristic rules to accelerate the solution process.

4.1   Overview

An upper bound z_{sup} is provided as an input to the optimization algorithm. This bound may be obtained using heuristic methods. A three-step procedure is executed at each branching node.

The first step tightens time windows $[r_u, d_u]$ using the upper bound, the precedence constraints of the problem, those imposed by the branching procedure, and others that can be deduced from rules. Problem feasibility tests are carried out. If such tests conclude that the problem is infeasible, the node is abandoned.

The second step computes a solution to the relaxed job shop problem obtained from the Dantzig-Wolfe decomposition. Two cases are possible:

- The solution process is completed with the proof that no solution for the relaxation is possible. In this case, there is no solution to the job shop problem at the current node, and the node is abandoned.

- A solution is found for the relaxation; the process is halted, although optimality is not necessarily obtained. While a solution of the relaxation satisfies time and precedence constraints of the job shop problem, it may violate the machine constraints. If the process is stopped prior to optimality, the cost of the solution is not a lower bound for the current node. A lower bound, however, is of little interest because its cost is necessarily inferior to the z_{sup} and is not sufficient to eliminate the node at this stage. In fact, any solution to the relaxation has a cost less than z_{sup} as it satisfies the time constraints that were tightened using the value of $z_{sup} - 1$.

The third step applies a heuristic to calculate a solution that satisfies all constraints of the job shop problem, using the solution obtained in step 2. Once again, two cases are possible:

- A solution is found for the job shop problem. In this case, the upper bound is adjusted. Furthermore, branching nodes that have been solved to optimality and that have a lower bound greater than or equal to the new upper bound are eliminated. If the current node cannot be eliminated, the three-step procedure for processing a branching node is restarted with the new value of z_{sup}.

- No solution to the job shop problem is found. In this case, a pair of operations that conflict on one machine is selected. Two new problems are created by imposing an order on these two operations.

The branching tree is explored depth-first to find feasible solutions as quickly as possible. The advantage of proceeding in this way is that the operation time windows can be tightened, reducing the number of dynamic programming states. The following sections contain further details on the steps of the algorithm.

4.2   Preprocessing of the branching nodes

Before starting the solution process at a branching node, rules are applied to find precedence constraints and tighten time windows. An efficient implementation of these rules is described in Brucker, Jurisch and Sievers (1994); Carlier and Pinson (1990). In addition, the feasibility of each single-machine sequencing problem is verified, using calls to subproblems if necessary.

Precedence constraints

Let $\text{Succ}(u)$ denote the set of operations that must be carried out on machine i_u after operation u, and $\text{Prec}(u)$ the set of operations that must be carried out on machine i_u before operation u. Operation $v \in \text{Prec}(u)$ if and only if $u \in \text{Succ}(v)$.

Precedence relations may be deduced from simple rules. In particular, $v \in \text{Succ}(u)$ if $u \neq v$, $i_u = i_v$ and if one of the following conditions holds:

- The relation $u \to v$ is imposed by the branching.

- By the time constraints, operation v cannot be carried out before operation u: $r_v + p_v + p_u > d_u$.

- The relation $u \to v$ may be obtained by transitivity: $\exists w \mid u \in \text{Succ}(w)$ and $w \in \text{Succ}(v)$.

Other precedence constraints are deduced from more complex rules involving blocks of operations carried out on the same machine. Let X

be a subset of operations to be carried out on the same machine. Let
$r_X = \min_{v \in X} r_v$, $d_X = \max_{v \in X} d_v$ and $p_X = \sum_{v \in X} p_v$.

- If there is a set X of operations that must be carried out on machine
  i_u, such that $u \notin X$ and

$$\min\{r_X, r_u\} + p_u + p_X > d_X,$$

  then all operations in X must precede operation u, that is, $u \in$
  Succ(v) for all $v \in X$.

- If there exists a set X of operations that must be carried out on
  machine i_u, such that $u \notin X$ and

$$r_X + p_u + p_X > \max\{d_X, d_u\},$$

  then operation u must precede all operations in X, that is, $v \in$
  Succ(u) for all $v \in X$.

The problem is not feasible if the precedence constraints induce a
cycle, that is, if there exists u, v such that $u \in$ Succ(v) and $v \in$ Succ(u).

Time constraints

The time intervals $[r_u, d_u]$ are tightened using the upper bound $z_{\sup}$,
the precedence constraints of the job shop problem and the precedence
constraints among operations carried out on the same machine.

The new earliest time r_u to begin operation u is the largest of the
following quantities:

- r_u,

- $d'_u - p_u - z_{\sup} + 1$, (JIT objective),

- $r_v + p_v$, $\forall v \colon (v, u) \in A$,

- $\min_{v \in X} r_v + \sum_{v \in X} p_v$, $\forall X \colon X \subseteq$ Prec(u).

The new latest time d_u to terminate operation u is the smallest of the
following quantities:

- d_u,

- $z_{\sup} - 1$, ($C_{\max}$ objective),

- $d'_u + z_{\sup} - 1$, (JIT, $T_{\max}$ objectives),

- $d_v - p_v$, $\forall v \colon (u, v) \in A$,

- $\max_{v \in X} d_v - \sum_{v \in X} p_v$, $\forall X \colon X \subseteq$ Succ(u).

The problem is infeasible if an operation u can be found such that $r_u + p_u > d_u$.

4.3   Processing at a branching node

Dantzig-Wolfe decomposition is applied to the job shop problem associated with the current node.

Master problem

At each iteration of the Dantzig-Wolfe algorithm, the master problem calls the subproblems to receive columns with negative marginal cost. In the case at hand, subproblems are solved using a computationally intensive dynamic programming algorithm. It is not necessary to solve subproblems exactly to obtain columns with negative marginal cost, especially during the initial iterations. Subproblems are solved heuristically by limiting the number of dynamic programming states.

The limit on the number of states is controlled by a parameter passed to the subproblems from the master problem. The subproblem returns a boolean value indicating whether the state space has been explored completely or only partially. The master problem increases the limit if no further columns are generated or if the objective does not increase sufficiently. The optimal solution is found when all subproblems are solved exactly and generate no further columns.

As discussed in Section 4.1, the problem is not necessarily solved to optimality. Before raising the limit on the number of dynamic programming states, the feasibility of the relaxed master problem is verified. If the problem is feasible, column generation terminates and the heuristic search for a feasible solution to the job shop problem begins immediately (Section 4.4).

Subproblem

The subproblem is a sequencing problem on a single machine, $(n \mid r_u, d_u, \text{prec} \mid \sum g'_u(C_u))$, and is solved by dynamic programming. States are eliminated using both exact and heuristic criteria.

Exact criteria ensure that eliminated states cannot lead to an optimal solution. Only states that satisfy the precedence constraints are constructed. Several of these states are eliminated using rules based on the time constraints. These rules are given in Gélinas and Soumis (1997). Other states are eliminated using bounds. The dual variable λ_i of the master problem provides an upper bound on the cost of a schedule on machine i that may improve the solution to the master problem. A lower bound is computed for the cost of schedules constructed from a dynamic programming state. The state is eliminated if the lower bound is not promising.

Finally, states are eliminated using a heuristic criterion if their number exceeds the limit passed from the master problem. This criterion is based on the quality and feasibility of a state. States with a good lower bound and those that appropriately place operations that must terminate early are retained. When states are eliminated using the heuristic criterion, the subproblem solution may not be optimal, and the master problem is so notified.

4.4   Branching node post-processing

The Dantzig-Wolfe solution satisfies time and precedence constraints at a cost less than the upper bound z_{sup}. This solution will be used as a starting point for another solution that also satisfies the machine constraints.

The disjunctive graph $G = (V, C \cup D)$ associated with the job shop problem will be used in this regard. The nodes of the graph correspond to operations, including two fictitious operations representing the beginning and end of operations, $V = \{0, 1, \ldots, N, *\}$. Execution times p_u are associated with nodes of the graph. Arcs of the graph fall into two types. The set C of conjunctive arcs includes precedence arcs; arcs $(0, u)$ where u is the first operation of a job; and arcs $(u, *)$ where u is the last operation of a job. The set D contains disjunctive arcs representing pairs of operations processed on the same machine. The arc pair $\{(u, v), (v, u)\}$ is said to be resolved if one of the two arcs is selected and the other rejected. In selecting (u, v), we require that operation u be performed before operation v, which corresponds to the addition of a precedence constraint (conjunctive arc) in the graph G.

Some disjunctive arc pairs are resolved at the current node using rules stated in Section 4.2. To obtain a feasible solution, the rest of the disjunctive arcs are resolved temporarily, according to the order of the operations in the relaxed solution.

$$u \to v \quad \text{if} \quad \sum_{h \in \Omega_{i_u}} y_h C_u^h \leq \sum_{h \in \Omega_{i_v}} y_h C_v^h.$$

A longest path problem with time windows is then solved from node 0 to all other nodes. If the arrival time at each node is such that the operation may be carried out within the specified time interval, then a feasible solution has been found with a cost below the upper bound. The upper bound is then updated.

Local search

The solution obtained to the longest path problem is then improved by inverting a disjunctive arc that has been resolved temporarily. Of interest are the arcs that belong to the longest path (if the solution is feasible) or to an infeasible path (if the solution is infeasible). The time gain obtained locally by inverting the disjunctive arcs is calculated. Let (u^*, v^*) be the arc yielding the maximal gain. If this gain is positive, the arc is inverted and the longest path problem is solved on the new graph. The process terminates when there is no further local improvement.

JIT objective

A further stage occurs when a feasible solution is found for the objective JIT. The solution to the longest path algorithm places the operations as early as possible within the time windows $[r_u, d_u]$. A better solution can be obtained by delaying operations so that they end as close as possible to the desired termination time. The maximum tardiness $T_{\max}$ of an operation with respect to its desired termination time is calculated in the feasible solution. The time windows are then temporarily tightened in such a way that this maximum tardiness is not exceeded: $[r_u, \min\{d_u, d'_u + T_{\max}\}]$; and a longest path problem is solved by pulling node $*$ back toward the other nodes in the graph. The upper bound for the job shop problem is adjusted using this new solution.

4.5   Branching strategies

If no solution is found using the procedure described in Section 4.4, branching occurs on a pair of operations that are carried out on the same machine and in conflict in the relaxed solution.

When there are many candidates when selecting a pair, we use the following rules:

- reduce the set of candidates to a set of pairs in conflict on the longest path found in Section 4.4 if this set is not empty,

- select the earliest scheduled pair in the set of candidates.

5.   Experimentation

Numerical experiments were conducted using the Dantzig-Wolfe algorithm implemented in C on a HP9000/735 computer. The following sections describe the test problems and present the results obtained.

5.1   Test problems

Problems with ten machines and up to 500 operations were generated and solved using $C_{\max}$, $T_{\max}$ and JIT objectives. Problem sizes are described in Table 10.1.

Table 10.1. Sizes of the job shop problem instances.

Number of machines	Number of jobs	Number of operations per job	Total number of operations
10	250	2	500
10	100	3	300
10	30	5	150
10	10	10	100

A problem is constructed as follows. First, the number of machines, the number of jobs, and the number of operations per job are established. For each operation, a machine is selected at random in such a way that no job has two operations on the same machine. The length of an operation is generated uniformly in the interval $[1, 100]$. The times r_u are initialized to zero; the times d_u are initialized to a large value (∞). To select the desired completion times for jobs and their mutually compatible operations, a feasible schedule is constructed using decision rules. The operation that can begin earliest is placed first. In the case of a tie, the operation in that job having the most outstanding work is selected. The completion time T for this schedule is used to generate times d'_u. Let $u_1, u_2, \ldots, u_{n_j}$ be the operations in job j, in order. Times $d'_{u_{n_j}}$ are generated in the interval $[\sum_{k=1}^{n_j} p_{u_k}, T]$; in addition, we set

$$d'_{u_k} = d'_{u_{k+1}} - p_{u_{k+1}}, \quad k = n_j - 1, \ldots, 1.$$

The data d'_u are ignored for the objective $C_{\max}$. If $T_{\max}$ is to be minimized, we desire that the processing of job j terminate no later than time $d'_{u_{n_j}}$. If JIT is to be minimized, we desire that the processing begin at time $d'_{u_{n_j}} - \sum_{k=1}^{n_j} p_{u_k}$ and continue without stopping until time $d'_{u_{n_j}}$.

Ten problems are generated for each problem size, for a total of 40 job shop problems. Of particular interest are the job shop problems with many jobs and few operations per job. Such problems are easy to solve with the objective $C_{\max}$ because machines can operate without stopping. The schedule constructed using decision rules is optimal for all problems having 30 jobs or more. Therefore, we only present results for 10-job problems using the objective $C_{\max}$.

An upper bound is provided for the optimization algorithm. The schedule obtained with decision rules provides an upper bound for the objective $C_{\max}$. This bound is from 7% to 21% greater than the cost of the optimal solution for 10-job problems. For $T_{\max}$ and JIT, the upper bound is taken to be the optimal value plus 20%. (The optimal value is known because the algorithm has already been executed once using a large value as upper bound.) Future applications of the algorithm will require a heuristic procedure to produce a feasible solution for the objectives $T_{\max}$ and JIT. The cost of this solution will be an upper bound that should be no more than 20% from the optimal solution.

5.2   Numerical results

This section first presents results for specific steps of the algorithm. It then presents results for the job shop problems and analyzes the behavior of the algorithm with different initial upper bounds.

Solution of the Dantzig-Wolfe decomposition

The relaxed job shop problem is generally not solved to optimality for the first branching node. The value of the lower bound that could be obtained from Dantzig-Wolfe decomposition was obtained in a separate calculation. Table 10.2 gives the lower bound and the cost of the optimal solution for 10- and 30-job problems. Since the dynamic programming algorithm requires too much memory for the 100- and 200-job problems, optimal solutions are not obtained for them.

The lower bound is fairly distant from the optimal solution at the top of the branching tree, which provides justification for the solution approach presented here. We don't use much effort to get exact solutions to the Dantzig-Wolfe relaxation at the top of the tree. The job shop problem becomes more highly constrained at the lower level of the tree, as branching decisions are taken and feasible solutions are found. So, lower bounds become easier to get by exactly solving the Dantzig-Wolfe relaxation and are of better quality. This bound eliminates nodes associated with these more constrained problems.

On the other hand, the Dantzig-Wolfe solution is very useful for finding feasible solutions at each node of the branching tree, and is used to establish the order of operations in a schedule constructed heuristically.

Dynamic programming

Subproblems are solved using a dynamic programming algorithm. Two statistics are particularly germane as measures of the problem difficulty: the number of states and the number of labels. A state is asso-

Table 10.2. Lower bound from Dantzig-Wolfe decomposition.

Oper	C_{max}		T_{max}		JIT	
	DW	Opt	DW	Opt	DW	Opt
30× 5	-	-	164.0	174	135.4	235
	-	-	127.0	156	109.3	197
	-	-	99.6	205	117.7	220
	-	-	285.1	346	285.2	346
	-	-	67.0	67	133.2	196
	-	-	159.1	188	159.1	188
	-	-	72.7	121	87.1	152
	-	-	112.7	199	114.5	199
	-	-	91.6	218	127.6	278
	-	-	134.0	146	134.0	153
10×10	717.9	792	14.5	80	69.2	172
	792.0	867	53.1	69	106.6	172
	750.0	810	0.0	6	79.0	155
	742.8	845	2.4	99	75.7	167
	825.0	885	78.6	221	97.2	266
	625.6	728	14.8	90	60.9	162
	689.3	811	10.9	109	68.6	166
	743.2	840	26.1	108	70.5	190
	841.0	855	24.9	95	79.6	207
	747.0	766	36.6	119	102.7	208

ciated with a set of operations. A cost function is associated with each state. The cost function is piecewise linear and represented by a list of labels, one label per piece. At iteration k of the dynamic programming algorithm, all states associated with sets of k operations are considered. In a job shop problem with 10 machines, 100 jobs and 3 operations per job, there are a total of 300 operations and an average of 30 operations per machine. Consider a subproblem with 30 operations. At iteration 10 of the dynamic programming algorithm, there are a possible C_{10}^{30} states, that is more than 30 million states. While exact criteria can eliminate states, the number of them that remain to be considered in an exact procedure may be very large. The proposed algorithm uses heuristic criteria to eliminate states.

Table 10.3 gives the average and maximum number of states constructed in one iteration of the dynamic programming algorithm during the solution of the job shop problems. In all dynamic programming iter-

Table 10.3.   Dynamic programming statistics.

Oper	C_{max} States Avg	Max	Labels Avg	Max	T_{max} States Avg	Max	Labels Avg	Max	JIT States Avg	Max	Labels Avg	Max
250× 2	-	-	-	-	11.8	16	2.8	51	11.3	16	2.6	59
100× 3	-	-	-	-	11.0	233	2.7	42	9.3	120	2.7	47
30× 5	-	-	-	-	6.5	133	2.0	27	5.2	109	2.2	34
10×10	3.9	15	2.1	23	3.6	16	2.1	24	3.5	18	2.4	31

ations in all problems solved, no more than 233 states (i.e., a very small number) were constructed. This was sufficient, however, to find a feasible solution to the relaxed job shop problem or prove that no solutions exists.

To prove that there are no solutions to the relaxed problem, the subproblems must be solved exactly. The implementation of the algorithm increases the limit on the number of states until no further elimination occurs using heuristic criteria. While such a procedure may require that a large number of states be considered, this did not occur in the numerical experiments conducted for this study. Two reasons may explain this. First, the relaxed job shop problem is almost always feasible when Dantzig-Wolfe decomposition is applied. Rules applied in the preprocessing stage help to identify infeasible job shop problems. If the node is not eliminated using these rules, a solution is usually found to the decomposition. Second, when the relaxation has no solution, the subproblems are highly constrained and states are eliminated using exact criteria that are highly effective under the circumstances.

The table also gives the mean and maximum number of labels required to represent the cost function attached to a state. The number of labels increases with the number of operations and with the width of the time windows (Gélinas and Soumis, 1997). Larger numbers of labels imply greater calculations and manipulations in the dynamic programming algorithm. The average number of labels per state was low in problems solved for this study.

Elimination of nodes in branching tree

Figure 10.4 illustrates the branching tree obtained for one of the 250-job problems using the objective T_{max}. Nodes are numbered in the order in which they were explored. The value of the initial upper bound is 432. The relaxed job shop problem is not solved to optimality for the initial nodes; a lower bound of 360 is obtained at the fifth node. Going down

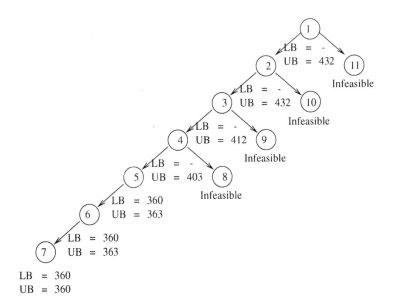

Figure 10.4. Branching tree.

the tree, feasible solutions are found with respective costs of 412, 403, 363 and finally 360. Nodes 7, 6 and 5 are then eliminated using the lower bound. While exploration continues from node 4, the new nodes are found to be infeasible when the rules from the pre-processing stage are applied. This tree contains 11 explored nodes, four nodes eliminated in the preprocessing stage and three nodes eliminated by the Dantzig-Wolfe solution.

Table 10.4 gives the percentage of nodes eliminated during solution of the various job shop problems using the three objectives.

Tot: percentage of the nodes eliminated.

Pre: percentage of the nodes eliminated in the pre-processing stage.

Table 10.4. Percentage of eliminated branching nodes.

Oper	$C_{\max}$			$T_{\max}$			JIT		
	Tot	Pre	DW	Tot	Pre	DW	Tot	Pre	DW
250×2	-	-	-	82	21	61	57	43	14
100×3	-	-	-	78	21	57	53	41	12
30×5	-	-	-	63	35	28	54	43	11
10×10	51	47	4	51	48	3	51	48	3

DW: percentage of the nodes eliminated by the Dantzig-Wolfe solution.

Percentages of the eliminated nodes, both overall and by the Dantzig-Wolfe solution, were observed to be higher for problems having few operations per job. The decomposition proposed here is more appropriate for problems of this type. The master problem selects convex combinations of schedules for a single machine so as to satisfy precedence constraints. It ignores disjunctive constraints for problems of sequencing on a single machine. A solution to the relaxed problem may differ significantly from a feasible solution to the job shop problem if there are many precedence constraints.

Results for the job shop problems

The algorithm was used to solve all job shop problems using the three objectives. Only one problem, with objective JIT, was not solved to optimality. Results are presented in Tables 10.5, 10.6 and 10.7. These tables contain the following information:

MaxWork: maximum total processing time on a single machine
$$= \max_{i=1,...,m} \left\{ \sum_{u|i_u=i} p_u \right\}.$$

UB: upper bound.

Opt: cost of the optimal solution, or of the best solution if optimality is not attained.

UtilMach: percentage utilization of machines. Let T be the value of $C_{\max}$ in the optimal solution. The mean percentage utilization (Avg) and maximum percentage utilization (Max) of the machines are calculated for the interval $[0, T]$.

IterDW: total number of iterations in the Dantzig-Wolfe algorithm.

BB: total number of nodes explored in the branching tree.

Cpu: CPU time in seconds. It indicates the time required to solve the master problem (TM), the subproblems (TS), and the total time (TT).

All problems were solved in less than 20 minutes for the objectives $C_{\max}$ and $T_{\max}$; most problems were solved in less than one hour for the objective JIT. Two 100-job problems required much more CPU time for the JIT objective; one of these problems was not solved to optimality. In fact, the JIT objective is more difficult to optimize as it is sensitive

Table 10.5. Numerical results for the objective $C_{\max}$.

| Oper | MaxWork | UB | Opt | UtilMach | | IterDW | BB | Cpu (secs) | | |
				Avg	Max			TM	TS	TT
10×10	627	960	792	61.9	79.2	775	169	107	7	120
	652	982	867	60.4	75.2	1340	310	194	12	216
	750	960	810	62.1	92.6	1218	286	164	10	184
	673	937	845	62.8	79.6	2568	622	529	28	574
	679	1016	885	56.8	76.7	1195	248	235	16	258
	599	780	728	66.1	82.3	555	122	73	4	82
	610	896	811	61.5	75.2	2138	530	329	20	366
	706	975	840	63.8	84.0	6048	1555	927	50	1026
	677	945	855	63.9	79.2	3039	852	387	20	435
	682	873	776	64.0	87.9	311	66	38	3	42

to operations that end late or begin too early in the schedule. The cost of the solution using $T_{\max}$ is a lower bound for JIT.

Table 10.5 indicates the maximum operation time on a single machine (MaxWork). In fact, this number is a lower bound for $C_{\max}$. The precedence constraints cause waiting times on the machines, so that the cost of the optimal solution is well above this bound. The percentage utilization of the machines decreases when changing the objective from $C_{\max}$ to $T_{\max}$ and to JIT. The objective $C_{\max}$ produces better machine utilization because the optimal schedule compresses operations on the bottleneck machine as much as possible. When the objective JIT is used, it may be advantageous to create waiting times on machines, so that operations begin and end at the desired times.

The algorithm spends most of its time solving the master problem. Little time is spent solving subproblems because the number of dynamic programming states is restricted. The difference between the total time (TT) and time spent solving the master problem (TM) and the subproblems (TS) is accounted for by pre- and post-processing at the branching nodes.

Behavior of the algorithm using different starting values

Table 10.8 presents results obtained using various starting values of the upper bound for the objective JIT. The algorithm was executed once using an upper bound of 1000. Later executions used tighter upper bounds, at 20%, 10% and 5% of the optimal value. With an improved

Table 10.6.   Numerical results for the objective $T_{\max}$.

Oper	UB	Opt	UtilMach Avg	UtilMach Max	IterDW	BB	Cpu (secs) TM	Cpu (secs) TS	Cpu (secs) TT
250× 2	594	495	77.7	87.2	17	1	8	10	18
	269	224	72.8	98.4	10	1	4	3	8
	129	107	78.0	99.2	45	11	46	15	65
	225	187	77.3	97.2	62	9	62	21	86
	410	341	77.4	99.9	45	3	40	17	61
	432	360	76.5	95.8	91	11	84	48	138
	219	182	75.7	99.4	61	10	106	22	136
	250	208	78.0	95.5	202	37	723	90	835
	542	451	76.7	98.2	28	1	26	13	41
	480	400	80.1	92.0	71	1	53	32	87
100× 3	250	208	72.7	96.9	339	103	639	50	715
	236	196	69.9	93.1	470	163	839	61	938
	276	230	74.6	92.7	297	77	474	45	532
	300	250	73.3	91.3	91	18	170	18	190
	128	106	83.5	98.1	693	318	878	60	1031
	255	212	76.0	98.0	161	37	393	29	429
	249	207	75.2	92.6	105	18	153	17	174
	203	169	72.1	96.3	89	29	182	14	200
	254	211	66.8	94.2	35	2	28	6	35
	98	81	68.6	98.8	148	31	346	23	375
30× 5	209	174	71.5	90.1	417	177	113	12	136
	188	156	68.9	91.6	406	173	103	10	122
	246	205	65.2	94.4	123	42	61	5	68
	416	346	52.9	77.0	132	51	58	5	65
	81	67	64.9	95.6	377	126	66	7	83
	226	188	70.1	97.2	2175	877	315	41	469
	146	121	74.9	97.8	406	138	101	9	120
	239	199	74.0	85.4	221	67	69	7	81
	262	218	70.8	90.3	1001	280	244	31	293
	176	146	61.5	92.5	391	102	89	10	107
10×10	96	80	53.7	68.6	2206	442	481	22	518
	83	69	53.3	66.3	82	16	8	1	10
	8	6	54.3	81.0	116	15	10	1	12
	119	99	52.3	66.4	2204	375	513	26	550
	266	221	50.8	68.6	1925	364	514	31	556
	108	90	58.6	73.0	1426	279	247	12	269
	131	109	54.8	67.0	753	149	147	8	159
	130	108	51.8	68.2	3184	703	586	33	641
	114	95	55.5	68.8	1478	326	240	14	264
	143	119	55.7	76.5	303	50	51	3	56

Table 10.7. Numerical results for the objective JIT.

| Oper | UB | Opt | UtilMach | | IterDW | BB | Cpu (secs) | | |
			Avg	Max			TM	TS	TT
250× 2	594	495	76.3	85.6	286	30	749	176	959
	269	224	70.5	95.3	271	9	107	70	242
	329	274	72.6	92.3	88	13	99	29	149
	321	267	73.5	92.3	366	41	392	117	579
	410	341	70.3	90.8	105	5	137	43	193
	432	360	72.1	90.3	380	66	1052	206	1333
	255	212	72.2	94.8	824	183	448	137	960
	305	254	75.8	92.8	992	163	1234	306	1753
	542	451	69.1	88.4	171	10	367	88	475
	480	400	76.9	88.4	293	25	726	153	911
100× 3	320	266	66.6	88.7	8453	2156	16629	1397	18926
	294	245	66.8	89.0	723	176	1339	106	1506
	305	254	72.1	89.5	665	139	1203	100	1353
	326	271	70.8	88.1	423	79	992	83	1096
	280	233	75.6	88.8	485	121	680	64	774
	255	216*	70.2	90.5	21148	5000	23159	1958	30542
	249	207	73.9	91.1	466	111	793	64	897
	249	207	68.3	91.2	2341	645	2240	202	2758
	254	211	65.3	92.0	363	46	241	36	298
	336	280	62.6	90.2	357	7	328	65	404
30× 5	282	235	64.2	80.8	11310	3281	2903	322	3437
	237	197	66.2	88.0	976	300	200	25	247
	264	220	58.7	85.0	219	50	117	8	128
	416	346	53.8	78.3	349	103	102	14	121
	236	196	58.0	85.5	429	108	61	7	77
	226	188	62.0	86.0	1338	503	242	28	329
	183	152	71.7	93.6	755	172	163	14	200
	239	199	68.9	79.5	642	165	133	14	162
	334	278	63.5	80.9	937	280	237	27	283
	184	153	58.7	88.3	721	157	134	13	163
10×10	172	143	48.5	62.0	4240	766	1102	44	1173
	172	143	47.9	59.6	723	131	125	6	137
	155	129	46.6	69.5	59	3	38	1	40
	167	139	50.3	63.9	722	129	132	7	144
	266	221	41.8	56.5	1766	313	495	26	531
	162	135	53.1	66.0	2136	416	507	21	542
	166	138	48.2	59.0	601	91	126	5	135
	190	158	48.0	63.2	2187	338	549	29	590
	207	172	50.5	62.6	5054	1081	1157	59	1253
	208	173	48.1	66.0	671	106	88	6	98

* : Not optimal

Table 10.8. Behavior of the algorithm using different starting values for the objective JIT.

Oper	$z_\text{sup} = 1000$ BB	Cpu	$z_\text{sup} = 1.20z_\text{opt}$ BB	Cpu	$z_\text{sup} = 1.10z_\text{opt}$ BB	Cpu	$z_\text{sup} = 1.05z_\text{opt}$ BB	Cpu
250×2	43	1368	30	959	15	443	45	591
	15	388	9	242	59	429	19	100
	7	343	13	149	5	46	5	29
	67	1286	41	579	45	343	19	137
	3	499	5	193	1	68	1	28
	134	2337	66	1333	55	746	51	535
	176	4082	183	960	113	701	49	288
	163	7073	163	1753	231	1976	98	774
	4	294	10	475	10	279	6	143
	1	773	25	911	24	524	15	339
100×3	401	4375	2156	18926	346	1642	273	1784
	276	5478	176	1506	97	748	329	4308
	216	3385	139	1353	170	990	70	486
	99	1752	79	1096	104	770	75	715
	107	1073	121	774	25	245	135	486
	921	7434	-	-	811	5235	165	1002
	189	2336	111	897	69	457	51	407
	332	3606	645	2758	220	1223	199	967
	96	1346	46	298	58	375	21	123
	25	1150	7	404	27	305	18	150
30×5	418	482	3281	3437	226	258	217	240
	247	372	300	245	108	100	121	66
	158	324	50	128	111	221	36	57
	59	147	103	121	44	76	11	28
	91	173	108	77	57	55	148	103
	600	528	503	329	151	105	178	121
	572	679	172	200	241	262	43	55
	289	319	165	162	65	79	109	86
	305	430	280	283	233	258	144	142
	305	664	157	163	66	61	15	15

bound, time windows may be narrower. Furthermore, the branching tree is smaller and solution time is faster.

Some exceptions, however, can be observed. The algorithm presented here searches the branching tree using a depth first strategy and com-

Table 10.8. continued

Oper	$z_{sup} = 1000$ BB	Cpu	$z_{sup} = 1.20z_{opt}$ BB	Cpu	$z_{sup} = 1.10z_{opt}$ BB	Cpu	$z_{sup} = 1.05z_{opt}$ BB	Cpu
10×10	706	1266	766	1173	648	1146	619	987
	91	135	131	137	85	96	33	36
	36	124	3	40	23	41	21	14
	145	233	129	144	140	184	119	162
	645	1112	313	531	383	691	125	243
	890	989	416	542	239	265	407	505
	158	242	91	135	114	198	64	75
	1332	1610	338	590	159	183	349	443
	1072	1243	1081	1253	904	968	1322	1326
	61	136	106	98	70	48	18	18

putes feasible solutions using a heuristic. If good solutions are found along the initial branches explored, the algorithm gives good results regardless of the starting value. The converse is true as well: even with a good starting value, it is possible to select a bad search direction and explore many nodes before finding good solutions. This seems to have been the case for the two 100-job problems that are very difficult to solve starting from a value situated 20% from the optimal value. These two problems were solved much more easily using a different starting value.

Such exceptions apart, better bounds generally yield better results. One possible fruitful approach could be to develop effective heuristic methods for finding good upper bounds before the optimization methods are called upon. Interestingly, however, the algorithm managed to solve all problems in reasonable times, even with very poor bounds.

6.   Conclusion

This study has presented a formulation of the job shop problem that uses Dantzig-Wolfe decomposition. This approach breaks down the problem of coordination between machines and procedures to construct a schedule for each machine. In this way, an efficient algorithm may be applied to each problem component. Exchange of information between the master problem and the subproblems produces better lower bounds than approaches that treat each machine independently.

The algorithm presented here uses Dantzig-Wolfe decomposition and a branching strategy based on conflict solution procedure. We measure the effort provided at each stage of the resolution. While the optimal value of the Dantzig-Wolfe decomposition is a lower bound for the job shop problem, this bound is not efficient (especially for the initial branching nodes). Importantly, good bounds for the job shop problem are difficult to obtain, even with other formulations. Therefore, rather than putting considerable effort into finding lower bounds, this approach settles for a solution whose cost falls below the upper bound. Such a solution forms the basis of a heuristic schedule construction method for the job shop problem.

Subproblems in the decomposition are solved using dynamic programming. This technique is rarely used in sequencing problems because the number of states grows too quickly. The approach presented here applies this technique successfully by controlling the size of the state space to be explored. Even when only a small number of states are explored, the dynamic programming algorithm produces good schedules for the master problem.

The algorithm has been tested on 10-machine problems using three objectives: $C_{\max}$, $T_{\max}$ and an objective consistent with the Just-In-Time philosophy. Interesting problems for the objective $C_{\max}$ have as many jobs as machines. There exist methods that are better than ours at solving such problems. Nevertheless, the present algorithm has solved problems involving 10 jobs and 10 operations per job in less than 20 minutes each.

This algorithm is particularly efficient for problems involving many jobs and few operations per job. Such problems arise frequently in industry, as machines are increasingly versatile and jobs are processed with few changes of machine. Objectives other than $C_{\max}$ (such as minimization of delivery delays or storage periods) in fact appear to be more interesting in practice. Most existing methods consider the objective $C_{\max}$ and are poorly adapted to other objectives, especially when these objectives involve irregular functions of the operation completion times. The algorithm described here can handle such objectives. Finally, problems of up to 500 operations were solved using an objective consistent with a Just-In-Time approach.

References

Adams, J., Balas, E., and Zawack, D. (1988). The shifting bottleneck procedure for job shop scheduling. *Management Science*, 34(3):391–401.

Applegate, D. and Cook, W. (1991). A computational study of the job-shop scheduling problem. *Journal on Computing*, 3(2):149–157.

Balas, E., Lenstra, J.K., and Vazacopoulos, A. (1995). One machine scheduling with delayed precedence constraints. *Management Science*, 41(1):94–109.

Barker J.R. and McMahon, G.B. (1985). Scheduling the general job-shop. *Management Science*, 31(5):594–598.

Blazewicz, J., Domscke, W., and Pesch, E. (1996). The job shop scheduling problem: Conventional and new solution techniques. *European Journal of Operational Research*, 93:1–33.

Brinkkotter, W. and Bruckner, P. (2001). Solving open benchmark instances for the job-shop by parallel head-tail adjustment. *Journal of Scheduling*, 4(1):53–64.

Brucker, P. and Jurisch, B. (1993). A new lower bound for the job-shop scheduling problem. *European Journal of Operational Research*, 64:156–167.

Brucker, P., Jurisch, B., and Sievers, B (1994). A branch and bound algorithm for the job-shop scheduling problem. *Discrete Applied Mathematics*, 49:107–127.

Carlier, J. (1982). The one-machine sequencing problem. *European Journal of Operational Research*, 11:42–47.

Carlier, J. and Pinson, E. (1989). An algorithm for solving the job-shop problem. *Management Science*, 35(2):164–176.

Carlier, J. and Pinson, E. (1990). A practical use of Jackson's preemptive schedule for solving the job shop problem. *Annals of Operations Research*, 26:269–287.

Chen, Z.-L. and Powell W.B. (1999a). A column generation based decomposition algorithm for parallel machine just-in-time scheduling problem. *European Journal of Operational Research*, 116:220–232.

Chen, Z.-L. and Powell W.B. (1999b). Solving parallel machine scheduling problems by column generation. *INFORMS Journal on Computing*, 11(1):79–94.

Chen, Z.-L. and Powell W.B. (2003). Exact algorithms for scheduling multiple families of jobs on parallel machines. *Naval Research Logistics*, 50(7):823–840.

Dantzig, G.B. and Wolfe, P. (1960). Decomposition principle for linear programs. *Operations Research*, 8:101–111.

Desrochers, M., Desrosiers, J., and Solomon, M. (1992). A new optimization algorithm for the vehicle routing problem with time windows. *Operations Research*, 40:342–354.

Dorndorf, V. Pesch, E., and Phan-Huy, T. (2000). Constraint propagation techniques for the disjunctive scheduling problem. *Artificial Intelligence*, 122(1–2):189–240.

Dorndorf, V. Pesch, E., and Phan-Huy, T. (2002). Constraint propagation and problem decomposition: A preprocessing procedure for the job-shop problem. *Annals of Operations Research*, 115(1–4):125–145.

Fisher, M.L. (1973). Optimal solution of scheduling problems using Lagrange multipliers—Part 1. *Operations Research*, 21:1114–1127.

Fisher, M.L., Lageweg, B.J., Lenstra, J.K., and Rinnooy Kan, A.H.G. (1983). Surrogate duality relaxation for job shop scheduling. *Discrete Applied Mathematics*, 5:65–75.

French, S. (1982). *Sequencing and Scheduling: An Introduction to the Mathematics of the Job-Shop*. Wiley, New York.

Garey, M.R. and Johnson, D.S. (1979). *Computers and Intractability*. W.H. Freeman and Co., San Francisco.

Gélinas, S. and Soumis, F. (1997). A dynamic programming algorithm for single machine scheduling with ready times and deadline to minimize total weighted completion time. MIS Collection in the *Annals of Operation Research*, 69:135–156.

Giffler, B. and Thompson, G.L. (1960). Algorithms for solving production scheduling problems. *Operations Research*, 8:487–503.

Hoitomt, D.J., Luh, P.B., and Pattipati, K.R. (1993). A practical approach to job-shop scheduling problems. *IEEE Transactions on Robotics and Automation*, 9(1):1–13.

Join, A.S. and Meeran, S. (1999). Deterministic job-shop scheduling: Past, present and future. *European Journal of Operational Research*, 113(2):390–434.

Lageweg, B.J., Lenstra, J.K., and Rinnooy Kan, A.H.G. (1977). Jobshop scheduling by implicit enumeration. *Management Science*, 24(4): 441–450.

Martin, P. and Shmoys, D.B. (1996). A new approach to computing optimal schedule for the job-shop scheduling problem. *Proceedings of the 5th International IPCO conference*, pp. 389–403.

McMahon, G. and Florian, M. (1975). On scheduling with ready times and due dates to minimize maximum lateness. *Operations Research*, 23(3):475–482.

Muth, J.F. and Thompson, G.L. (1963). *Industrial Scheduling*. Englewood Cliffs, New Jersey, Prentice-Hall.

Rinnooy Kan, A.H.G. (1976). *Machine Scheduling Problems: Classification, Complexity and Computations*. The Hague, The Netherlands, Martinus Nijhoff, 39.

Roy, B. and Sussman, B. (1964). Les problèmes d'ordonnancement avec contraintes disjonctives. *NoteDS* No.9 bis, SEMA, Paris.

van den Akker, J.M., Hoogeveen, J.A., and van de Velde, S.L. (1995). Parallel machine scheduling by column generation. *Technical Report*, Center for Operations Research and Econometrics, Université Catholique de Louvain, Belgium.

Chapter 11

APPLYING COLUMN GENERATION TO MACHINE SCHEDULING

Marjan van den Akker
Han Hoogeveen
Steef van de Velde

Abstract   The goal of a scheduling problem is to find out when each task must be performed and by which machine such that an optimal solution is obtained. Especially when the main problem is to divide the jobs over the machines, column generation turns out to be very successful. Next to a number of these 'partitioning' problems, we shall discuss a number of other problems that have successfully been tackled by a column generation approach.

1. Introduction

In this chapter we discuss the application of the technique of column generation to scheduling problems. Scheduling problems appear in for instance production facilities, where the goal is to find an optimal allocation of scarce resources to activities over time (Lawler, Lenstra, Rinnooy Kan and Shmoys, 1993). The scarce resources are usually called machines, and the activities are called tasks or jobs; the basic scheduling problem is then to find for each of the tasks an execution interval on one of the machines that are able to execute it, such that all side-constraints are met. Obviously, this should be done in such a way that the resulting solution, which is called a schedule, is best possible, that is, it minimizes the given objective function.

Although the technique of column generation has been around since the early sixties, its first application to a machine scheduling problem is of a much later date. As far as we know, it was first used by Marjan van den Akker. She wanted to solve single-machine scheduling problems with an additive objective function by using polyhedral combinatorics.

To this end, she formulated such a problem as an integer linear programming problem using a time-indexed formulation, which is based on binary variables x_{jt} that indicate whether a job j $(j = 1, \ldots, n)$ starts at time t $(t = 0, \ldots, T)$. The solution to the LP-relaxation turned out to give a very strong lower bound for the small-sized problem instances, but at that time (1994), it was not possible to solve the LP-relaxation for a larger sized problem using the conventional techniques. To alleviate this problem, she reformulated the problem by Dantzig-Wolfe decomposition and solved it by column generation. She presented these results at a conference in Giens (France) and compared the performance of column generation to solving the time-indexed formulation with CPLEX. Robert Bixby, the man behind CPLEX, was in the audience, and he obviously was not happy hearing that CPLEX was not able to solve all the instances. Therefore, he offered to try to solve some of the instances using CPLEX. The outcome was as follows: Bixby could solve the LP instances using the CPLEX barrier method on a very fast computer but still this required more computation time than column generation. Bixby sportingly admitted defeat and wrote to Marjan that she did not even need to buy a better computer to solve these instances.

After this initial success of column generation, attention shifted towards parallel machine scheduling problems with an additive objective function. A parallel machine scheduling problem consists of two parts: Determine for each job which machine should execute it, and given this decision, find an optimal schedule for each of the resulting single-machine problems. If the difficulty of the problem lies in dividing the jobs over the machines, that is, if it is easy to find an optimum solution given the set of jobs to be executed by each machine, then the problem can be modelled as an integer linear programming problem in the following way. First, enumerate all relevant subsets of jobs that can be assigned to a single machine and then formulate it as a set covering problem, where each job should be contained in exactly one subset of jobs. Obviously, column generation seems to be an attractive technique to solve the LP-relaxation, as explicitly enumerating all relevant subsets is not a feasible approach. This was discovered by three groups independently. After these initial results, many follow-up papers have been written, dealing with extensions.

Finally, we discuss a number of problems that have been tackled by using a combination of constraint satisfaction and column generation. Here we solve a linear programming problem using column generation to find out whether there can exist a feasible schedule. The problems discussed here have an objective function of the min max type, whereas the problems discussed in the other sections are of the sum type.

Since we are dealing with a large variety of problems, we introduce the necessary notation in the section in which the problem is described. Here we use the three-field notation scheme introduced by Graham, Lawler, Lenstra and Rinnooy Kan (1979), to denote scheduling problems. The remainder of this chapter is organized as follows. Since the parallel machine scheduling problem leads to a more 'traditional' column generation model than the single machine scheduling problem, in Section 2 we discuss the parallel machine scheduling problem with the objective of minimizing total weighted completion time. We describe here how to formulate it as an integer linear program and how to solve the LP-relaxation using column generation. We also show how to use this in a branch-and-bound algorithm. We discuss and compare the implementations by Van den Akker, Hoogeveen and Van de Velde (1999); Chen and Powell (1999a). In Section 3 we briefly discuss a number of similar problems that can be solved in a similar way. In Section 4 we look at the time-indexed formulation for single-machine scheduling problems and show how the LP-relaxation of this formulation can be solved using column generation. Moreover, we study a column generation formulation for a flow shop scheduling problem, which can also be obtained from the time-indexed formulation. Finally, in Section 5 we discuss how column generation can be combined with constraint satisfaction.

2.   Column generation for $P\|\sum w_j C_j$

We start with describing the archetypal result in the area of parallel machine scheduling that is solved using column generation. In Section 3 we will see that similar results can be derived for many more problems.

2.1   Problem description

The problem $P\|\sum w_j C_j$ is described as follows. There are m identical machines, $M_1, \ldots, M_m$, available for processing n independent jobs, $J_1, \ldots, J_n$. Job J_j $(j = 1, \ldots, n)$ has a processing requirement of length p_j and a weight w_j. Each machine is available from time zero onwards and can handle no more than one job at a time. Preemption of jobs is not allowed, that is, once started, the execution of the job must continue on the same machine until it has been completed. Hence, a *schedule* can be specified by just specifying the completion times, which we denote by $C_1, \ldots, C_n$; it is *feasible* if no job starts before time zero and if there are at most m jobs being executed at the same time. Given these completion times, we can then find a feasible machine allocation, such that no machine processes more than one job at a time and no job starts processing before time zero. The objective is to find a schedule with minimum total weighted completion time $\sum_{j=1}^{n} w_j C_j$. The problem is $\mathcal{NP}$-hard in the strong sense when the number of machines is part of the problem instance. Smith (1956), has shown that the single-machine problem is solvable in $O(n \log n)$ time. This algorithm is based on the observation that there exists an optimal schedule with the following two properties:

(i) The jobs are processed contiguously from time zero onwards.

(ii) The jobs are sequenced in order of non-increasing ratios w_j/p_j.

Since in the parallel machine case the machines are independent of each other as soon as the jobs have been partitioned, we know that there exists an optimal schedule in which each machine executes the jobs that have been assigned to it in order of non-increasing order of w_j/p_j without any unnecessary idle time. This implies that the difficulty of solving this problem comes from finding an optimal division of the jobs over the machines. One way to achieve this is by applying dynamic programming; this can be implemented to run in $O\big(n(\sum_{j=1}^{n} p_j)^{m-1}\big)$ time and space. Especially the space requirement becomes unmanageable when m or $\sum_{j=1}^{n} p_j$ increase.

To overcome this problem, we present a branch-and-bound procedure, where the lower bound comes from solving the LP-relaxation of a suitable integer linear programming problem through column generation. This approach has been applied by two groups of researchers: Van den Akker, Hoogeveen and Van de Velde (1999); Chen and Powell (1999a). The only difference is in the branch-and-bound procedure; we will discuss both procedures in Section 2.5.

Anticipating the branch-and-bound procedure by Van den Akker et al., we derive an execution interval for each job. The execution interval of a job J_j is $(j = 1, \ldots, n)$ is specified by a release date r_j, before which it cannot be started, and a deadline $\bar{d}_j$, at which it has to be finished. We can derive initial release dates and deadlines from the following two characteristics that an optimal schedule is known to possess:

(iii) No machine becomes idle before the latest start time of a job. Therefore, the last job on any machine is completed between time $H_{\min} = \sum_{j=1}^{n} p_j/m - (m-1)p_{\max}/m$ and $H_{\max} = \sum_{j=1}^{n} p_j/m + (m-1)p_{\max}/m$, where $p_{\max} = \max_{1 \leq j \leq n} p_j$.

(iv) If $w_j \geq w_k$ and $p_j \leq p_k$, where at least one of the inequalities is strict, then there exists an optimal schedule in which job J_j is started no later than job J_k.

Property (iii) yields for each job an initial deadline $\bar{d}_j = \sum_{k=1}^{n} p_k/m + (m-1)p_j/m$. We use Property (iv) to derive tighter release dates and deadlines in the root node, and more effectively, in the nodes of the branch-and-bound tree (see Subsection 2.5.1). Since Property (ii) decrees that each machine executes the jobs in order of non-increasing w_j/p_j ratios, we reindex the jobs such that

$$\frac{w_1}{p_1} \geq \frac{w_2}{p_2} \geq \cdots \geq \frac{w_n}{p_n};$$

to avoid trivialities, we assume that $n > m$. We define $\mathcal{P}_j$ as

$$\mathcal{P}_j = \{J_k \mid k < j, w_k \geq w_j, p_k \leq p_j\}$$

and $\mathcal{S}_j$ as

$$\mathcal{S}_j = \{J_k \mid k > j, w_k \leq w_j, p_k \geq p_j\}.$$

Accordingly, there is an optimal schedule in which all jobs in the set $\mathcal{P}_j$ start no later than job J_j. Hence, if $|\mathcal{P}_j| \geq m$, then we may conclude that at least $|\mathcal{P}_j| - m + 1$ jobs belonging to $\mathcal{P}_j$ are completed at or before the starting time of J_j, from which we can derive an earliest starting time. In a similar spirit, we can tighten the deadline of J_j.

2.2 ILP formulation

As we have argued above, we have solved the $P\|\sum_{j=1}^{n} w_j C_j$ problem when we have found the optimal assignment of the jobs to the machines. We define a *machine schedule* as a string of jobs that can be assigned together to any single machine. Let a_{js} be a constant that is equal to 1 if job J_j is included in machine schedule s and 0 otherwise. Accordingly, the column $(a_{1s}, \ldots, a_{ns})^T$ represents the jobs in machine schedule s. Let $C_j(s)$ be the completion time of job J_j in s; $C_j(s)$ is defined only if $a_{js} = 1$. Because of Property (ii) above and the reindexing of the jobs, they appear in s in order of their indices without any idle time between their execution. Hence, we have that $C_j(s) = \sum_{k=1}^{j} a_{ks}p_k$. So the cost c_s of machine schedule s is readily computed as

$$c_s = \sum_{j=1}^{n} w_j C_j(s) = \sum_{j=1}^{n} w_j a_{js} \left[\sum_{k=1}^{j} a_{ks}p_k \right].$$

We call a machine schedule s *feasible* if $r_j + p_j \leq C_j(s) \leq \bar{d}_j$ for each job J_j included in s. Let S be the set containing all feasible machine schedules. We introduce variables x_s ($s = 1, \ldots, |S|$) that assume value 1 if machine schedule s is selected and 0 otherwise. The problem is then to select m machine schedules, one for each machine, such that together they contain each job exactly once and minimize total cost.

Mathematically, we have to determine values x_s that solve the problem

$$\min \sum_{s \in S} c_s x_s$$

subject to

$$\sum_{s \in S} x_s = m, \tag{11.1}$$

$$\sum_{s \in S} a_{js} x_s = 1, \text{ for each } j = 1, \ldots, n, \tag{11.2}$$

$$x_s \in \{0, 1\}, \text{ for each } s \in S. \tag{11.3}$$

We obtain the linear programming relaxation by replacing conditions (11.3) by the conditions $x_s \geq 0$ for all $s \in S$; we do not need to enforce the upper bound of 1 for x_s, since this follows immediately from the conditions (11.2). We solve the LP-relaxation using column generation. The initial set $\overline{S}$ of columns can be generated heuristically, after which we add columns with negative reduced cost, if these exist.

2.3   The pricing algorithm

We describe next the approach by Van den Akker, Hoogeveen and Van de Velde (1999). This is identical to the approach by Chen and Powell (1999a), except that Van den Akker et al. take the additionally generated release dates and deadlines into account.

In our problem, the reduced cost c'_s of any machine schedule s is given by

$$c'_s = c_s - \lambda_0 - \sum_{j=1}^{n} \lambda_j a_{js},$$

where λ_0 is the given value of the dual variable corresponding to condition (11.1) and $\lambda_1, \ldots, \lambda_n$ are the given values of the dual variables corresponding to conditions (11.2). To test whether the current solution is optimal, we determine if there exists a machine schedule $s \in S$ with negative reduced cost. To that end, we solve the *pricing problem* of finding the machine schedule in S with minimum reduced cost. Since λ_0 is a constant that is included in the reduced cost of each machine schedule, we essentially have to minimize

$$c_s - \sum_{j=1}^{n} \lambda_j a_{js} = \sum_{j=1}^{n} \left[w_j \left(\sum_{k=1}^{j} a_{ks} p_k \right) - \lambda_j \right] a_{js}$$

subject to the release dates and the deadlines of the jobs and Properties (i) and (ii).

Our pricing algorithm is based on dynamic programming. It uses a forward recursion, where we add the jobs in order of increasing index. Let $F_j(t)$ denote the minimum reduced cost for all feasible machine schedules that consist of jobs from the set $\{J_1, \ldots, J_j\}$ in which the last job is completed at time t. Furthermore, let $P(j) = \sum_{k=1}^{j} p_k$. For the machine schedule that realizes $F_j(t)$, there are two possibilities: Either J_j is not part of it, or the machine schedule contains J_j. As to the first possibility, we must select the best machine schedule with respect to the first $j-1$ jobs that finishes at time t; the value of this solution is $F_{j-1}(t)$. The second possibility is feasible only if $r_j + p_j \leq t \leq \bar{d}_j$. If this is the case, then we add J_j to the best machine schedule for the first $j-1$ jobs that finishes at time $t - p_j$; the value of this solution is $F_{j-1}(t-p_j) + w_j t - \lambda_j$. The initialization is then

$$F_j(t) = \begin{cases} -\lambda_0, & \text{if } j = 0 \text{ and } t = 0, \\ \infty, & \text{otherwise.} \end{cases}$$

The recursion is then for $j = 1, \ldots, n$, $t = 0, \ldots, \min\{P(j), \max_{1 \leq k \leq j} \bar{d}_k\}$

$$F_j(t) = \begin{cases} \min\{F_{j-1}(t), & F_{j-1}(t - p_j) + w_j t - \lambda_j\}, \\ & \text{if } r_j + p_j \leq t \leq \bar{d}_j, \\ F_{j-1}(t), & \text{otherwise.} \end{cases} \tag{11.4}$$

After we have found $F_n(t)$, for $t = 0, \ldots, \sum_{j=1}^{n} p_j$, we can compute the optimal solution value as

$$F^* = \min_{H_{\min} \leq t \leq H_{\max}} F_n(t).$$

Accordingly, if $F^* \geq 0$, then the current linear programming solution is optimal. If $F^* < 0$, then it is not, and we need to introduce new columns to the problem. Candidates are associated with those t for which $F_n(t) < 0$; they can be found by backtracking.

2.4 A special type of fractional solution

Let now x^* denote the optimal solution to the linear programming relaxation of the set partitioning formulation and let S^* denote the set containing all columns s for which $x_s^* > 0$. If x^* is integral, then x^* constitutes an optimal solution for $P\|\sum_{j=1}^{n} w_j C_j$, and we are done.

There exists one further case in which we do not have to apply branch-and-bound to find an optimal integral solution; this case occurs if for all jobs J_j $(j = 1, \ldots, n)$ the completion time $C_j(s)$ is equal to C_j in each $s \in S^*$. This special case plays a crucial role in the branch-and-bound algorithm of Van den Akker et al.

THEOREM 11.1 *If $C_j(s) = C_j$ for each job J_j $(j = 1, \ldots, n)$ and for each s with $x_s^* > 0$, then the schedule obtained by processing J_j in the time interval $[C_j - p_j, C_j]$ $(j = 1, \ldots, n)$ is feasible and has minimum cost.*

If x^* is fractional and does not satisfy the conditions of Theorem 11.1, then we need a branch-and-bound algorithm to find an optimal solution. We present two such algorithms in the next section.

2.5 Two branch-and-bound algorithms

It is well known that applying branch-and-bound in combination with column generation makes sense only if it is possible to prevent a column from being generated by the pricing algorithm when this column does not satisfy the constraints issued by the branching strategy. Van den Akker, Hoogeveen and Van de Velde (1999); Chen and Powell (1999a), present two alternative branching strategies in which the solution space is partitioned by imposing constraints on the columns that can easily be incorporated into the pricing algorithm. We will first discuss their algorithms and then compare these.

2.5.1 Branching on the execution interval.
We start with the branch-and-bound algorithm by Van den Akker et al. If we have a fractional optimal solution that does not satisfy the conditions of Theorem 11.1, then there is at least one job J_j for which

$$\sum_{s \in S^*} C_j(s) x_s^* > \min\{C_j(s) \mid x_s^* > 0\} \equiv C_j^{\min};$$

we call such a job J_j a *fractional job*. Our branching strategy is based on partitioning the execution interval (see Carlier, 1987). In each node of the branch-and-bound tree, we first identify the fractional job with smallest index, and, if any, then create two descendant nodes: One for the condition that $C_j \leq C_j^{\min}$ and one for the condition that $C_j \geq C_j^{\min} + 1$. The first condition essentially specifies a new deadline for J_j by which it must be completed, which is smaller than its current deadline $\bar{d}_j$. For the problem corresponding to this descendant node, we therefore set $\bar{d}_j \leftarrow C_j^{\min}$. Since the jobs in $\mathcal{P}_j$ start no later than J_j and $\bar{d}_j - p_j$ is the latest possible start time of J_j, we must have that each $J_k \in \mathcal{P}_j$ is completed by time $\bar{d}_j - p_j + p_k$, which may be smaller than its current deadline $\bar{d}_k$. Hence, we let $\bar{d}_k \leftarrow \min\{\bar{d}_k, \bar{d}_j - p_j + p_k\}$ for each $J_k \in \mathcal{P}_j$. Since we have $\mathcal{P}_k \subset \mathcal{P}_j$ for each job $J_k \in \mathcal{P}_j$, updating the deadline of job J_k will not lead to any further updates.

The second condition specifies a release date $C_j^{\min} + 1 - p_j$ before which J_j cannot be started, which is larger than its current release date r_j. For the problem corresponding to this descendant node, we therefore set $r_j = C_j^{\min} + 1 - p_j$. Since the jobs in $\mathcal{S}_j$ start no earlier than J_j, we can possibly increase the release dates of these jobs as well. For each $J_k \in \mathcal{S}_j$, we let $r_k \leftarrow \max\{r_k, r_j\}$. Hence, this partitioning strategy not only reduces the feasible scheduling interval of J_j—It may also reduce the feasible scheduling intervals of the jobs in $\mathcal{P}_j$ and $\mathcal{S}_j$.

The nice thing of this partitioning strategy is that either type of condition can easily be incorporated in the pricing algorithm without increasing its time or space requirement: We can use exactly the same recursion as before.

The update of the execution intervals may result in an infeasible instance, which implies that we can prune this node. Since deciding whether there exists a schedule that respects all release dates and deadlines is $\mathcal{NP}$-complete, we only check a necessary condition for feasibility. If the instance corresponding to the current node fails this check, then we can prune the corresponding node and backtrack. To apply the check, we replace the deadlines $\bar{d}_j$ by due dates d_j $(j = 1, \ldots, n)$; there exists a schedule obeying the release dates and deadlines if and only if the outcome value of the optimization problem $P|r_j|L_{\max}$ is smaller than or equal to zero. We compute a lower bound (see Vandevelde, Hoogeveen, Hurkens and Lenstra, 2004, for an overview) for $P|r_j|L_{\max}$ and conclude infeasibility if this lower bound is positive.

It may also be that the current set of columns $\overline{S}$, which was formed while solving the linear programming relaxation in the previous node, does not constitute a feasible solution, due to the new release dates and deadlines. We work around this problem in the following way. We first remove the infeasible columns that are not part of the current solution to the linear programming relaxation. As to the infeasible columns that are currently part of the linear programming solution, we consider these as 'artificial variables', which we make unattractive by increasing their costs with some big value M, after which we can continue with our column generation algorithm.

2.5.2 Branching on the immediate successor.

The second branching rule that we discuss is due to Chen and Powell (1999a). Here we use that a schedule is fully characterized when we know for each job its immediate predecessor and its immediate successor on the same machine; we introduce two dummy jobs J_0 and J_{n+1} to model the case that a job is the first or last one on a machine. The number of machines that is used is then equal to the number of immediate successors of dummy job J_0.

Given a solution of the linear program, we can compute for each pair of jobs J_i and J_j, with $i < j$ the value y_{ij} as

$$y_{ij} = \sum_{s \in S_{ij}} x_s,$$

where S_{ij} denotes the subset of the columns in which job J_i is an immediate predecessor of job J_j. We have the following theorem.

THEOREM 11.2 *If the solution of the linear programming formulation in a node leads to an integral set of y_{ij} values, then the corresponding solution is optimal for this node.*

If we encounter fractional values y_{ij}, then we branch on the one that is closest to 0.5; we impose the constraint of $y_{ij} = 0$ in the first branch, and we require $y_{ij} = 1$ in the second branch. In both nodes, we remove the columns that do not satisfy the constraint. If the remaining set of columns does not constitute a feasible solution, then we can apply some heuristic to find a schedule that satisfies all branching conditions, after which we can add the corresponding machine schedules to the column set. Unfortunately, we need to adjust the pricing algorithm to find the columns with minimum reduced cost that satisfy the branching conditions. We change the dynamic programming algorithm by recording the identity of the job that completes at time t.

Let $F_j(t)$ denote the minimum reduced cost for all feasible machine schedules in which J_j is the last job, which is completed at time t. Furthermore, let $B(j)$ $(j = 1, \ldots, n, n+1)$ denote the index set that contains the indices of the jobs that are allowed as immediate predecessors of job J_j; here $B(n+1)$ refers to the jobs that can finish last. To find the value $F_j(t)$, we have to find the minimum of $F_i(t - p_j)$ over all $i \in B(j)$, to which we add the cost of completing job J_j at time t. After we have found $F_n(t)$ for all relevant values of t, we can compute the optimal solution value as

$$F^* = \min_{H_{\min} \leq t \leq H_{\max}, j \in B(n+1)} F_j(t).$$

Accordingly, if $F^* \geq 0$, then the current linear programming solution is optimal. If $F^* < 0$, then it is not, and we need to introduce new columns to the problem. Candidates are associated with those j, t for which $F_j(t) < 0$, where j is to restricted to the indices in the set $B(n+1)$; these can be found by backtracking.

Note that the adjustment of the dynamic programming algorithm has increased its running time by a factor $O(n)$.

2.6 Comparing the two algorithms

In this subsection, we compare the two algorithms with respect to their running times and with respect to implementation issues. Unfortunately, no study has been undertaken so far to compare the running times of both algorithms; therefore, we provide only general remarks with respect to this point.

An obvious advantage of the algorithm by Van den Akker et al. is that the running time of the pricing algorithm is not affected by the branching scheme, whereas it now requires $O(n^2 \sum_{j=1}^{n} p_j)$ time in the algorithm by Chen and Powell. A further advantage of the branching scheme used by Van den Akker et al. is that release dates and deadlines of other jobs can be strengthened using the sets $\mathcal{P}_j$ and $\mathcal{S}_j$, which implies that a node not only can get pruned because of the lower bound, but also because of its infeasibility with respect to the release dates and deadlines. Unfortunately, this yields the clear disadvantage that, if it is unclear whether a node corresponds to a feasible schedule, then it is required to work with infeasible machine schedules that are made unattractive by adding a penalty term to their cost.

An advantage of the algorithm by Chen and Powell is that it is easier to implement, since in each node of the branch-and-bound tree it is possible to find a set of columns that constitute a feasible solution.

Overall, we expect that on average the algorithm by Van den Akker et al. will be faster than the algorithm by Chen and Powell, but that it can be more sensitive with respect to 'bad' instances. We should keep in mind though, as reported by Van den Akker et al., that solving the LP-relaxation the first time is the bulk of the work. Hence, there is more to be gained by solving the initial LP-relaxation, than by speeding up the branch-and-bound algorithm.

3. Related results

A similar approach as used for the $P \| \sum w_j C_j$ problem can be applied to a number of related parallel machine scheduling problems. Van den Akker, Hoogeveen and Van de Velde (1999); Chen and Powell (1999a), extend the result of the previous section in two directions, which are

- problems with additive objective functions that have all jobs in the relevant part of the schedule sequenced according to some priority rule. Problems that fall in this category are the problem of minimizing the weighted number of late jobs, which is denoted by $P\|\sum w_j U_j$, and total weighted late work, which is denoted by $P\|\sum_{j=1}^n w_j V_j$.

- parallel machine problems with non-identical machines.

We discuss these extensions in the next two subsections. After that, we describe some other related results that have appeared in the literature since then. We only indicate the major differences with the approach used to solve the $P\|\sum w_j C_j$ problem.

3.1   Problems $P\|\sum w_j U_j$ and $P\|\sum w_j V_j$

These problems are defined as follows. Again, there are m parallel identical machines $M_1, \ldots, M_m$ to process n jobs $J_1, \ldots, J_n$, where job J_j $(j = 1, \ldots, n)$ has processing time p_j and weight w_j. Moreover, each job J_j has a *due date* d_j $(j = 1, \ldots, n)$ by which it ideally should be completed. In both cases jobs are not penalized as long as they are completed at or before their due dates. If a job is completed after its due date, that is, $C_j > d_j$, then it is called *late*. U_j $(j = 1, \ldots, n)$ is an indicator function that gets the value 1 if J_j is late, and 0, otherwise. The late work of J_j is defined as the portion of work of J_j that is performed after its due date d_j. Accordingly, we have that $V_j = \min\{p_j, \max\{0, C_j - d_j\}\}$.

For either problem, there is an optimal schedule in which each machine first performs the on-time jobs in order of non-decreasing due dates and then the late jobs in any sequence (Lawler and Moore, 1969; Potts and Van Wassenhove, 1992). The late jobs appear thus in the irrelevant part of the schedule, and in fact it does not matter if, when, and by what machine the late jobs are executed. These problems are therefore equivalent to *maximizing* $\sum_{j=1}^n w_j(1 - U_j)$, the weighted number of on-time jobs, and $\sum_{j=1}^n w_j(p_j - V_j)$, the total weighted on-time work. These problems lend themselves much better for the column generation approach, since the pricing algorithm needs to focus then only on the on-time jobs. These complementary problems then boil down to determining values x_s $s \in S$ that solve

$$\max \sum_{s \in S} c_s x_s$$

subject to

$$\sum_{s \in S} a_{js} x_s \leq 1, \text{ for each } j = 1, \ldots, n, \tag{11.5}$$

and conditions (11.1) and (11.3). The only difference, except for the computation of the cost c_s, is that only the (partially) on-time jobs will be present in one of the selected machine schedules. Relaxing the integrality constraints we are left with a linear programming problem that can be solved using column generation. The pricing algorithm maximizes the reduced cost, which can be achieved by applying dynamic programming, where the jobs are added in order of non-decreasing due date. In the dynamic programming algorithm, a job will only be included if it is (partially) on-time; we can incorporate release dates and deadlines.

To find an optimal integral solution, we can again use one of the two branch-and-bound algorithms described in Section 2.5. We refer to Chen and Powell (1999a); Van den Akker and Hoogeveen (2004), for details.

3.2   Non-identical machines

If the machines are not identical, then the processing time of J_j depends on the machine M_i that executes it. There are two variants: *Uniform* and *unrelated* machines. In the first case, each machine has a given speed s_i $(i = 1, \ldots, m)$, and the time needed to process job J_j on machine M_i is equal to $p_{ij} = p_j/s_i$ $(i = 1, \ldots, m; j = 1, \ldots, n)$. In the latter case the processing times can take any value. This implies, however, that in the latter case the priority rule is not equal for all machines for the objective of minimizing total weighted completion time. This can be easily overcome, though, by working with a different set of machine schedules for each machine. Let $S(i)$ denote the set of feasible machine schedules for machine M_i $(i = 1, \ldots, n)$. We need to adjust the formulation given in Section 2.2 only slightly in case of non-identical machines. The problem then becomes

$$\min \sum_{i=1}^{m} \sum_{s \in S(i)} c_s x_s$$

subject to

$$\sum_{s \in S(i)} x_s = 1, \text{ for each } i = 1, \ldots, m, \tag{11.6}$$

$$\sum_{i=1}^{m} \sum_{s \in S(i)} a_{js} x_s = 1, \text{ for each } j = 1, \ldots, n,$$

$$x_s \in \{0, 1\}, \text{ for each } s \in S(i), i = 1, \ldots, m.$$

For the pricing algorithm, we need to perform the recursion m times, one time for each machine separately. Accordingly, the pricing algorithm runs in $O\left(n \sum_{i=1}^{m} \sum_{j=1}^{n} p_{ij}\right)$ time. Obviously, it is worth-while to reduce the set of eligible machines for each job by preprocessing.

To find an integral solution, we must apply branch-and-bound. The branch-and-bound algorithm of Chen and Powell can still be applied. The branch-and-bound algorithm by Van den Akker et al. cannot be applied anymore, but this can now be replaced by a forward branching strategy where jobs are assigned to machines; this branching strategy can easily be combined with column generation.

3.3   Common due date problems

In case of a common due date problem jobs are penalized when they are not completed exactly at their due date, which is common to all jobs. This objective function is motivated by the 'just-in-time' philosophy. Van den Akker, Hoogeveen and Van de Velde (2002), consider the single-machine problem with the objective of minimizing the total weighted deviation of the completion times from the common due date, which is denoted by d. The cost of completing job J_j ($j = 1, \ldots, n$) at time C_j is then equal to $\alpha_j(d - C_j)$ if $C_j \leq d$; it amounts to $\beta_j(C_j - d)$ if $C_j \geq d$, where α_j and β_j are given. It is well-known that in case of a *large* common due date there exists an optimal schedule in which some job is completed exactly at time d; the common due date is considered to be large, if ignoring the unavailability of the machine before time zero does not affect the optimum solution. Under these circumstances, a schedule consists of an *early* and a *tardy* part, which comprises all jobs that are executed before and after the due date, respectively. It is easily proven that, given the jobs that are in the early part, it is optimal to execute these in order of non-decreasing α_j/p_j ratio, such that the last one is completed at time d. Similarly, the jobs in the tardy part must be executed in order of non-increasing β_j/p_j ratio, where the first job starts at time d. Van den Akker, Hoogeveen and Van de Velde (2002), report that a straightforward application of column generation, in which both pricing problems are solved independently through a dynamic programming algorithm like the one used to solve $P\|\sum w_j C_j$ is able to solve problems with up to 60 jobs. But this performance can be improved to solve problems with up to 125 jobs by combining column generation with Lagrangean relaxation. The main key to this success is not to solve both pricing problems independently, but to retain the constraint that the total processing time of the jobs in the early and tardy part must sum up to $\sum_{j=1}^{n} p_j$. Without this constraint we can only conclude that the LP-relaxation has been solved to optimality if there is neither a machine schedule for the early part nor for the tardy part with negative reduced cost. If we retain it, however, then we can stop column generation if there are no two machine schedules, one for the early and one for the tardy part, with total processing time equal to $\sum_{j=1}^{n} p_j$ with total reduced cost smaller than zero. Since the total processing of the machine schedule is kept in the state-variable used in the dynamic programming algorithm, the constraint that the total processing time must be equal to $\sum_{j=1}^{n} p_j$ is easily verified and provides no computational burden. Van den Akker et al. show how to derive an intermediate lower bound during the column generation process of solving the LP-relaxation. This can be used in a number of ways to speed up the process of solving the LP-relaxation; for details we refer to Van den Akker et al. Quite unexpectedly, the solution to the LP-relaxation was always integral in all of their computational experiments; they show by example that this is no structural result.

Chen and Powell (1999b), generalize the above problem to the case with $m \geq 2$ parallel identical machines. They solve the LP-relaxation through a standard implementation and apply branch-and-bound to find an optimal solution, where they branch on the immediate successor. In their computational experiments they report that instances with up to 60 jobs are solvable, where the hardest instances are the ones with 2 machines.

3.4   Common due window problems

Chen and Lee (2002), generalize the common due date problem to a *common due window* problem. In this case, the completion of a job is penalized only if it falls outside this window, which we denote by $[d_1, d_2]$. There are m parallel identical machines to execute the jobs. The objective is to minimize total cost, where the cost of completing job J_j is equal to $\alpha_j(d_1 - C_j)$ if $C_j \leq d_1$ and $\beta(C_j - d_2)$ if $C_j \geq d_2$, where α_j and β_j are given. In contrast to the previous subsection, we do not assume that d_1 is large. Therefore, we cannot guarantee that there exists an optimal schedule in which a job is completed at time d_1 or d_2, but in that case the first job must start at time zero. We partition each machine schedule into three parts:

- an early partial schedule, in which the last job completes at or before time d_1. The jobs in this part must be scheduled in non-decreasing order of α_j/p_j ratio;

- a fully tardy partial schedule, in which the first jobs start at or after time d_2. The jobs in this part must be scheduled in non-increasing order of β/p_j ratio;

- a middle partial schedule.

Here an early or fully tardy partial schedule can be empty. The ILP formulation is such that we search for at most m partial schedules of each type, such that for each early partial schedule that ends at time t_1 there exists a middle partial schedule that starts at time t_1, and such that for each middle partial schedule that ends at time t_2 there exists one fully tardy partial schedule that starts at time t_2.

We find a lower bound by solving the LP-relaxation through column generation. Here we need to find the early (fully tardy, middle) partial schedule with minimum reduced cost, which can be solved through a dynamic programming approach. Note that we do not have to solve a separate pricing problem for each possible value of t_1 and t_2. This lower bound is then used in a branch-and-bound algorithm where we branch on the immediate predecessor.

3.5   Scheduling with set-up times

Chen and Powell (2003), have investigated an extension of the problems $P \| \sum w_j C_j$ and $P \| \sum w_j U_j$ involving set-up times. The jobs are divided into several families, which roughly have the same characteristics. Before the start of a new job a set-up time is required, unless the consecutive jobs belong to the same family. The set-up times can be either sequence dependent or sequence independent.

We only work out the problem of minimizing total weighted completion time subject to sequence dependent set-up times; the other problems can be attacked in a similar fashion. Again, we work with machine schedules, which correspond to a given subset of the jobs and the order in which they are to be executed by a machine; we explicitly need the ordering, since finding an optimal order for a given set of jobs is a hard problem. Our ILP formulation reflects that we look for m machine schedules containing each job exactly once with minimum total cost.

If we solve the LP-relaxation using column generation in the standard way, then we need to generate machine schedules in which each job is contained at most once, and the constraint that 'each job is contained at most once' makes the problem intractable. Therefore, we disregard this constraint and generate machine schedules in which a job can be contained more than once; after all, these columns will not appear in an integral solution. We can find such a column through dynamic programming, where we use state-variables $f_j(t)$ that indicate the minimum reduced cost of completing job j at time t. In our branch-and-bound algorithm we branch on the immediate successor. A similar approach occurs in Subsection 4.1.

3.6   Scheduling jobs and maintenance activities

Lee and Chen (2000), have studied another extension of the problem $P\|\sum w_j C_j$ in which maintenance is required. Each machine needs to be maintained once in the period $[0, T]$, which requires t time units. Lee and Chen (2000), study two variants of this problem, depending on the capacity of the repair facility, which is either assumed to be one or infinite. Note that T is an upper bound on the completion of the maintenance; jobs can be completed after time T.

If the capacity of the repair facility is infinite, then the machine schedules are independent except for the jobs that they contain. Any machine schedule then consists of two subsets of jobs with a maintenance activity in between, where the jobs within one subset are in order of non-increasing w_j/p_j ratio. Just like in the previous subsection, we allow any job to be part of both subsets of the same machine schedule; we can then solve the LP-relaxation using column generation. We use this as a lower bound in our branch-and-bound algorithm, in which we branch on the immediate successor.

The constraint that just one machine can be maintained at a time yields an additional dependency between two machine schedules. Without loss of generality, we assume that the machines are maintained in order of their index. Since the machines are no longer identical, we use a different set of machine schedules for each machine, where we add a series of constraints in our ILP-formulation that decree that the completion times of the maintenance activities are at least t time units apart between two consecutive machines. We can then solve the LP-relaxation using column generation in the same fashion as for the case with an infinite capacity of the repair facility. When we apply a branch-and-bound algorithm in which we branch on the immediate successor, we run the risk that we do not end up with a feasible solution to the original problem, albeit that Lee and Chen (2000), report that this never happened in their computational experiments. If this might occur, though, then we can find an optimal, feasible solution by branching on the execution interval.

4.   Time-indexed formulation

We describe next the scheduling problems to which column generation was applied the first time, which are the problems that can be formulated using a time-indexed formulation. Time-indexed formulations for machine scheduling problems have received a great deal of attention; not only do the linear programming relaxations provide strong lower bounds, but they are good guides for approximation algorithms as well (see Hall,

Schulz, Shmoys and Wein, 1997). Unfortunately, a time-indexed formulation yields a large ILP-formulation. Van den Akker, Hurkens and Savelsbergh (2000), have shown how Dantzig-Wolfe decomposition techniques and column generation can be applied to alleviate to some extent the difficulties associated with the size of time-indexed formulations. In the next subsection, we discuss their approach. After that we study a Dantzig-Wolfe reformulation for the flow-shop scheduling problem with earliness, tardiness and intermediate inventory holding cost developed by Bülbül, Kaminsky and Yano (2001).

4.1   Single-machine scheduling

A time-indexed formulation is based on time-discretization, i.e., time is divided into periods, where period t starts at time $t - 1$ and ends at time t. The planning horizon is denoted by T, which means that we consider the time-periods $1, 2, \ldots, T$. Consider any single-machine scheduling problem in which the jobs are independent and the objective function is of the sum type. Using binary variables x_{jt} that indicate whether job job j $(j = 1, \ldots, n)$ starts in time period t $(t = 1, \ldots, T - p_j + 1)$, we get the following ILP-formulation:

$$\min \sum_{j=1}^{n} \sum_{t=1}^{T-p_j+1} c_{jt} x_{jt}$$

subject to

$$\sum_{t=1}^{T-p_j+1} x_{jt} = 1 \text{ for each } j = 1, \ldots n, \qquad (11.7)$$

$$\sum_{j=1}^{n} \sum_{s=t-p_j+1}^{t} x_{js} \leq 1 \text{ for each } t = 1, \ldots, T, \qquad (11.8)$$

$$x_{jt} \in \{0, 1\} \text{ for each } j = 1, \ldots, n; \quad t = 1, \ldots, T - p_j + 1,$$

where c_{jt} is equal to the cost that job j contributes to the objective function when it starts in time period t. The assignment constraints (11.7) state that each job has to be started exactly once, and the capacity constraints (11.8) state that the machine can handle at most one job during any time period.

We apply Dantzig-Wolfe decomposition to the LP-relaxation of the above model. We obtain the reformulation by representing the polytope P given by the capacity constraints (11.8) and the nonnegativity constraints as the convex hull of its extreme points. Since the constraint

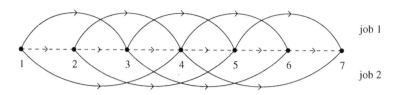

Figure 11.1. The network N for a 2-job example.

matrix of P is an *interval matrix*, the extreme points are integral, and
therefore, they represent schedules. As the assignment constraints (11.7)
are not involved in the description of P, jobs do not have to be started exactly once in such a schedule. Therefore, we call these *pseudo-schedules*.

Let x^k $(k = 1, \ldots, K)$ be the extreme points of P. Any $x \in P$
can be written as $\sum_{k=1}^{K} \lambda_k x^k$ for some nonnegative values λ_k such that
$\sum_{k=1}^{K} \lambda_k = 1$. Using this, we can reformulate the LP-relaxation of the
time-indexed formulation with decision variables λ_k. The cost coefficient
of λ_k is then equal to the cost of the pseudo-schedule x^k; the constraint
that each job j $(j = 1, \ldots, n)$ is started exactly once then becomes
$\sum_{k=1}^{K} n_{jk} \lambda_k = 1$, where n_{jk} denotes the number of times that job j is
started in pseudo-schedule x^k. Finally, we have the convexity constraint
stating that the sum of the lambda's is equal to one, and of course, the
nonnegativity constraints.

We solve the LP-relaxation through column generation, where we determine interesting pseudo-schedules. Pseudo-schedule x^k is characterized by the binary variables x_{jt}^k, which indicate whether job j starts at
time t. Given the dual variables π_j $(j = 1, \ldots, n)$, corresponding to the
reformulated assignment constraint for job j, and α, corresponding to
the convexity constraint, we find that the reduced cost $\bar{c}_k$ of the variable
λ_k is given by

$$\bar{c}_k = \sum_{j=1}^{n} \sum_{t=1}^{T-p_j+1} (c_{jt} - \pi_j)x_{jt}^k - \alpha. \tag{11.9}$$

Ignoring the constant α, we see that the pricing problem can be solved as
finding a shortest path in a network N. This network N has a node for
each of the time periods $1, 2, \ldots, T+1$ and two types of arcs: Process arcs
and idle time arcs. For each job j and each period t, with $t \leq T - p_j + 1$,
there is a process arc from t to $t + p_j$ representing that the machine
processes job j during the time periods $t, \ldots, t + p_j - 1$. Moreover, there
is an idle time arc from t to $t + 1$ $(t = 1, \ldots, T)$.

If we set the length of the arc referring to starting job j in time period
t equal to $c_{jt} - \pi_j$, for all j and t, and we set the length of all idle time

arcs equal to 0, then the reduced cost of the variable λ_k is precisely the length of path P_k minus the value of dual variable α. Therefore, solving the pricing problem corresponds to finding the shortest path in the network N with arc lengths defined as above. Since the network is directed and acyclic, the shortest path problem, and thus the pricing problem, can be solved in $O(nT)$ time by dynamic programming.

Van den Akker, Hurkens and Savelsbergh (2000), show that this column generation approach can be combined with *polyhedral combinatorics*. Here the goal is to strengthen the lower bound obtained from the LP-relaxation by adding inequalities that are violated by the current, fractional optimum but that are valid for the integral optimum. Van den Akker, Hurkens and Savelsbergh (2000), show the following general result for any LP-problem obtained through Dantzig-Wolfe decomposition.

THEOREM 11.3 *If for an* LP-*problem obtained by Dantzig-Wolfe decomposition the pricing problem can be solved for arbitrary cost coefficients, i.e., if the algorithm for the solution of the pricing problem does not depend on the structure of the cost coefficients, then the addition of a valid inequality $dx \leq d_0$ in terms of the original variables does not complicate the pricing problem.*

The computational results are somewhat disappointing, in the sense that improving the quality of the lower bound by adding valid inequalities takes quite some. To find an optimal integral solution, we can apply branch-and-bound, where we can add valid inequalities in each node. Here we use a forward branching strategy, that is, we build a partial schedule from left to right, such that we can find a lower bound in each node by solving the corresponding LP-relaxation through column generation.

4.2   Flow shop scheduling

Bülbül, Kaminsky and Yano (2004), consider a flow shop scheduling problem with earliness, tardiness, and intermediate inventory holding cost. In a flow shop each job must be processed by all of the m machines, starting at machine 1, after which it moves to machine 2, etc.; the time needed to move from one machine to the next is negligible, but in the application studied by Bülbül et al. intermediate products have to be conserved, which cost is measured by the intermediate inventory holding cost. The earliness and tardiness cost are defined on basis of the completion time on machine m. Finally, each job j has a release date r_j at which it becomes available; the holding cost is computed from this time on.

For a given schedule σ we use C_{ij} to denote the time at which job j ($j = 1, \ldots, n$) finishes on machine i ($i = 1, \ldots, m$), and we use q_{ij} to denote the time that job j spends in the queue before machine i. We compute the cost of schedule σ as

$$C(\sigma) = \sum_{j=1}^{n} h_{1j}[C_{1j} - p_{1j} - r_j] + \sum_{i=2}^{m}\sum_{j=1}^{n} h_{ij}q_{ij} + \sum_{j=1}^{n}[\alpha_j E_j + \beta_j T_j],$$

where h_{ij} is the holding cost per time unit for job j in the queue of machine i, and where α_j and β_j are the earliness and tardiness cost of job j.

We formulate this problem as an LP with variables C_{ij} and q_{ij}. To this formulation we apply Dantzig-Wolfe decomposition, where we put the constraints connecting the completion times of job j ($j = 1, \ldots, m$) on the different machines in the master problem, whereas we put the constraints involving one machine at a time in the subproblem. Another way to formulate the problem is to use a time-indexed formulation. Since this time-indexed formulation has the same Dantzig-Wolfe reformulation as the formulation in terms of completions times, we can show that the optimal value of the LP-relaxation of the Dantzig-Wolfe reformulation is at least equal to the value of the LP-relaxation of the time-indexed formulation. For small instances, we can solve the latter to optimality using CPLEX, and it turns out that the heuristic approach that we use to solve the former yields comparable bounds much faster.

To apply the Dantzig-Wolfe reformulation for the model defined in C_{ij} and q_{ij} variables, we must for each machine i ($i = 1, \ldots, m$) enumerate all feasible schedules σ_i^k, which are expressed in completion times C_{ij}^k, ($j = 1, \ldots, n$). We then have to select one schedule total per machine (which can be a convex combination of a number of schedules σ_i^k), such that the constraints stating that the jobs visit the machines in the right order are satisfied. Obviously, we apply column generation to avoid the full enumeration of the single-machine schedules. For machines 1 to $m-1$ the pricing problem boils down to a problem of the form $1|r_j|\sum w_j C_j$, whereas for machine m the pricing problem has the structure of problem $1|r_j|\sum(\alpha_j E_j + \beta_j T_j)$. Hence, all pricing problems are strongly $\mathcal{NP}$-hard. To solve the subproblems we apply a heuristic by Bülbül, Kaminsky and Yano (2001), which is based on a time-indexed formulation that is relaxed to an assignment problem. We further determine additional columns in the column generation step by applying local improvement to the already generated single-machine schedules. To speed up the process, a whole number of other engineering tricks is applied. To obtain a good heuristic solution, we use the information present in the near-optimal fractional solutions.

5.   Column generation and constraint satisfaction

In this section we discuss a number of problems that are solved by applying a constraint satisfaction approach; this is suitable if the objective function is of the min max kind, like the makespan (which is the case in the problems we discuss below). In a first step, the minimization problem is transformed into a feasibility problem by asking the question whether there exists a feasible solution with value no more than T; the minimization problem is then solved by applying binary search over T. Because of the upper bound T on the makespan we can then derive execution intervals $[r_j, \bar{d}_j]$ for each job j $(j = 1, \ldots, n)$, which then are narrowed by applying techniques from constraint satisfaction (we refer to Baptiste, Le Pape and Nuijten (2001), for details). If it is not possible to decide the feasibility problem now, then we check another necessary condition, which is based on linear programming.

5.1   Resource constraint project scheduling

The main difference between scheduling problems and resource constrained project scheduling problems is that there are no machines but resources, which can be *renewable* or *non-renewable*, where a resource is called renewable, if it is required during the execution of a task, which does not 'consume' it. Brucker and Knust (2000), look at a standard problem in which n jobs have to be scheduled. Each job j has a processing time p_j $(j = 1, \ldots, n)$, and preemption is not allowed. Executing job j requires r_{jk} units of each renewable resource k $(k = 1, \ldots, r)$ of which there are R_k units available at any point in time. Moreover, there are precedence constraints, which decree that a job cannot start until a given set of predecessors has been completed. The goal is to find a schedule with minimum makespan, which is equal to the completion time of the job that finishes last. Brucker and Knust (2000), solve this problem by first applying constraint satisfaction, like described above. If a decision has not been reached yet, the authors check another necessary condition. This condition checks whether it is possible to get all the work done while respecting the execution intervals and, partly, the precedence constraints; it can be checked by solving a linear programming problem. Hereto, we sort all release dates and deadlines in a combined list and partition the time horizon accordingly into a set of intervals. For each interval we derive all sets of jobs that can be executed simultaneously in this time interval, and for each such set we include a variable that indicates the amount of time during which this set is executed. Then we combine all intervals into one LP. Here we have for each job a constraint that decrees that the total amount of work done on this job is at least equal to the processing time, and for each interval we have a constraint that the total time spent on processing the subsets corresponding to this interval is no more than the size of the interval; we check whether this LP admits a feasible solution. Note that a 'yes' answer does not necessarily lead to a feasible, preemptive schedule, because of the precedence constraints between jobs that are executed in different sets. We turn this LP-formulation into an equivalent optimization problem in which we minimize the total penalty, where a penalty term is added to measure the amount of work on a job that is not executed. We solve this problem through column generation. Here, we have to find for each interval the subsets of independent jobs with negative reduced cost. Since this is an independent set problem, which is intractable, we apply branch-and-bound to solve it (see Baar, Brucker and Knust, 1998, for details). If the outcome of the LP-formulation does not decide the feasibility problem, then we can apply a branch-and-bound algorithm with a branching rule in which the execution interval of a job is split (see Carlier, 1987); we can apply the combination of constraint satisfaction and linear programming in each node.

Note that it is not necessary to apply constraint satisfaction at all to solve this problem. Brucker, Knust, Schoo and Thiele (1998), present a branch-and-bound algorithm for the optimization version of the resource constrained project scheduling problem defined above, where the lower bound is determined by solving a similar LP like the one above, without execution intervals, and with the objective of minimizing the total time that all subsets are processed; this LP-problem is again solved using column generation. The clear advantage of applying constraint satisfaction first is that the execution interval gets split up, which makes it easier to solve the pricing algorithm. Brucker and Knust (2000), do not present an elaborate comparison of both methods, but they remark that 'it confirms the conclusions of Klein and Scholl (1999), who reported that destructive methods are often superior to constructive approaches'.

In a follow-up paper Brucker and Knust (2003), apply their approach of combining constraint satisfaction and linear programming to an extended version of the basic resource-constrained project scheduling problem. In this more general problem activities can be executed in different *modes*, which may result in a different use of resources and a different processing time. Furthermore, there are minimum and maximum time-lags associated with each precedence constraint, which bound the difference in starting time between the two corresponding activities. Finally, there are also non-renewable resources involved, which makes sense in this setting, since jobs can be carried out in different modes. The objective is again to minimize the makespan. It turns out that the same approach can be applied, but we need to pay special attention to the possibility of processing activities in a different mode. In our LP-formulation for a given time interval, we conclude that a given subset of operations can be executed at the same time if they satisfy the precedence constraints, the capacity constraints, if no job occurs in two or more different modes, and if the remaining quantity of each of the non-renewable resources is sufficient for processing the remaining activities in the mode that requires the minimum of that resource. Brucker and Knust (2003), report that this approach leads to very promising results.

5.2   Scheduling a transportation robot

A further contribution by Brucker and Knust (2002), in this field concerns their work on the problem of scheduling a transportation robot in a job shop scheduling problem. In a job shop problem each job consists of a number of operations that have to be carried out in a given order by a prespecified machine. When an operation in a job has been completed, then this job has to be moved to the machine that executes the next operation. This moving around is carried out by a robot, which can transport only one job at a time. Brucker and Knust look at the problem that remains when the order in which each machine executes the operations of the different jobs is given: We have to find an optimal schedule for the robot. We model this by considering the robot as a single machine. The *tasks* that it has to execute correspond to the moving of a job in between the execution of two consecutive operations of this job; if the corresponding operations are O_i and O_j, then we call this task T_{ij}. The processing time of T_{ij} is put equal to the time needed to transport the job between the corresponding machines. There is a generalized precedence constraint between each pair of tasks T_{ij} and T_{jk} and associated with it is a minimum time-lag of size p_j, such that O_j can be processed in between. Similarly, if O_i is a direct predecessor of O_j on the same machine, where O_i and O_j belong to different jobs, then we introduce a generalized precedence constraint between the tasks T_{hi} and T_{jk}, where O_h is the direct predecessor of O_i and O_k is the direct successor of O_j in the job, with a minimum time-lag of size $p_i + p_j$. Since the robot may have to travel empty to a machine to pick up some job, we have a *sequence-dependent set-up time*: If task T_{jk} follows some task T_{hi}, then a minimum time s_{ij} has to elapse between the completion of T_{hi} and the start of T_{jk}, where s_{ij} is equal to the time needed to travel between the machines executing O_i and O_j. Finally, to cope with the beginning O_b and ending operation O_e of a job, we introduce a release date (tail) for the first (last) task corresponding to this chain. The release date is equal to the earliest possible completion time of O_b, and the tail has size p_e. Our goal is to minimize the maximum *delivery time*, which is equal to the completion time of a task on the machine plus its tail (which is zero, if undefined); this maximum delivery time then corresponds to the makespan of the job shop schedule.

We start by applying the constraint satisfaction techniques, which gives us execution intervals that must be satisfied to get a schedule with maximum delivery time no more than T. We then proceed with an additional feasibility check using column generation; the approach used here resembles the approach used in the time-indexed formulation. We split up the precedence constraints such that we obtain a number of chains (which have at most one predecessor and successor), which contain each task, and a remaining set of precedence constraints. Our ILP formulation is based on identifying for each chain a feasible schedule for this set of jobs (taking into account the time-lags and the execution intervals) that can be combined into a feasible schedule for the whole instance, that is, these so-called chain-schedules must respect the remaining precedence constraints and the capacity of the machine. Each chain-schedule is encoded by denoting for each time-interval $[t - 1, t]$ whether a job is being processed and by denoting the completion times of the jobs in the chain-schedule. We apply column generation to solve the LP-relaxation, where we have to identify interesting chain-schedules in each iteration. Brucker and Knust (2002), show that the pricing problem can be solved to optimality using dynamic programming.

The partitioning of the precedence constraints into chains is not unique. In fact, if we let each chain consist of only one task (and keep all precedence constraints explicitly), then we obtain a time-indexed formulation. It is easy to show that the LP-relaxation of the above formulation dominates the LP-relaxation of the time-indexed formulation.

References

Van den Akker, J.M., Hoogeveen, J.A., and Van de Velde, S.L. (1999). Parallel machine scheduling by column generation. *Operations Research*, 47:862–872.

Van den Akker, J.M., Hoogeveen, J.A., and Van de Velde, S.L. (2002). Combining column generation and Lagrangean relaxation to solve a single-machine common due date problem. *INFORMS Journal on Computing*, 14:37–51.

Van den Akker, J.M., and Hoogeveen, J.A. (2004). Minimizing the number of tardy jobs. In: *Handbook of Scheduling* (J.Y.-T. Leung,ed.),

Algorithms, Models, and Performance Analysis, pp. 227–243, CRC Press, Inc. Boca Raton, Fl, USA.

Van den Akker, J.M., Hurkens, C.A.J., and Savelsbergh, M.W.P. (2000). Time-indexed formulations for single-machine scheduling problems: column generation. *INFORMS Journal on Computing*, 12:111–124.

Baar, T., Brucker, P., and Knust S. (1998). Tabu-search algorithms and Lower bounds for the resource-constrained project scheduling problem, In: *Meta-heuristics: Advances and Trends in Local Search Paradigms for Optimization*, (S. Voss, S. Martello , I. Osman, and C. Roucairol, eds.), pp. 1–18, Kluwer, Dordrecht.

Baptiste, P., Le Pape, C., and Nuijten, W. (2001). Constraint-Based Scheduling: Applying Constraint Programming to Scheduling Problems. Kluwer Academic Publishers, Dordrecht, The Netherlands.

Brucker, P., Knust, S., Schoo, A., and Thiele, O. (1998). A branch-and-bound algorithm for the resource-constrained project scheduling problem. *European Journal of Operational Research*, 107:272–288.

Brucker, P. and Knust, S. (2000). A linear programming and constraint propagation-based lower bound for the RCPSP. *European Journal of Operational Research*, 127:355–362.

Brucker, P. and Knust, S. (2002). Lower bounds for scheduling a single robot in a job-shop environment. *Annals of Operations Research*, 115:147–172.

Brucker, P. and Knust, S. (2003). Lower bounds for resource-constrained project scheduling problems. *European Journal of Operational Research*, 149:302–313.

Bülbül, K., Kaminsky, P, and Yano, C. (2001). Preemption in single-machine earliness/tardiness scheduling. Submitted to: *Naval Research Logistics* for publication. Department of IEOR, University of California at Berkeley.

Bülbül, K., Kaminsky, P, and Yano, C. (2004). Flow Shop Scheduling with Earliness, Tardiness and Intermediate Inventory Holding Costs. *Naval Research Logistics*, 51:1–39.

Carlier, J. (1987). Scheduling jobs with release dates and tails on identical machines to minimize makespan. *European Journal of Operational Research*, 29:298–306.

Chen, Z.L. and Lee, C.-Y. (2002). Parallel machine scheduling with a common due window. *European Journal of Operational Research*, 136:512–527.

Chen, Z.L. and Powell, W.B. (1999). Solving parallel machine scheduling problems by column generation. *INFORMS Journal on Computing*, 11:78–94.

Chen, Z.L. and Powell, W.B. (1999). A column generation based decomposition algorithm for a parallel machine just-in-time scheduling problem. *European Journal of Operational Research*, 116:220–232.

Chen, Z.L. and Powell, W.B. (2003). Exact algorithms for scheduling multiple families of jobs on parallel machines. *Naval Research Logistics*, 50:823–840.

Graham, R.L., Lawler, E.L., Lenstra, J.K., and Rinnooy Kan A.H.G. (1979). Optimization and approximation in deterministic sequencing and scheduling: A survey. *Annals of Discrete Mathematics*, 5:287–326.

Hall, L.A., Schulz, A.S., Shmoys, D.B., and Wein, J. (1997). Scheduling to minimize average completion time: Off-line and on-line approximation algorithms. *Mathematics of Operations Research*, 22:513–544.

Klein, R. and Scholl, A. (1999). Conmputing lower bounds by destructive improvement—an application to resource-constrained project scheduling. *European Journal of Operational Research*, 112:322–346.

Lawler, E.L., Lenstra, J.K., Rinnooy Kan, A.H.G., and Shmoys D.B. (1993). Sequencing and scheduling: Algorithms and complexity. In: *Logistics of Production and Inventory* (S.C. Graves, P.H. Zipkin, and A.H.G. Rinnooy Kan, eds.), Handbooks in Operations Research and Management Science, Volume 4, pp. 445–522, North-Holland, Amsterdam.

Lawler, E.L. and Moore, J.M. (1969). A functional equation and its application to resource allocation and sequencing problems. *Management Science*, 16:77–84.

Lee, C.-Y. and Chen, Z.L. (2000). Scheduling jobs and maintenance activities on parallel machines. *Naval Research Logistics*, 47:145–165.

Potts, C.N. and Van Wassenhove, L.N. (1992). Single machine scheduling to minimize total late work. *Operations Research*, 40:586–595.

Smith, W.E. (1956). Various optimizers for single-stage production. *Naval Research Logistics Quarterly*, 3:325–333.

Vandevelde, A.M.G., Hoogeveen, J.A., Hurkens, C.A.J., and Lenstra, J.K. (2004). Lower bounds for the head-body-tail problem on parallel machines: A computational study of the multiprocessor flow shop. Forthcoming in: *INFORMS Journal on Computing*.

Chapter 12

IMPLEMENTING MIXED INTEGER COLUMN GENERATION

François Vanderbeck

Abstract   We review the main issues that arise when implementing a column generation approach to solve a mixed integer program: setting-up the Dantzig-Wolfe reformulation, adapting standard MIP techniques to the context of column generation (branching, preprocessing, primal heuristics), and dealing with issues specific to column generation (initialization, stabilization, column management strategies). The description of the different features is done in generic terms to emphasize their applicability across problems. More hand-on experiences are reported in the literature in application specific context, f.i., see Desaulniers et al. (2001) for vehicle routing and crew scheduling applications. This paper summarizes recent work in the field, in particular that of Vanderbeck (2002, 2003).

Introduction

The Dantzig-Wolfe reformulation approach is an application of a decomposition principle: one chooses to solve a large number of smaller size and typically well-structured subproblems instead of being confronted to the original problem whose size and complexity are beyond what can be solved in reasonable time. The approach is well suited for problems whose constraint set admits a natural decomposition into sub-systems representing well known combinatorial structure.

Consider a mixed integer problem of the form:

$$Z^{\text{MIP}} = \min \quad c(x,y) \tag{12.1}$$
$$[\text{MIP}] \quad s.t. \quad A(x,y) \geq a \tag{12.2}$$
$$B(x,y) \geq b \tag{12.3}$$
$$x \geq 0 \tag{12.4}$$
$$y \in \mathbb{N}^p \tag{12.5}$$

where $A \in \mathbb{Q}^{l \times (n+p)}$, $B \in \mathbb{Q}^{m \times (n+p)}$, are rational matrices and $c \in \mathbb{Q}^{(n+p)}$, $a \in \mathbb{Q}^l$, and $b \in \mathbb{Q}^m$ are rational vectors. This structure encompasses cases in which the problem has

1 **Difficult Constraints:** $A(x, y) \geq a$ represents *difficult constraints*. $B(x, y) \geq b$ represents a *more tractable* combinatorial subproblem that can be solved much more efficiently than the global problem. A classic example is the Traveling Salesperson Problem (TSP) where sub-system B represents 1-tree constraints while sub-system A stands for degree-2 constraints at each node. However, the word "efficiently" needs not necessarily mean polynomially solvable.

2 **Linking Constraints:** $B(x, y) \geq b$ has a block diagonal structure while $A(x, y) \geq a$ represents *linking constraints*. An example is the Cutting Stock Problem (CSP) where B defines a feasible way to cut each wide paper roll into products of smaller width (a knapsack subproblem for each wide roll) and A defines product demand covering constraints.

3 **Multiple Sub-Systems:** $A(x, y) \geq a$ and $B(x, y) \geq b$ represent each a *more tractable* sub-system (possibly having its own block diagonal structure) and the difficulty arises from having them simultaneously. An example is the Capacitated Multi-Item Lot-Sizing (CMILS) problem, where there is a subsystem associated with each item defining a single-item lot-sizing problem and a subsystem associated with each period defining a knapsack capacity constraint.

They are alternative ways to take advantage of such problem structure to compute efficiently strong dual bounds around which an efficient exact optimization approach can be developed: Lagrangian relaxation (Geoffrion, 1974), cutting plane approach (Crowder et al., 1983), and variable redefinition (Martin, 1987) are such techniques.

1.   Dantzig-Wolfe reformulation

The Dantzig-Wolfe reformulation approach is a special form of variable redefinition. It can be presented using the concept of generating sets: For each sub-system on which the decomposition is based, one defines a finite set of generators from which each subproblem solution can be generated. The variables of the reformulation shall be the weights of the elements of these generating sets.

Saying that G^B is a *generating set* for subproblem

$$X^B \equiv \{(x, y): B(x, y) \geq b, x \geq 0, y \in \mathbb{N}^p\}. \tag{12.6}$$

means that

$$X^B = \left\{ (x,y) = \sum_{g \in G^B} g\lambda_g : \lambda \in R^B \right\}.$$

Set G^B is typically large but finite even if X^B is not. R^B represents specific restrictions on the weights λ_g, including cardinality and integrality restrictions.

This framework encompasses various definitions of D-W reformulation. In the classic *convexification* approach, the generating set is defined as the set of extreme points and rays of conv(X^B). An alternative approach is to base the decomposition on a property that integer polyhedral can be generated from a finite set of feasible integer solutions plus a non-negative integer linear combination of integer extreme rays (Nemhauser and Wolsey, 1988). This is the *discretization* approach of Vanderbeck (2002). It generalizes to Mixed Integer Programs, by applying discretization in the integer variables (in the y-space) while convexification is used for the continuous variables. In Table 12.1, we specify the definition of G^B and R^B for each of the above cases. Vanderbeck (2002) discusses other possible definitions of generating sets. In Table 12.1, we assume boundedness of the subproblem X^B and a single block in matrix B to simplify the notation (the generalization to the unbounded case is straightforward, the case of a block diagonal subsystem is treated below).

The reformulation based on subproblem B takes the form:

$$M(B) \equiv \min \left\{ \sum_{g \in G^B} (cg)\lambda_g : \sum_{g \in G^B} (Ag)\lambda_g \geq a, \lambda \in R^B \right\}. \qquad (12.7)$$

Due to its large number of variables the linear relaxation of the reformulation is solved using a delayed column generation technique. Then, the reformulation is commonly called the master program while the slave is the pricing subproblem. For $M(B)$, the pricing subproblem takes the form

$$\zeta^B(\pi) \equiv \min\{(c - \pi A)g : g \in G^B\}. \qquad (12.8)$$

where π are dual variables associated with constraints $\sum_{g \in G^B} (Ag)\lambda_g \geq a$. The procedure starts by solving a master LP restricted to a subset of columns. While the subproblem returns negative reduced cost columns, they are added to the linear program that is re-optimized. The intermediate master LP values are not valid dual bounds. However, a Lagrangian dual bound can readily be computed from the subproblem value: applying Lagrangian relaxation to M, dualizing the A constraints

Table 12.1. Definitions of G^B and R^B under convexification and discretization approaches.

Convexification:

$$\begin{cases} G^B = \{(x,y) \in \mathbb{R}^n_+ \times \mathbb{N}^p \colon (x,y) = \text{ extreme point of } \text{conv}(X^B)\} \\ R^B = \left\{ \lambda \geq 0 \colon \sum_{g \in G^B_c} \lambda_g = 1, \sum_{g \in G^B} y^g \lambda_g \in \mathbb{N}^p \right\} \end{cases}$$

Discretization for an IP:

$$\begin{cases} G^B = \{y \in \mathbb{N}^p \colon By \geq b\} \\ R^B = \left\{ \lambda \colon \sum_{g \in G^B} \lambda_g = 1, \lambda_g \in \mathbb{N} \; \forall g \in G^B \right\} \end{cases}$$

Discretization for a MIP:

$$\begin{cases} G^B_p = \text{proj}_y X^B = \{y \in \mathbb{N}^p \colon B(x,y) \geq b, x \geq 0\} \\ S^B(y) = \{x \in \mathbb{R}^n_+ \colon x \text{ is an extreme point of } B(x,y) \geq b\} \\ G^B = \{(x,y) \in \mathbb{R}^n_+ \times \mathbb{N}^p \colon y \in G^B_p, x \in S^B(y)\} \\ G^B(y) = \{g = (x^g, y^g) \colon y^g = y \in G^B_p, x^g \in S^B(y)\} \\ R^B = \left\{ \lambda \colon \sum_{g \in G^B} \lambda_g = 1, \sum_{g \in G^B(y)} \lambda_g \in \mathbb{N} \; \forall y \in G^B_p \right\} \end{cases}$$

with weights $\pi \geq 0$ yields a valid bound of the form

$$L(\pi) = \pi a + f\big(\zeta^B(\pi)\big) \tag{12.9}$$

where function $f(.)$ takes a form that varies with the definition of the cardinality restrictions in R^B (Vanderbeck, 2003). In the case of a single subproblem taking the form of a bounded integer polyhedron, $f\big(\zeta^B(\pi)\big)$ simply equals $\zeta^B(\pi)$. Upon completion of the column generation procedure, this Lagrangian bound equals the master LP value. This column generation algorithm is embedded in a branch-and-bound procedure to solve the integer problem. The overall algorithm is known as *Branch-and-Price*.

The interest of the reformulation is twofold. First, it leads to a solution method implementing the desired decomposition of the problem and exploiting our ability to solve the subproblem. Second, it often leads to strong dual bounds: Lagrangian duality theory (Geoffrion, 1974) tells us that the master LP (under a standard definition of the generating set) is equivalent to the Lagrangian dual defined by dualizing the master constraints, itself equivalent to solving an LP on the intersection of the master constraints and the convex hull of the MIP subproblem polyhedron. For $M(B)$, this writes

$$Z_{\text{LP}}^{M(B)} = \min\{c(x,y) \colon A(x,y) \geq a, (x,y) \in \text{conv}(X^B)\}. \qquad (12.10)$$

Thus, $Z_{\text{LP}}^{\text{MIP}} \leq Z_{\text{LP}}^{M(B)} \leq Z^{\text{MIP}}$. The first inequality is typically strict unless $X_{\text{LP}}^B = \text{conv}(X^B)$ (that is when the subproblem has the "integrality property"). In the latter case, although the Dantzig-Wolfe decomposition approach gives a dual bound equal to the standard LP relaxation of the MIP, it might be worthwhile: beyond the motivation of working with a specialized subproblem solver, it may allow to avoid symmetry (as developed below in the case of multiple identical subsystems); it also provides a good strategy for introducing variables dynamically (if the original MIP formulation is itself large scale, as in Briant and Naddef, 2004).

In a standard D-W reformulation, the definitions of G^B and R^B ensure that

PROPERTY 1 *The master linear programming relaxation offers a dual bound which is equal to the Lagrangian dual value* (12.10): *i.e., we must have*

$$\left\{ (x,y) = \sum_{g \in G^B} g\lambda_g \colon \lambda \in R_{\text{LP}}^B \right\} = \text{conv}(X^B);$$

PROPERTY 2 *The master program* (12.7) *is a valid mixed integer reformulation of the original* MIP *problem* (12.2)–(12.1).

However, in extensions of the D-W reformulation principle, one might consider generating set definitions where these properties are relaxed as discussed in Section 3.

When matrix B is block diagonal

$$B = \begin{pmatrix} B^1 & 0 & \cdots & 0 \\ 0 & B^2 & \cdots & 0 \\ \vdots & & \ddots & \vdots \\ 0 & 0 & \cdots & B^K \end{pmatrix} \qquad (12.11)$$

with K non-identical blocks, one defines a generating set G^{B^k} for each $k = 1, \ldots, K$. But, when the blocks are identical, i.e. $B^1 = B^2 = \cdots =$

B^K, one only needs to define one generating set $G^B = G^{B^1}$. The weights associated with the generators can then be aggregated,

$$\lambda_g = \sum_{k=1}^{K} \lambda_g^k \quad \forall g \in G^B \tag{12.12}$$

and the cardinality constraint in the associated restriction set R^B be written as

$$\sum_{g \in G^B} \lambda_g = K. \tag{12.13}$$

Then, there is a third benefit in working with the D-W reformulation: it does not suffer from the drawback of symmetry present in the original formulation. Indeed, in the original formulation, variables must be indexed by k and a given solution can be represented in several ways by permuting the k indexing. This is quite harmful when it comes to enforcing integrality through branching constraints.

A distinction is to be made between convexification and discretization approaches that relates to the expression of the integrality restrictions. With the former one must translate the master solution in the space of the original variables to express and enforce integrality. As a consequence, when there are multiple identical sub-systems, one cannot aggregate weights using (12.12) and the reformulation suffers from the same symmetry drawback as the original formulation. The discretization approach, offers a true integer programming reformulation with which it is easier to implement some branching schemes as we shall see.

In most applications, cardinality constraints (12.13) can be relaxed to either a $\leq$ or a $\geq$ constraint (whether $K = 1$ or $K > 1$) as the combinatorial relaxation yields the same optimum value. This is in particular the case when $0 \in G^B$. A generalization is to write the cardinality constraint in the form:

$$L^B \leq \sum_{g \in G^B} \lambda_g \leq U^B \tag{12.14}$$

where L^B and U^B define a lower and an upper bound on the number of generators that can be used from subproblem B. They have an economical interpretation: for instance in a vehicle routing problem they stand for the minimum and maximum number of vehicles that must/can be used.

Formulation (12.7) assumes a simple *Lagrangian relaxation* of a subsystem A while B are being kept as hard constraints. However, one can treat both sub-systems as hard constraints by duplicating variables and dualizing constraints that enforce equality of the copies: this *variable splitting* approach implements a *Lagrangian decomposition* (Guignard, 2004). Then, the master takes the form

$$M(A, B) \equiv \min \left\{ \sum_{g \in G^A} \left(\frac{c}{2}g\right) \lambda_g^A + \sum_{g \in G^B} \left(\frac{c}{2}g\right) \lambda_g^B \right.$$

$$\left. : \sum_{g \in G^A} g\lambda_g^A = \sum_{g \in G^B} g\lambda_g^B, \lambda^A \in R^A, \lambda^B \in R^B \right\} \quad (12.15)$$

and the dual bound is

$$Z_{\text{LP}}^{M(A,B)} = \min\{c(x, y) \colon (x, y) \in \text{conv}(X^A) \cap \text{conv}(X^B)\}. \quad (12.16)$$

In the sequel, unless said otherwise, we assume a simple Lagrangian relaxation with B as the sub-system.

2.   Choosing a solver for the master

There are alternatives to solving the master LP with the revised simplex algorithm. Seen in the dual space, the latter amounts to applying a cutting plane algorithm known as Kelley's method (Kelley, 1960; Cheney and Goldstein, 1959); the column generation subproblem is then to be seen as a separation routine generating cuts. Other approaches are known to have better performance: the bundle method (Lemaréchal, 1998) and ACCPM (Goffin and Vial, 2002) are cutting plane algorithms for the dual of the master that differ from Kelley's cutting plane method by the choice of proposal that is made to the subproblem solver.

In Kelley's method, the dual solution that is returned to the subproblem is the point that optimizes the current restricted master linear program. In the bundle method the proposal is the point that optimizes a modified objective over the restricted master polyhedron: a quadratic term that penalizes the norm of the deviation to the current best dual solution, $\hat{\pi}$, is included in the objective to stabilize the search ($\hat{\pi}$ is the dual solution that gave the best Lagrangian bound value). In ACCPM, the proposal is the analytic center of the localization set defined by the restricted master polyhedron intersected with a cut on the objective value (defined by the best Lagrangian bound). The latest version of ACCPM (du Merle and Vial, 2002) includes a proximal term like in the bundle method to enhance stability.

Another alternative to the simplex algorithm is the interior point method that is readily available in most commercial MIP solvers. Applied to the dual it has the benefit of (approximately) computing dual prices centered in the optimal face when the dual admits many alternative optimum solutions (as it often happens)—Before translating it into an extreme solution. Such "centered dual solutions" may help in stabilizing the column generation procedure. Finally, a sub-gradient algorithm is an alternative that exhibits slow convergence but requires little computations at each iteration. It is often used when one is happy with an approximate dual solution of the master LP. The sub-gradient algorithm also provides a candidate primal solution in the convex hull of generated subproblem solutions, but it may violate the master constraints (Briant et al., 2004).

3.   Options in setting-up the D-W reformulation

Given an application, there are typically many ways of formulating it as a MIP and selecting sub-systems on which the D-W decomposition is based. The quality of the resulting dual bounds can be compared theoretically using Lagrangian duality theory (Geoffrion, 1974). But, for practical purposes, the important consideration is the trade-off between the duality gap observed on practical data sets and the computational time required to solve resulting master and subproblems.

The options are:

- To adopt a tighter or looser definition of a sub-system. Some constraints can go in either sub-system A or B or be duplicated in both. Implicit constraints (cuts) can also be added. In particular, one can enforce bounds on subproblem variables that are induced by constraints that are not in the subproblem. Columns whose coefficients take value within the bounds implied by master constraints are called *proper*. Vanderbeck (2002) also defines *strongly proper columns* whose component values satisfy bounds implied by using probing techniques in the master. Including extra restrictions in a subproblem may make it much harder to solve but may yield better dual bounds. The opposite trade-off consists in relaxing the subproblem to ease its solution while weakening the dual bounds. This can be viewed as a form of *state-space-relaxation*— A classical technique in dynamic programming—See Gamache et al. (1999) for an example of its use in a column generation context. Such relaxation can be exploited just to provide a warm start, or to implement a primal-dual heuristic for the subproblem.

- To base the reformulation on one or several sub-systems: a Lagrangian relaxation approach yields reformulation (12.7) while a Lagrangian decomposition approach yields reformulation (12.15). Although the latter yields a theoretically stronger dual bound, the master program is typically much harder to solve: the number of dualized constraints (12.15) is often large and each restricted mas-

ter typically admits many alternative dual solutions (a degeneracy
that causes a large number of iterations in the column generation
procedure). Moreover the theoretically better dual bound may
in practice be not significantly better than the bound given by
simple Lagrangian relaxation (Monneris, 2002). Guignard (2004)
proposes to aggregate variables in order to reduce the number of
dualized constraints (12.15); in some case the same quality dual
bound can be obtained with aggregate representation of the link
between the subsystems.

- To reformulate a sub-system (say A), improving its formulation, either instead of treating it as a subproblem (while decomposition is
  performed based on a second sub-system (B) as in Michel (2003))
  or before applying D-W decomposition (as in Thizy, 1991). One
  might have a formulation of the convex-hull of subproblem A solutions: for instance, as a network flow problem equivalent to solving
  the subproblem by dynamic programming (Martin, 1987). Or one
  might know cutting planes describing the integer polyhedron A.
  However, the size of such explicit reformulation is often too large.
  Then an approximate improved formulation can be used: for instance by aggregating variables (as in van Vyve, 2003) or by adding
  only a subset of cuts. The formulation can also be improved dynamically, using a cutting plane algorithm, which, combined with
  the column generation procedure, leads to a branch-and-price-and-
  cut algorithm.

- To disaggregate a sub-system, say B, that has a block diagonal
  structure or treat it as a single block. As discussed in Vanderbeck
  (2003), this does not affect the quality of the final dual bound as
  $\mathrm{conv}(X^B) = \bigotimes_{k=1}^{K} \mathrm{conv}(X^{B^k})$. But intermediate dual bounds can
  be better with the aggregated approach. (However, one can work
  with a disaggregate master while adopting the Lagrangian bound
  computation and the column generation strategy of the aggregate
  approach.) On the other hand, aggregation influences the efficiency of the master solver: the disaggregate formulation typically
  requires fewer iterations (because the disaggregate columns can
  be recombined in different implicitly defined aggregate columns);
  but solving a disaggregate master can be more costly computationally (in particular, when the master is solved using the bundle
  method, solving the disaggregate form of the quadratic master is
  much harder than the aggregate form).

4.   Branching scheme

In a branch-and-bound algorithm, the branching scheme consists in partitioning the solution space into sub-spaces so as to cut off the current fractional solution while ensuring that all feasible integer solutions are included in one of the sub-spaces. In a branch-and-price algorithm one can fall back to this standard scheme by using a convexification approach. Then, one returns to the space of the original variables (x, y) to check integrality and derive branching constraints (Villeneuve et al., 2003).

Alternatively, with the discretization approach, one has a mixed integer programming D-W reformulation and branching can be implemented by partitioning the generating sets, $G = G^1 \cup G^2$, and enforcing separate cardinality constraints (12.14) on each subset:

$$L^1 \leq \sum_{g \in G^1} \lambda_g \leq U^1 \quad \text{and} \quad L^2 \leq \sum_{g \in G^2} \lambda_g \leq U^2$$

so as to eliminate infeasible fractional solutions for which $\sum_{g \in G(y)} \lambda_g \notin \mathbb{N}$ for some $y \in G_p$. The recursive partition of the generating set shall be pursued until the integrality constraints $\sum_{g \in G^B(y)} \lambda_g \in \mathbb{N}$ are met for all $y \in G_p^B$. The scheme seems to call for associating a subproblem with each subset of generators G^1 and G^2. This, of course, quickly gets out of hand due to the exponentially increasing number of subsets. Instead, the pricing subproblem associated with G is amended to model correctly the reduced cost of generators from any subsets. The resulting modifications to the pricing subproblem range from amending the objective coefficients to adding new variables and constraints that may destroy the initial combinatorial structure and make the subproblem less tractable.

In practice, the separation of the generating set shall define a true partition $G = \widehat{G} \cup (G \setminus \widehat{G})$. Branching is implemented by choosing $\widehat{G}$ such that $\sum_{g \in \widehat{G}} \lambda_g = \alpha \notin \mathbb{N}$ and by adding a disjunctive branching constraint

$$\sum_{g \in \widehat{G}} \lambda_g \leq \lfloor \alpha \rfloor \quad \text{or} \quad \sum_{g \in \widehat{G}} \lambda_g \geq \lceil \alpha \rceil. \tag{12.17}$$

An indicator variable $1_{\widehat{G}}$ saying whether or not the generated column belongs to $\widehat{G}$ is added to the pricing subproblem. Its cost equals the dual value of the branching constraint. MIP constraints can be necessary to define $1_{\widehat{G}}$ in terms of the existing subproblem variables while in the simplest case $1_{\widehat{G}}$ is equal to an existing subproblem variable. Thus, sets $\widehat{G}$ must be carefully chosen, making a trade-off between the efficiency of the branching scheme and the difficulties resulting from the subproblem modifications (Vanderbeck, 2000).

Branching on a single λ_g variable corresponds to choosing $\widehat{G}$ to be a singleton. This is typically a poor choice on both regards: it yields an unbalanced branch-and-bound tree and requires heavy modifications to the subproblem. Instead, the most aggregate definitions of $\widehat{G}$ that yield the most balanced tree are also the easiest to accommodate in the subproblem. The simplest definition is $\widehat{G} = G$, enforcing integrality of the number of generators used from a sub-system:

Rule 1: $\sum_{g \in G} \lambda_g \in \mathbb{N}$.

For instance, in a column generation approach to the facility location problem, this rule enforces $0 - 1$ decisions on opening facilities. The next easiest rule that does not yield any subproblem modification is to branch on existing binary variables: say $y_i \in \{0, 1\}$ is a subproblem variable then define $\widehat{G} = \{g \in G : y_i = 1\}$ and enforce

Rule 2: $\sum_{g \in G : \, y_i=1} \lambda_g \in \mathbb{N}$.

Then, $1_{\widehat{G}} = y_i$.

Next consider an integer variable of the subproblem, say $y_k \in \mathbb{N}$, with bounds $l_k \leq y_k \leq u_k$. It admits a binary decomposition:

$$y_k = \sum_{j=0}^{\lfloor \log u_k \rfloor} 2^j y_{kj} \quad \text{with} \quad y_{kj} \in \{0, 1\} \quad \forall j.$$

One can branch on these binary components starting with the heaviest:

Rule 3: $\sum_{g \in G : \, y_{kj}=1} \lambda_g \in \mathbb{N}$.

The modification to the subproblem consists in introducing the binary components either dynamically as they arise in branching constraints (one replaces y_k using $y_k = y_k' + \sum_{l=j}^{\lfloor \log u_k \rfloor} 2^l y_{kl}$ with $y_k' \in \mathbb{N}$, $y_k' \leq 2^j - 1$ and bounds $l_k \leq y_k \leq u_k$ must be explicitly enforced on the new expression of y_k); or statically by reformulating the subproblem as a $0-1$ program.

Both Rule 2 and 3 can sometimes be enforced directly in the subproblem rather than being set as a master constraint. This results in a stronger dual bound. When the upper bound $\lfloor \alpha \rfloor$ of branching constraint (12.17) is equal to zero, no generators can be chosen in $\widehat{G}$ and the subproblem solution space can be redefined to yield $G \setminus \widehat{G}$. Symmetrically, when the lower bound $\lceil \alpha \rceil$ of (12.17) equals the upper bound U

of the cardinality constraint (12.14), all the generators must be chosen in $\widehat{G}$ and the subproblem solution space can be redefined accordingly.

When the master solution remains fractional beyond these simplest rules, one can branch on pairs (s, t) of binary variables or binary components introduced for Rule 3:

Rule 4: $\sum_{g \in G:\ y_s = y_t = 1} \lambda_g \in \mathbb{N}$.

This rule generalizes a branching scheme attributed to Ryan and Foster (1981). It admits special cases where it can be implemented directly in the subproblem while it normally requires to introduce a new binary variable in the subproblem, say $1_{\widehat{G}} = y_{st}$. Specifically, there are four distinct cases:

1 When $\lfloor \alpha \rfloor = 0$, the constraint $y_s + y_t \leq 1$ is added to the subproblem.

2 When $\lfloor \alpha \rfloor > 0$, y_{st} is introduced in the subproblem along with the constraint $y_{st} \geq y_s + y_t - 1$.

3 When $\lceil \alpha \rceil = U$, the constraint $y_s = y_t$ is added to the subproblem.

4 When $\lceil \alpha \rceil < U$, y_{st} is introduced in the subproblem along with the constraints $y_{st} \leq y_s$ and $y_{st} \leq y_t$.

Columns present in the master that do not satisfy the new subproblem constraints must be deleted. Vanderbeck (2003) presents this rule in a more general context and defines two more special cases.

The above rules may not suffice to generate a solution that obey the integrality restrictions. Then one shall pursue branching by implementing rules based on subset $\widehat{G}$ defined by fixing more than 2 column components. Worst case results and subproblem modifications are given in Vanderbeck (2000). However, for many applications, the above rules suffice to ensure integer solutions: either theoretically—For instance when the master is a set partitioning problem—Or in practice—For example, most instances of the cutting stock problems can be solved by combining these basic branching rules with primal heuristics (Perrot, 2004).

Choosing specific components (or pairs) on which to branch is done according to branching priorities associated with variables of the original formulation. Which of the disjunctive branch is explored first depends on the priorities defined for original variables. A tree search strategy (best bound first, depth first, least discrepancy, ...) must be selected. The above hierarchy of rules along with a best bound first strategy focuses on improving dual bounds. However, if the goal is to obtain incumbent integer solution quickly, branching rules yielding an unbalanced tree may be appropriate along with a depth first or a least discrepancy search strategy. One might even consider branching by rounding-up λ_g if backtracking is not expected (or in a rounding heuristic).

The distinction between convexification and discretization approaches is mostly a matter of presentation framework. The discretization has an apparent drawback: it requires

$$\sum_{g \in G^B(y)} \lambda_g \in \mathbb{N} \quad \forall y \in G_p^B \tag{12.18}$$

although an integer solution can be obtained for which this isn't true: the current solution in λ may be fractional, while its translation in the original variables y is not. This is in particular the case in applications where the subproblem is a shortest path problem: as we know from the flow decomposition theorem, the arc flow y can be integer while the path flows λ are fractional. Thus one could say that the condition (12.18) is too strict and might lead to continue on branching when it is not necessary. This misfortune can easily be avoided by checking integrality using the convexification conditions $\sum_{g \in G^B} y^g \lambda_g \in \mathbb{N}^p$ while implementing branching with the goal of enforcing (12.18).

On the other hand, the convexification approach has its drawbacks too. With the plain implementation defined in Table 12.1, branching constraints are always implemented in the master and are therefore dualized. While, with the discretization approach, special cases can be detected automatically where branching can be enforced directly in the subproblem. Then, the branching constraints are part of the polyhedron that is convexified yielding a better Lagrangian dual bound. However, this apparent drawback of the convexification approach can be circumscribed in an application specific context by recognizing, in the master, constraints that concern the subproblem variables only and can therefore be placed in the subproblem. It is equivalent to redefining the D-W reformulation from the original formulation at each node of the branch-and-bound tree (Villeneuve et al., 2003). We could call this approach the "dynamic convexification approach", it reproduces the subproblem modifications of the discretization approach. Thus, the discretization approach can be understood as a way to implement implicitly/automatically the redefinition of the D-W reformulation at each branch-and-bound node with no need to do it explicitly/manually. While the convexification approach is normally static but can be reseted at each node.

The second drawback of the convexification approach arises when the sub-system on which the decomposition is based has a block diagonal structure, as it is often the case. Even if these blocks are identical, the convexification approach requires to associate different variables with

each block to enforce $\sum_{g \in G^{Bk}} y^g \lambda_g^k \in \mathbb{N}^p \ \forall k$, which results in symmetry. Again, this drawback can be overcome: by introducing auxiliary
aggregate variables, $y = \sum^k \sum_{g \in G^{Bk}} y^g \lambda_g^k$ on which to branch. Again,
this remark reveals what the discretization approach does implicitly: it
permits to branch on auxiliary variables. Even when diagonal blocks
are not identical but are similar, it is worthwhile branching on sets
$\widehat{G} \subset G = \cup_{k=1}^K G^k$ to avoid symmetry.

Note that when these two drawbacks of the convexification approach
cumulate, i.e. when you have a branching constraint that could go in the
subproblem in the presence of multiple blocks, the manipulations that
need to be done under convexification to compensate for them become
difficult to implement. For instance, consider implementing Rule 4 in
the presence of multiple identical blocks. Thus, our view is that the
discretization approach provides a theoretical framework in which it is
more natural to formulate branching constraints. In particular, any
branching rule implemented in the convexification framework can be
implemented in the discretization framework in the same way but the
converse is not true. The same conclusion holds for adding cutting planes
for the master.

5.   Preprocessing and variable fixing

Standard preprocessing techniques (Savelsbergh, 1994) can also be
used in a column generation approach. Not directly on the column generation re-formulation that is only known implicitly through the solution of subproblems, but in the solution space of the original formulation
(12.2)—(12.1), which is also that of the subproblem variables. Bounds
on subproblem variables can be tightened not only based on subproblem
constraints but also on the basis of their input in the master constraints.
Tighter bounds might permit to eliminate redundant constraints (in the
subproblem or in the master) or to prove infeasibility. The reduced cost
fixing techniques permit to strengthen subproblem variable bounds further. Preprocessing can also be used to estimate the range of dual price
values: this dual information is helpful for a warm start of the column
generation procedure.

When the sub-system is made of blocks (12.11), preprocessing can
be used to derive bounds on aggregate variables $(x, y) = \sum_{k=1}^K (x^k, y^k)$,
which in turn implies bounds on the disaggregate variables (x^k, y^k). The
useful formulation where standard preprocessing can be applied is therefore:

$$\min c(x, y) \qquad\qquad (12.19)$$

$$A(x, y) \geq a \tag{12.20}$$

$$B^k(x^k, y^k) \geq b \quad \forall k = 1, \ldots, K \tag{12.21}$$

$$(x, y) = \sum_{k=1}^{K} (x^k, y^k) \tag{12.22}$$

$$0 \leq l^k \leq (x^k, y^k) \leq u^k \quad \forall k = 1, \ldots, K$$

$$0 \leq l \leq (x, y) \leq u. \tag{12.23}$$

Using constraints (12.20) to strengthen bounds (l, u) and constraints (12.21) to strengthen bounds (l^k, u^k), one can then combine these strengthening through constraints (12.22).

Formulation (12.19)–(12.23) does not need to be generated explicitly, but these implicit relations can be exploited for preprocessing. As it is standard, the lower bound on a variable x_j with positive coefficient a_{ij} in $\geq$ constraints A_i can be improved if it violates

$$l_j \geq \left(a_i - \sum_{v \neq j:\, a_{iv} > 0} a_{iv} u_v - \sum_{v \neq j:\, a_{i,v} < 0} a_{iv} l_v \right) \Big/ a_{ij}$$

and similarly for other coefficient sign and constraint sense and for (x^k, y^k) in the B^k constraints. Then, combining aggregate and disaggregate bounds can result in further strengthening if one of the following relations is violated:

$$l_i^k \geq l_i - \sum_{t \neq k} u_i^t, \quad u_i^k \leq u_i - \sum_{t \neq k} l_i^t,$$

$$l_i \geq \sum_{k=1}^{K} l_i^k, \qquad u_i \leq \sum_{k=1}^{K} u_i^k.$$

When the diagonal blocks are identical, $(l^k, u^k) = (\hat{l}, \hat{u}) \; \forall k = 1, \ldots, K$, the above procedure simplifies: one only needs to preprocess one sub-block and the strengthening through combining aggregate and disaggregate bounds takes the form

$$\hat{l}_i \geq l_i - (U - 1)\hat{u}_i, \quad \hat{u}_i \leq u_i - (L - 1)\hat{l}_i,$$

$$l_i \geq L\hat{l}_i, \qquad u_i \leq U\hat{u}_i,$$

where (L, U) are the bounds of the cardinality constraint (12.14). These relations are tested iteratively until no variable bounds can be improved. For integer variables, bounds are rounded to their integer parts.

If an incumbent integer solution INC is available, it can be shown to dominate solutions where a variable takes value within restricted range:

if a dual bound conditional to such restriction is available that takes value greater than INC. Lagrangian bound (12.9) can be made conditional to restricted range (l', u') on subproblem variables:

$$L(\pi | l', u') = \pi a + \min\{(c - \pi A)(x, y) \colon B(x, y) \geq b,$$
$$l' \leq (x, y) \leq u'\}. \quad (12.24)$$

Thus, further bound strengthening for integer variables y_i can be achieved as follows:

$$\text{while } L\big(\pi | l' = l + e^i(u_i - l_i), u\big) \geq \text{INC}, \quad \text{reset } u_i := u_i - 1,$$
$$\text{while } L\big(\pi | l, u' = u - e^i(u_i - l_i)\big) \geq \text{INC}, \quad \text{reset } l_i := l_i + 1,$$

where e^i is the ith unit vector. The reduced cost indicates which variable bound is candidate for such strengthening (Vanderbeck, 2003).

This so-called "variable fixing" scheme is computationally costly since it requires solving a subproblem for each test. However, a dual bound on the subproblem value is sufficient to define a valid test. Thus, in practice, one may solve a relaxed subproblem: for instance, a combinatorial relaxation of constraints $B(x, y) \geq b$ yields a trivial bound. A linear relaxation of the subproblem can yield a stronger bound (if the subproblem has the integrality property this bound has the same quality as (12.24)—This is the case when the subproblem is a shortest path problem). Yet another strategy is to dualize constraints $B(x, y) \geq b$ and apply a dual heuristic (such as several steps of a sub-gradient method) to estimate the Lagrangian prices.

When variable bounds cannot be further strengthened, constraint redundancy and infeasibility can be tested. The improved bounds on subproblem variables shall be used in the definition of the generating set. They guarantee that only *proper* columns shall be generated. However, this can make the subproblem much harder to solve. Take the two dimensional knapsack problem for example. The unbounded version can be solved in pseudo-polynomial time by dynamic programming while the bounded version is strongly NP-hard. For the single-item lot-sizing problem the uncapacitated version is polynomial while the capacitated case is NP-hard. Nevertheless it might be worth enforcing subproblem variable bounds since the D-W reformulation yields a better dual bound then.

Applying preprocessing to the dual of (12.19)–(12.23) allows to estimate the dual prices. This LP dual is a valid dual problem for the mixed integer program (12.2)–(12.1) or (12.7). However, the bounds on dual prices are not valid bounds for the Lagrangian dual (12.10). They merely provide estimates of dual price values. Take the example of the cutting stock problem. The primal LP takes the form:

$$\min\left\{\sum_k y_k : \sum_k x_{ik} \geq d_i \ \forall i, \sum_i w_i x_{ik} \leq W y_k \ \forall k, x_{ik} \geq 0, y_k \in [0,1]\right\}$$

where x_{ik} is the number of pieces of product i cut in wide-roll k, while y_k is 1 if wide-roll k is used. Its dual writes

$$\max\left\{\sum_i d_i \pi_i - \sum_k \nu_k : \sigma_k W - \nu_k \leq 1 \ \forall k, \pi_i - w_i \sigma_k \leq 0\right.$$
$$\left. \forall ik, \pi_i, \nu_k, \sigma_k \geq 0 \right\}.$$

It can be solved trivially through preprocessing: $\sigma_k \leq 1 + \nu_k/W \Rightarrow \pi_i \leq w_i/W(1 + \nu_k)$; the objective is maximized by setting the π_i's equal to those upper bounds and $\nu_k = 0 \ \forall k$ because $K \geq \sum_i w_i d_i/W$. Thus $\pi_i = w_i/W$ solves the dual LP. These values provide good estimates for starting up the column generation procedure but do not define valid upper bounds on the Lagrangian dual solution.

6.   Initialization

The generalized simplex algorithm used to solve the column generation formulation must be started with a feasible primal solution. It can be constructed heuristically. Alternatively, one can introduce artificial columns and combine phase 1 and phase 2 of the simplex method: at the end of phase 2, if artificial columns remain in the primal solution, their cost is increased and one returns to phase 2. Artificial columns remain useful to restart column generation after adding branching constraints. They also stabilize the column generation procedure. Dual methods need not be started with a primal feasible solution but they can benefit anyway from a warm start.

There are various ways to set artificial variables in the D-W reformulation from the most aggregate to the least: one can define a single artificial column, or one for each subproblem (representing the sort of solutions expected from this subproblem) or one for each master constraint (which offers an individual control on dual prices). In defining artificial column coefficients, one tries to feed the model with relevant information concerning the order of magnitude of the data: after the pre-processing phase, one has estimates of the maximum and minimum value that can be assumed by master constraint LHS, either globally or for a single subproblem solution; these values can be used to set artificial column coefficients.

In the primal, artificial columns act as slack variables that transform the hard constraints into soft ones: minimizing their cost amounts to minimizing the constraints' violations. In the dual of the master program, artificial columns define cuts (in the same way as regular columns do). Let us look for instance at the primal dual pair for the LP relaxation of (12.7) augmented with artificial columns associated with individual master constraints (indexed by i). The artificial columns are defined by their cost, constraint coefficient and upper bound: $(\hat{c}, \hat{a}, \hat{u})_i$.

$$\text{Primal:} \begin{cases} \min \sum_{g \in G} c_g \lambda_g + \sum_i \hat{c}_i \rho_i \\ \sum_{g \in G} a_{ig} \lambda_g + \hat{a}_i \rho_i \geq a_i \quad \forall i \\ \sum_{g \in G} \lambda_g = 1 \\ \lambda_g \geq 0 \quad \forall g \in G \\ 0 \leq \rho_i \leq \hat{u}_i \quad \forall i \end{cases}$$

$$\text{Dual:} \begin{cases} \max \sum_i a_i \pi_i - \sigma + \sum_i \hat{u}_i \nu_i \\ \sum_i a_{ig} \pi_i - \sigma \geq c_g \quad \forall g \in G \\ \hat{a}_i \pi_i \leq \hat{c}_i + \nu_i \quad \forall i \\ \pi_i, \nu_i \geq 0 \quad \forall i. \end{cases}$$

It appears that the artificial column cost over its constraint coefficient, $\hat{c}_i / \hat{a}_i$, defines an upper bound on the dual variable π_i that can be overruled (when $\nu_i > 0$) at a cost defined by the upper bound $\hat{u}_i$. In the same way, more aggregate definitions of artificial columns define generalized upper bounds on dual prices. Modifying their definition dynamically is a way to control dual variables. A general presentation of the primal-dual interpretation of modifications done to the master can be found in Briant et al. (2004)—It relies on Fenchel duality.

At a branch-and-price node other than the root, the pool of already generated columns offers a natural set for initialization, once the columns that do not satisfy branching constraints are eliminated, but it might be too large. Selecting only the columns that were potentially active at the parent node (i.e. with zero reduced cost) offers a good filter.

An intelligent initialization of the restricted master program offers a *warm start*. It might be worthwhile working harder at initialization. One can generate an initial set of columns based on the above dual price estimates. In turn, dual price estimates can be refined further using a truncated version of the master solver: for instance, perform a few

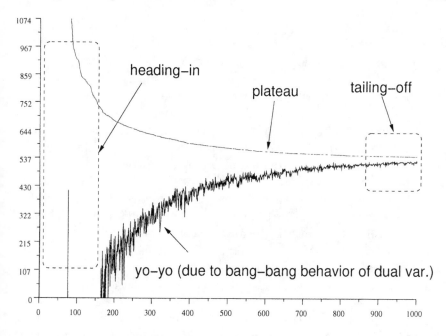

Figure 12.1. Illustration of the convergence of a simplex-based column generation approach on the TSP instance eil76 of the TSPLIB: the upper (resp. lower) curve gives the primal (resp. dual) bound at each iteration.

pivots of the simplex or several iterations of a sub-gradient algorithm with a subproblem optimized over the existing columns. The spirit is to avoid using the computationally costly procedure (whether it is the exact solution of the subproblem or that of the restricted master) while the problem knowledge is approximative. This strategy has sometimes proved helpful (Caprara et al., 2000). However, the exact method (for the master or the subproblem) is often very quick at the outset (because of the simplicity of initial master formulation or subproblem objective) and using approximate procedures does not always translate in big time savings.

7.   Stabilization

The standard method for solving the LP relaxation of D-W reformulation, i.e. using the simplex algorithm with dynamic pricing of variables, suffers from several drawbacks as illustrated in Figure 12.1: (i) a slow convergence (a phenomenon called the *"tailing-off effect"* in reference to the long convergence tail); (ii) the first iterations produce irrelevant columns and dual bounds due to poor dual information at the outset (let us call this the *heading-in effect*), (iii) *degeneracy* in the primal and hence multiple optimal solutions in the dual: one can observe that the restricted master solution value remains constant for several iterations (which can be called the *plateau effect*); (iv) instability in the dual solutions that are jumping from one extreme value to another (the *bang-bang effect*); the distance between the dual solution at an intermediate iteration k, π^k, and an optimum dual solution, π^*, does not necessarily decrease at each iteration, i.e., one can have $\|\pi^k - \pi^*\| > \|\pi^{k-1} - \pi^*\|$; (v) likewise, the intermediate Lagrangian dual bounds (12.9) do not converge monotonically (the *yo-yo effect*).

Stabilization techniques have been developed to reduce these drawbacks (a review is given in Lübbecke and Desrosiers (2004)). Vignac (2003) classifies them into 4 categories according to the mechanism that is used: (i) defining bounds on dual prices (the dynamic boxes of Marsten et al., 1975), (ii) smoothing dual prices that are returned to the subproblem (Neame, 1999) takes a convex combination of the current master dual solution and that of the previous iterate, Wentges (1997) takes a convex combination of the current dual solution and the dual solution that yielded the best Lagrangian bound), (iii) penalizing deviation from a stability center, usually defined as the dual solution that gave the best Lagrangian bound value (as in the bundle method of Lemaréchal (1998)), (iv) working with interior dual solution rather than extreme dual solutions (as in the solution method ACCPM of Goffin and Vial, 2002, or in Rousseau et al. (2003), where the center of the dual optimal face is computed approximately). These can be combined into hybrid methods (du Merle et al., 1999; Neame, 1999; du Merle and Vial, 2002).

These techniques often have an intuitive interpretation in the primal space of the λ variables (Briant et al., 2004). The basic techniques (dynamic boxstep and smoothing) can be implemented within the linear programming framework. Observe that using simplex re-optimization after adding a column rather than optimization from scratch can bring some form of stabilization toward the previous primal solution (whether this is desirable or not). Other techniques, such as penalizing deviation from a stability center, can be tuned to remain usable within the linear programming context (by choosing the L_1 norm for instance). Otherwise, an alternative master solver (as presented in Section 2) must be used. It is only by using a different master solver (such as bundle or ACCPM) that one can get theoretical convergence rate better than that of the simplex-method.

There are basic tips to reduce instability. One should avoid including redundant constraints in the master as they create degeneracy. In particular, one should use inequality constraints rather than equalities when this combinatorial relaxation yields the same solution. This also bounds the dual variables to positive values. Including the original variable in the master problem, as proposed in Poggi de Aragão and Uchoa (2003), can yield harmful redundancy. A more aggregate formulation of master constraints can bring stability (Perrot, 2004).

Basic stabilization can be obtained by managing artificial columns dynamically to control dual variables along with an intelligent initialization of the restricted master program that allows to alleviate the heading-in effect (Briant et al., 2004). The next easy step is to use a form of smoothing of dual prices. The more advanced stabilization techniques, in particular the hybrid methods, require intensive parametrization (which may hide the logic). They can be seen as barely attempting to mimic the behavior of alternative dual price updating methods such as an interior point solver, the bundle method or ACCPM. They are attempt at getting the benefits of these alternative approaches while staying in the Simplex world. Therefore, their theoretical convergence rate is that of the simplex method. When instability is problematic beyond the use of basic procedures, one should consider methods based on non-differentiable convex optimization that really bring a different logic.

In a branch-and-price algorithm however the exact solution of the master LP might not always be needed. A node can sometimes be cut by bound before reaching optimality of the master LP. The convergence tail can also be truncated by branching early.

8. Column re-optimization and post-processing

Re-optimizing existing columns before calling on the oracle subproblem to generate new columns, has proved to be a successful strategy: for instance, Savelsbergh and Sol (1995) apply a local search heuristic to active columns. It can be used as an alternative to a multiple column generation strategy: instead of introducing several columns of negative reduced cost at a given iteration, return only one but scan the others in the re-optimization phase after updating the dual prices.

Re-optimization can be somewhat formalized using the concept of *base-pattern* of Vanderbeck (2002). If the generating set defines our universe, *base-patterns* are the aggregate states resulting from some sort of projection/mapping onto a smaller size universe. In a sense a base-pattern summarizes the main properties of the columns/generators that are mapped onto it. It can be viewed as a seed for the generation of all the columns that are represented by it. To re-optimize a column, one can compute the best reduced cost column that can be obtained from its base-pattern, assuming that this re-optimization is much cheaper than solving the subproblem.

For example for the capacitated multi-item lot-sizing problem, the setup solution y to the discrete setup subproblem represents a base pattern for all associated production levels x. Then, re-optimizing x for a fixed y involves solving a shortest path subproblem. For the cutting stock problem, a cutting pattern defines a partition of the wide-roll where cut pieces can be replaced by an assortment of smaller items. The re-optimization can be heuristic.

Both examples call upon the concept of *exchange vectors*. An exchange vector is a valid perturbation: added to a feasible column, it gives rise to another feasible column (see Vanderbeck, 2002, for a formal definition). Exchange vectors define dual cuts (Valério de Carvalho, 2002); handling exchange vectors implicitly through re-optimization is the equivalent of using a cutting plane procedure rather than setting all cuts a priori in the dual master formulation. It can be called after a few master simplex iterations and not wait until the restricted master LP is re-optimized.

The column generation literature includes discussions on what could be *dominant/redundant* columns (a review is given in Lübbecke and Desrosiers (2004)). Since columns for the master primal are cuts for its dual, one can turn to the well-defined concepts of redundant dual constraints versus facet defining constraints to characterize columns. A more pragmatic concept that can be imported in column generation is that of *lifting*: the idea is to apply post-processing to subproblem solutions to increase column coefficients in master constraints (assuming "$\geq$" constraints) to their maximum feasible value (feasibility being defined by the subproblem constraints).

A sequential lifting procedure consists in enumerating each master constraint i with zero dual price $\pi_i = 0$ and solve the auxiliary subproblem

$$\max\{A_i(x,y)\colon B(x,y) \geq b, (c - \pi A)(x,y) \leq \zeta(\pi), x \geq 0, y \in \mathbb{N}\}$$

where $\zeta(\pi)$ is defined in (12.8). Alternatively, it can be formulated in terms of variables representing variations from the current subproblem solution. Lifting is a practical approach when this auxiliary subproblem is trivial: for instance, in the cutting stock application, it consists in filling-up the knapsack with zero profit items.

9.   Primal heuristics

The column generation approach is also a natural setting for deriving good primal solutions to a combinatorial problem: the decomposition principle can be exploited in greedy, local search or math-programming based heuristics. The price coordination mechanism of the Dantzig-Wolfe approach brings the global view that may be lacking when local search or constructive heuristics are applied directly to the original problem formulation (these latter approaches are often qualified as myopic).

There are examples of decomposition based heuristics in the literature that range from the simplest to the latest heuristic paradigm: a standard heuristic is to initialize the master with a set of heuristically generated columns and to solve the MIP master problem restricted to that initial set. Alternatively, on can use the columns generated while solving the master LP (approximately or exactly) to define the restricted master IP. However, there is typically no guarantee that such a static restricted set of columns holds a feasible integer solution. Methods that dynamically generate further columns needed for the integer solution avoid such drawback.

A rounding heuristic can be implemented by iteratively selecting in the master LP solution the branching variable that is closest to its integer part and rounding it. Variables that are currently integer can be fixed first. It is important to use pre-processing to adapt the subproblem to the residual master problem. The definition of branching variable is not tuned to improving dual bounds as in Section 4: on can round λ_g variables as in Vanderbeck (1999) or the underlying original variables as in Elhedhli and Goffin (2004).

A generic greedy heuristic for the master consists in generating iteratively the columns that have the smallest ratio of cost per unit of constraint satisfaction, take it in the solution and reiterate for the residual master problem (this requires generating columns that are proper for the residual problem).

The local search paradigm can be implemented by exchanging columns in a restricted pool over which the restricted integer master program is solved (heuristically or exactly). Examples of meta-heuristic based on decomposition can also be found in the literature. Taillard (1999) proposed an Adaptive Memory Programming heuristic for the vehicle routing problem based on managing a pool of columns representing feasible routes: columns with a history of being present in good solutions are iteratively selected that are suitable for the residual problem; if no more columns are available, remaining customers are being introduced in an existing route if feasible or new routes are generated for them.

10. Automated reformulation

Implementations of column generation based solution methods that are reported in the literature have been developed for a specific application. These algorithms are building on general purpose features such as the ideas reviewed herein. They are therefore amenable to a generic solver for decomposable MIPs. If such generic solvers were not proposed, it is because of the belief that there remain mandatory manual steps in the process of implementing a column generation code: the user must define the reformulation, the form taken by a column and its reduced cost, the user must say how the subproblem is modified after branching; he must define the form of intermediate Lagrangian bounds for its problem. Once the user has all that defined then he can use the *tool-box* offered by Abacus (Thienel, 1995), BCP (Ladányi et al., 1998), Maestro (Chabrier, 2003), or Minto (Savelsbergh and Nemhauser, 1991) to ease the implementation. Then, generic parts of the code can easily be adapted for re-use in other applications.

However, it is possible to automatize the reformulation, letting a code apply the Dantzig-Wolfe decomposition based on the original formulation and the user indication of what constraints must be dualized. All further modifications resulting from branching or adding cuts in the master for instance can be taken into account automatically too. Vanderbeck (2003) presents a framework for representing a MIP in an aggregate/implicit way from which the column generation reformulation can easily be generated automatically along with the subproblem and Lagrangian bound expressions. A prototype "black-box" column generation generic solver for MIP, named *BaPCod*, was developed along these principles. The underlying idea is simple: start with the most general form assumed by a decomposable MIP, write the formulas for its D-W reformulation, and code them. All simpler problems will be special cases.

Acknowledgments

The author received support from INRIA via a Collaborative Research Action (ARC) between INRIA Rhones-Alpes, HEC Geneva and the University of Bordeaux 1. We are very grateful for the patience of the anonymous referee in listing the typos of the original draft and making constructive remarks.

References

Briant, O., Lemaréchal, C., Meurdesoif, Ph., Michel, S., Perrot, N., and Vanderbeck, F. (2004). Comparison of bundle and classical column generation. *Working paper*, in preparation.

Briant, O. and Naddef, D. (2004). The optimal diversity management problem. *Operations Research*, 52:515–526.

Caprara, A., Fischetti, M., and Toth, P. (2000). Algorithms for the set covering problem. *Annals of Operations Research*, 98:308–319.

Chabrier, A. (2003). Génération de colonnes et de coupes utilisant des sous-problèmes de plus court chemin. *Ph.D Thesis*, Université d'Angers, France.

Cheney, E. and Goldstein, A. (1959). Newton's method for convex programming and Tchebycheff approximations. *Numerische Mathematik*, 1:253–268.

Crowder, H.P., Johnson, E.L., and Padberg, M.W. (1983). Solving large-scale zero-one linear programming problems. *Operations Research*, 31: 803–834.

Desaulniers, G., Desrosiers, J., and Solomon, M.M. (2001). Accelerating strategies in column generation methods for vehicle routing and crew scheduling problems. In: *Essays and Surveys in Metaheuristics* (C.C. Ribeiro and P. Hansen, eds.), pp. 309–324, Kluwer, Boston.

du Merle, O. and Vial, J.-Ph. (2002). Proximal ACCPM, a cutting plane method for column generation and Lagrangian relaxation: application to the *p*-median. HEC/Logilab, *Technical Report*, 2002.23.

du Merle, O., Villeneuve, D., Desrosiers, J., and Hansen, P. (1999). Stabilized column generation. *Discrete Mathematics*, 194, 229–237.

Elhedhli, S. and Goffin, J.-L. (2004). The integration of interior-point cutting plane methods within branch-and-price algorithms. *Mathematical Programming*, 100(2):267–294.

Gamache, M., Soumis, F., Marquis, G., and Desrosiers, J. (1999). A column generation approach for large scale aircrew rostering problems. *Operations Research*, 47:247–263.

Geoffrion, A.M. (1974). Lagrangian relaxation for integer programming. *Mathematical Programming Studies*, 2:82–114.

Goffin, J.-L. and Vial, J.-Ph. (2002). Convex non-differentiable optimization: a survey focused on the analytic center cutting plane method, *Optimization Methods and Software*, 17:805–867.

Guignard, M. (2004). Lagrangean Relaxation. In: *Handbook of Applied Optimization* (M. Resende and P. Pardalos, eds.), Oxford University Press.

Kelley, J.E. (1960). The cutting plane method for solving convex programs. *Journal of the SIAM*, 8:703–712.

Kim, S., Chang, K.N., and Lee, J.Y. (1995). A descent method with linear programming subproblems for nondifferentiable convex optimization. *Mathematical Programming*, 71:17–28.

Ladányi, L., Ralphs, T.K., and Trotter Jr., L.E. (1998). Branch, cut, and price: sequential and parallel. In: *Computational Combinatorial Optimization* (M. Junger and D. Naddef, eds.), Springer.

Lemaréchal, C. (1998). Lagrangian Relaxation. In: *Computational Combinatorial Optimization* (M. Junger and D. Naddef,eds.), Springer.

Lübbecke, M.E. (2001). Engine scheduling by column generation. *Ph.D Thesis*, Braunschweig University of Technology, Cuvillier Verlag, Göttingen.

Lübbecke, M.E. and Desrosiers, J. (2004). Selected topics in column generation. Forthcoming in: *Operations Research*.

Marsten, R.E., Hogan, W.W., and Blankenship, J.W. (1975). The BOXSTEP method for large-scale optimization. *Operations Research*, 23:389–405.

Martin, R.K. (1987). Generating alternative mixed-integer programming models using variable redefinition. *Operations Research*, 35(6):820–831.

Michel, S. (2003). Optimisation des tournées de véhicules avec gestion des stocks. *Rapport de stage*, DEA, Université Bordeaux 1, France.

Monneris, K. (2002). Problème de production par lots: Méthode de branch-and-price. *Rapport de stage*, DEA, Université Bordeaux 1, France.

Neame, P.J. (1999). Nonsmooth dual methods in integer programming. *Ph.D Thesis*, Department of Mathematics and Statistics, The University of Melbourne, Australia.

Nemhauser, G.L. and Wolsey, L.A. (1988). Integer and Combinatorial Optimization. John Wiley & Sons, Inc.

Perrot, N. (2004). Advanced IP column generation strategies for the cutting stock stock problem and its variants. *Ph.D Thesis*, University Bordeaux 1, France.

Poggi de Aragão, M. and E. Uchoa (2003). Integer program reformulation for robust branch-and-cut-and-price. In: *Proceedings of the Conference Mathematical Program in Rio: A Conference in Honour of Nelson Maculan*, pp. 56-61.

Rousseau, L.-M., Gendreau, M., and Feillet, D. (2003). Interior point stabilization for column generation. *Working paper*.

Ryan, D. M. and Foster, B.A. (1981). An integer programming approach to scheduling. In: *Computer Scheduling of Public Transport Urban Passenger Vehicle and Crew Scheduling* (A. Wren, ed.), pp. 269–280, North-Holland, Amsterdam,

Savelsbergh, M.W.P. (1994). Preprocessing and probing techniques for mixed integer programming problems. *ORSA Journal on Computing*, 6:445–454.

Savelsbergh, M.W.P. and Nemhauser, G.L. (1991). Functional description of MINTO, a Mixed INTeger Optimizer (Version 3.0). *Report COC-91-03D*, Georgia Institute of Technology.

Savelsbergh, M.W.P. and Sol, M. (1995). The general pickup and delivery problem. *Transportation Science*, 29(1):17–29.

Taillard, E.D. (1999). A heuristic column generation method for the heterogeneous VRP. *Recherche-Opérationnelle*, 33:1–14.

Thienel, S. (1995). ABACUS—A Branch-and-cut system. *Ph.D Thesis*, Universität zu Köln.

Thizy, J.-M. (1991). Analysis of Lagrangian decomposition for the multi-item capacitated lot-sizing problem. *INFOR*, 29(4):271-283.

Valério de arvalho, J.M. (2002). Using extra dual cuts to accelerate column generation. Forthcoming in: *INFORMS Journal on Computing*.

Vanderbeck, F. (1999). Computational study of a column generation algorithm for bin packing and cutting stock problems, *Mathematical Programming*. Serie A, 86:565–594.

Vanderbeck, F. (2000). On Dantzig-Wolfe decomposition in integer programming and ways to perform branching in a branch-and-price algorithm. *Operations Research*, 48(1):111–128.

Vanderbeck, F. (2002). A generic view at the Dantzig-Wolfe decomposition approach in mixed integer programming, *Working Paper*, no. U-02.22, MAB, Université Bordeaux 1, France.

Vanderbeck, F. (2003). Automated Dantzig-Wolfe re-formulation or how to exploit simultaneously original formulation and column generation re-formulation. *Working Paper*, no. U-03.24, MAB, Université Bordeaux 1, France.

van Vyve, M. (2003). A solution approach of production planning problems based on compact formulations of lot-sizing models. *Ph.D Thesis*, Université Catholique de Louvain, Belgique.

Vignac, B. (2003). Accélération de la convergence de l'algorithme de génération de colonne. *Rapport de stage*, DEA, Université Bordeaux 1, France.

Villeneuve, D., Desrosiers, J., Lübbecke, M.E., and Soumis, F. (2003). On compact formulations for integer programs solved by column generation. Forthcoming in: *Annals of Operations Research*.

Wentges, P. (1997). Weighted Dantzig-Wolfe decomposition for linear mixed-integer programming. *Transactions in Operational Research*, 4(2):151–162.